A Concise Introduction to Geometric Numerical Integration

"This book is highly recommended for graduate students, postgraduate researchers, and researchers interested in beginning study in the field of geometric numerical integration."

—*Mathematical Reviews*

A Concise Introduction to Geometric Numerical Integration, Second Edition, presents the main themes, techniques and applications of geometric integrators for researchers in mathematics, physics, astronomy and chemistry who are already familiar with numerical tools for solving differential equations. It also offers a bridge from traditional training in the numerical analysis of differential equations to understanding recent, advanced research literature on numerical geometric integration.

Accessible to researchers and post-graduate students from diverse backgrounds, this introductory book gets readers up to speed on the most significant ideas, methods and applications. Readers can reproduce the figures and results given in the text using the MATLAB® programs and model files available online.

New to the second edition

- Focus on single-step methods for solving differential equations.
- Full-scale rewrite of Chapter 3, with a new introduction and more comprehensive treatment of non-autonomous systems and the processing technique.
- New section dedicated to multi-product methods in Chapter 4, and a mention of the class of Crouch–Grossman and positivity-preserving integrators.
- A fresh introduction and a more detailed discussion of splitting and composition methods with complex coefficients in Chapter 6.
- A brand-new Chapter 7, showcasing some compelling applications where geometric integrators are now being used.
- Redesigned figures throughout, with improved clarity and readability. The codes for generating them are provided on the website accompanying the book.
- New exercises, as well as an enhanced list of bibliographic references.

A Concise Introduction to Geometric Numerical Integration

2nd Edition

Sergio Blanes and Fernando Casas

CRC Press
Taylor & Francis Group
Boca Raton London New York

CRC Press is an imprint of the
Taylor & Francis Group, an **informa** business
A CHAPMAN & HALL BOOK

Designed cover image: Sergio Blanes and Fernando Casas

Second edition published 2026
by CRC Press
2385 NW Executive Center Drive, Suite 320, Boca Raton FL 33431

and by CRC Press
4 Park Square, Milton Park, Abingdon, Oxon, OX14 4RN

CRC Press is an imprint of Taylor & Francis Group, LLC

First edition published by CRC Press 2016

ISBN: 978-1-032-86246-0 (hbk)
ISBN: 978-1-032-86460-0 (pbk)
ISBN: 978-1-003-52762-6 (ebk)

DOI: 10.1201/9781003527626

Typeset in CMR10 font
by KnowledgeWorks Global Ltd.

For those who are no longer here

To Marina,
In memoria aeterna

Contents

List of Figures

List of Tables

Preface to the First Edition

"A basic idea behind the design of numerical schemes is that they can preserve the properties of the original problems as much as possible.... Different representations for the same physical law can lead to different computational techniques in solving the same problem, which can produce different numerical results..." (Kang Feng, cited in [142]).

Differential equations play an important role in applied mathematics and are omnipresent in the sciences and in technical applications. They appear in many different fields such as chemical reaction kinetics, molecular dynamics, electronic circuits, population dynamics, control theory and astrodynamical problems, to name just a few. However, since the early days of the subject, it has become evident that very often finding closed solutions is either simply impossible or extremely difficult. Therefore, computing or approximating solutions of differential equations, partial as well as ordinary, linear or nonlinear, constitutes a crucial ingredient in all mathematical sciences.

Very often in applications, the differential equation modeling the physical phenomenon one aims to study possesses qualitative (geometric) properties that are absolutely essential to preserve under discretization. Hamiltonian systems constitute a paradigmatic example. These appear in many different contexts (classical, statistical and quantum mechanics, molecular dynamics, celestial mechanics, etc.) and have a number of features that are not shared by generic differential equations. These specific traits may be traced back to the fact that Hamiltonian flows define symplectic transformations in the underlying phase space.

The numerical integration of Hamiltonian systems by a conventional method results in discrete dynamics that are not symplectic since there is *a priori* no reason whatsoever as to why numerical schemes should respect this property. If the time interval is short and the integration scheme provides reasonable accuracy, the resulting violation of the symplectic character may be tolerable in practice. However, in many applications one needs to consider large time intervals so that the computed solution is useless due to its lack of symplecticity. One has then to construct special-purpose integrators that when applied to a Hamiltonian problem preserve the symplectic structure at the discrete level. These are known as *symplectic integration algorithms*, and they not only outperform standard methods from a qualitative point of view but also the numerical error accumulates more slowly. This of course becomes very important in long-time computations.

Starting from the case of symplectic integration, the search for numerical integration methods that preserve the geometric structure of the problem was generalized to other types of differential equations possessing a special structure worth being preserved under discretization. Examples include volume-preserving systems, differential equations defined in Lie groups and homogeneous manifolds, systems possessing symmetries or reversing symmetries, etc. Although diverse, all these differential equations have one important common feature, namely that they all preserve some underlying geometric structure which influences the qualitative nature of the phenomena they produce. The design and analysis of numerical integrators preserving this structure constitute the realm of *Geometric Numerical Integration*. In short, in geometric integration one is not only concerned with the classical accuracy and stability of the numerical algorithm, but the method must also incorporate into its very formulation the geometric properties of the system. This gives the integrator not only an improved qualitative behavior but also allows for a significantly more accurate long-time integration than with general-purpose methods. In the analysis of the methods a number of techniques from different areas of mathematics, pure and applied, come into play, including Lie groups and Lie algebras, formal series of operators, combinatorics, differential and symplectic geometry, etc.

In addition to the construction of new numerical algorithms, an important aspect of geometric integration is the explanation of the relationship between the preservation of the geometric properties of a numerical method and the observed favorable error propagation in long-time integration.

Geometric numerical integration has been an active and interdisciplinary research area since the 1990s and continues nowadays to be the subject of intensive development. The monograph [293] played a substantial role in spreading the interest of symplectic integration within the international community working in the numerical analysis of ordinary differential equations. A more recent book on numerical Hamiltonian dynamics is [218], whereas the most authoritative monograph on geometric numerical integration at present is [170], with two editions and more than 3000 citations in Google Scholar.

Although books [142, 170, 218, 293] constitute invaluable references in the field of geometric numerical integration, it is the authors' belief that there is still a gap to be filled. On the one hand, books [142, 218, 293] are devoted almost exclusively to symplectic integration and numerical Hamiltonian dynamics. On the other hand, the monograph [170] is without any doubt the standard reference on the subject, but as such it might be too advanced for researchers or postgraduate students with different backgrounds wishing to initiate themselves in the field. The present book thus has a double goal. First, it is intended as a (necessarily concise) introduction to the main themes, techniques and applications of geometric integrators for researchers in mathematics, physics, astronomy or chemistry already familiar with numerical tools for solving differential equations. Second, it might constitute a bridge from the traditional training in the numerical analysis of differential equations to the most recent and advanced research literature in the field of numerical geometric integration.

It is assumed only that the reader has a basic undergraduate training in differential equations, linear algebra and numerical methods for differential equations. More advanced mathematical material necessary in several parts of the book is collected in the Appendix. Also, for the reader's convenience (and also for self-consistency) we have distributed within the book some of the basic features of Hamiltonian dynamical systems, in order to facilitate the discussion and understanding of their numerical treatment. Much emphasis is put on illustrating the main techniques and issues involved by using simple integrators on well-known physical examples. These illustrations have been developed with the help of MATLAB, and the corresponding codes and model files (including updates) can be downloaded from the website accompanying the book

`https://www.gicas.uji.es/GNIBook.html`

Readers can reproduce the figures given in the text with the programs provided. In fact, it is not difficult to change parameters or even the numerical schemes to investigate how they behave on the systems provided. Although we have used MATLAB, there are, of course, other widely available free alternatives to this commercial package. We mention in particular OCTAVE, which is distributed under the Free Software Foundation's GNU Public License.

The book is based on a set of lectures delivered over the years by the authors to postgraduate students and also to an audience composed of mathematicians and physicists interested in numerical methods for differential equations. The actual project was suggested to one of the authors by Prof. Goong Chen (Texas A&M University at Qatar) during one of those lectures in Doha, Qatar, and has been made possible by NPRP Grant No. #5-674-1-114 from the Qatar National Research Fund (a member of Qatar Foundation) and by Project MTM2013-46553-C3 from the Ministerio de Economía y Competitividad (Spain).

We would like to thank all those who have helped us during the different stages in the elaboration of this book with their insights, corrections and suggestions, and in particular to Anna Arnal, Siu A. Chin, Cristina Chiralt, David Cohen and Ander Murua. The assistance of Laura and Marina Casas is greatly acknowledged.

We are especially grateful to our editors at CRC Press, Bob Stern and Sarfraz Khan, for their encouragement and help during this process.

Castellón and Valencia, October 2015

Sergio Blanes
Fernando Casas

Preface to the Second Edition

The positive reception of the previous edition of this book, as illustrated by the number of citations, has confirmed what the authors stated in the preface: that a gap existed in the literature on geometric numerical integration, one that could be filled by an introductory book offering a broad overview of the topic, while also presenting some of the latest advances in the field. Now, nearly 10 years after its initial publication and following the suggestion of the editors, we present this second edition, with some major modifications with respect to the first, while maintaining a more or less similar length.

In this new version we have corrected all the misprints and errors identified by us and also kindly communicated by several colleagues, in particular Alejandro Escorihuela-Tomàs. We would like to express our sincere gratitude to all of them for their interest and invaluable support, and especially to our friend and long-time collaborator Ander Murua, for stimulating discussions on most of the topics treated here and to José Luis Ruiz for his assistance during the editing process. We are also grateful to the Spanish (and the rest of the EU) citizens for supporting our work through the research project PID2022-136585NB-C21, MCINN/AEI/10.13039/501100011033/FEDER, UE.

The final stages of this project were completed while the first author was on sabbatical from his home institution, hosted by the Department of Applied Mathematics and Theoretical Physics, University of Cambridge, thanks to a mobility grant provided by the Ministerio de Ciencia, Innovación y Universidades (Spain).

Besides minor details and changes in the presentation throughout the book, the main differences with respect to the previous edition concern the following aspects:

- We have focused on single-step methods for solving differential equations. While we acknowledge the importance of multi-step methods, we believe they are already well-covered in other references from the perspective of geometric numerical integration.

- Chapter 3 has been almost entirely rewritten, with a new introduction and a more comprehensive treatment of non-autonomous systems and the processing technique.

- A new section dedicated to multi-product methods has been added to Chapter 4. We also mention the class of Crouch–Grossman and positivity-preserving integrators.

- Chapter 6 has been thoroughly revised, with a new introduction and a more detailed discussion of splitting and composition methods with complex coefficients.
- A new Chapter 7 has been added, showcasing some compelling applications where geometric integrators are being used nowadays.
- Most of the figures have been redesigned, with the aim of improving their clarity and readability. The codes for generating them are provided on the website accompanying the book.
- New exercises have been introduced, as well as an enhanced list of bibliographic references.

We hope this revised edition will be as well received by readers as the previous one.

> *Leer, por lo pronto, es una actividad posterior a la de escribir: más resignada, más civil, más intelectual.* (J. L. Borges)

May 2025

Sergio Blanes
Fernando Casas

Chapter 1

What is geometric numerical integration?

Differential equations constitute one of the most important tools for modeling the evolution in time of natural phenomena since Newton introduced them in his treatise on differential calculus. One of the first references to differential equations can be found in the following Newton's epigram: *Data aequatione quotcunque fluentes quantitae insolvent fluxiones in venire et vice versa*, which, in V.I. Arnold's free translation, can be stated as: *It is useful to solve differential equations* [15].

Modern applications of differential equations encompass such diverse areas as planetary motion, particle accelerators, fluid mechanics, population dynamics, electrical networks and molecular dynamics. From a mathematical point of view, the theory of differential equations comprises many different mathematical fields, and the problems arising in that theory have made fundamental contributions to linear algebra, the theory of Lie groups and functional analysis. Since their very introduction, different classical methods and techniques were developed to find solutions: series expansions, quadratures and elementary functions, principles of least actions, etc., by such luminaries as Newton himself, Euler, members of the Bernoulli family, Lagrange, Hamilton and others in the eighteenth and nineteenth centuries, whereas the theoretical analysis of properties of the solutions (existence, uniqueness, stability and differentiability with respect to initial values and parameters) started with Cauchy around 1820 and was virtually complete a century later.

In the meantime, at the end of the nineteenth century, Poincaré initiated what is now called the qualitative theory of differential equations and dynamical systems. The emphasis was then shifted to describe the qualitative behavior of the solution set of a given system of differential equations, including invariant sets and properties of the corresponding flow. Still, in many practical situations one aims to obtain not only qualitative information about the nature of the flow but also accurate representations of the corresponding solutions. It very often happens, however, that differential equations appearing in applied mathematics and the natural sciences do not have solutions which can be expressed in closed form, and thus one is compelled to seek approximate solutions by employing numerical methods. In fact, many numerical methods currently in use to solve differential equations possess a long history that may also be traced back to Euler, Cauchy, Adams, Heun and Runge, among

DOI: 10.1201/9781003527626-1

others. Those methods, though, were seldom used due to the limited computational resources available before the advent of digital electronic computers. It is worth remarking that the first system of differential equations solved by the ENIAC (Electronic *Numerical Integrator* and Calculator) computer was integrated using Heun's method [160]. The mathematical framework to analyze numerical integrators was developed during the 1950s, once numerical simulations in computers started to be perceived, rather than an oddity, as a "third leg" of physical research, complementing theory and experiment. As a result of these analyses, the new area of numerical analysis of differential equations emerged, and highly tuned and thoroughly tested software packages for general use were available by the 1960s and 1970s.

There are types of problems arising in many fields of science and applied mathematics that possess an underlying geometric structure which influences the qualitative character of their solutions and so one aims naturally to construct numerical approximations that preserve this geometry. However, many numerical integrators included in standard software packages do not take into account these distinctive features of the equations to be solved, and the question is whether it is possible to design new schemes providing approximate solutions that share one or several geometric properties with the exact solution. This is precisely the realm of *Geometric Numerical Integration*, a terminology introduced by Sanz-Serna in [286].

According to [244], *"Geometric integration" is the term used to describe numerical methods for computing the solution of differential equations while preserving one or more physical/mathematical properties of the system* exactly *(i.e., up to round-off error)*. Thus, rather than taking primarily into account prerequisites such as consistency and stability, the aim is to reproduce the qualitative features of the solution of the differential equation being discretized, in particular its geometric properties, such as the symplectic character (for Hamiltonian systems) and unitarity (for problems in quantum mechanics), but also the phase-space volume (for divergence-free vector fields), time-reversal symmetries, first integrals of motion (energy, linear and angular momentum), Casimirs, Lyapunov functions, etc. In these structure-preserving methods, one tries to incorporate as many of these properties as possible and, as a result, they exhibit an improved qualitative behavior. In addition, they typically allow for a significantly more accurate integration for long-time intervals than with general-purpose methods [170].

In this first chapter we take a glance at some of the issues involved in geometric integration in contrast with the standard procedure in numerical integration. We will focus mainly on the paradigmatic class of Hamiltonian systems and more particularly on three examples: the simple harmonic oscillator, the mathematical pendulum and the gravitational two-body problem. This will help us to introduce concepts and techniques that will be analyzed more thoroughly in later chapters.

1.1 First elementary examples and numerical methods

1.1.1 Simple harmonic oscillator

The one-dimensional harmonic oscillator is perhaps the simplest and most studied model in physics and computational mathematics. The system describes, in particular, the one-dimensional motion of a particle of mass m attached to an elastic helical spring with stiffness constant k in the linear approximation, and the corresponding equation of motion is

$$m\,\frac{d^2y}{dt^2} = m\,\ddot{y} = -k\,y, \qquad y \in \mathbb{R}. \tag{1.1}$$

Here and in the sequel we adopt Newton's notation for representing the time derivative. Alternatively, in terms of the position coordinate $q = y$ and the linear momentum $p = mv = m\dot{q}$, equation (1.1) can be written as a first-order system

$$\begin{aligned} \dot{q} &\equiv \frac{dq}{dt} = \frac{p}{m} = \frac{\partial H}{\partial p}(q,p) \\ \dot{p} &\equiv \frac{dp}{dt} = -kq = -\frac{\partial H}{\partial q}(q,p), \end{aligned} \tag{1.2}$$

where the *Hamiltonian*

$$H(q,p) = \frac{1}{2m}p^2 + \frac{1}{2}kq^2$$

represents the mechanical energy of the system, which is an invariant or *first integral* of the motion: $H(q(t),p(t)) = H(q(0),p(0)) \equiv E$ for all times. Here $T(p) = \frac{1}{2m}p^2$ is the kinetic energy, whereas $V(q) = \frac{1}{2}kq^2$ represents the potential energy. The simple harmonic oscillator constitutes then an example of a *Hamiltonian system* and the pair (q,p) are called *conjugate variables.* Although here q represents the configuration and p is a physical momentum, there are other examples where this is not necessarily the case.

System (1.2) can be expressed in matrix form as

$$\frac{d}{dt}\begin{pmatrix} q \\ p \end{pmatrix} = \begin{pmatrix} 0 & \frac{1}{m} \\ -k & 0 \end{pmatrix}\begin{pmatrix} q \\ p \end{pmatrix} \equiv A\begin{pmatrix} q \\ p \end{pmatrix}. \tag{1.3}$$

Introducing the vector $x = (q,p)^T$, we have

$$\dot{x} = f(x), \tag{1.4}$$

where $x \in \mathbb{R}^2$ describes the state of the system at a given time t and $f(x) = Ax = (p/m, -kq)^T$ is a *vector field* defined at each point x. The *flow* of the system, φ_t, maps $\mathbb{R}^2$ in $\mathbb{R}^2$ for each value of time t in such a way that $\varphi_t(\alpha)$ is the value $x(t)$ of the solution of (1.4) with initial condition $x(0) = \alpha$, whereas for fixed x_0 and varying t, $\varphi_t(x_0)$ provides the solution of the initial value problem defined by (1.4) and $x(0) = x_0$. More specifically, by explicitly solving the linear system (1.3), we get

$$\begin{pmatrix} q_0 \\ p_0 \end{pmatrix} \xrightarrow{\varphi_t} \begin{pmatrix} q(t) \\ p(t) \end{pmatrix} = \begin{pmatrix} \cos\omega t & \frac{1}{\omega}\sin\omega t \\ -\omega\sin\omega t & \cos\omega t \end{pmatrix} \begin{pmatrix} q_0 \\ p_0 \end{pmatrix} \equiv M_t \begin{pmatrix} q_0 \\ p_0 \end{pmatrix}, \tag{1.5}$$

where $\omega = \sqrt{k/m}$. Notice that $\det M_t = 1$. In consequence, the flow φ_t is an *area-preserving transformation*. Moreover, the system is *time reversible*, $\varphi_t^{-1} = \varphi_{-t}$: inverting the direction of the initial velocity does not change the trajectory, only the direction of motion along this trajectory. In this way $(q_0, -p_0) = \varphi_t(q(t), -p(t))$.

1.1.2 Some elementary numerical methods

The simple harmonic oscillator is a very particular dynamical system, since it can be explicitly solved and its flow possesses many distinctive qualitative properties not shared by other systems. This makes it an excellent test bench for studying the behavior of numerical integration methods. Now we will describe the approximations furnished by some of them.

Explicit Euler method. Without any doubt, the explicit Euler method is the simplest of all numerical methods for the differential equation (1.4). With a constant step size h, the approximation x_{n+1} to $x(t_{n+1})$, with $t_{n+1} = (n+1)h$, is obtained explicitly from x_n as

$$x_{n+1} = x_n + hf(x_n). \tag{1.6}$$

Starting with the initial value $x(0) = x_0$, the scheme computes the sequence of approximations x_1, x_2, etc. to the solution using one evaluation of f per step. Formula (1.6) induces a one-parameter family of maps $\psi_h : x_n \longmapsto x_{n+1}$ called the *numerical flow*.

For the simple harmonic oscillator formulated in the (q, p) variables, equation (1.2), the numerical solution provided by the explicit Euler method reads

$$\begin{pmatrix} q_{n+1} \\ p_{n+1} \end{pmatrix} = \begin{pmatrix} q_n + h\frac{p_n}{m} \\ p_n - hkq_n \end{pmatrix} = \begin{pmatrix} 1 & \frac{h}{m} \\ -hk & 1 \end{pmatrix} \begin{pmatrix} q_n \\ p_n \end{pmatrix} = (I + hA) \begin{pmatrix} q_n \\ p_n \end{pmatrix}, \tag{1.7}$$

where I is the 2×2 identity matrix and A is given by (1.3).

Implicit Euler method. In this scheme, the approximation x_{n+1} is computed implicitly from the relation

$$x_{n+1} = x_n + hf(x_{n+1}). \tag{1.8}$$

In general, a nonlinear algebraic system of equations needs to be solved at each step to determine x_{n+1}. For the harmonic oscillator (1.2), however, one has, after some algebra,

$$\begin{pmatrix} q_{n+1} \\ p_{n+1} \end{pmatrix} = \frac{1}{1+h^2\frac{k}{m}} \begin{pmatrix} 1 & \frac{h}{m} \\ -hk & 1 \end{pmatrix} \begin{pmatrix} q_n \\ p_n \end{pmatrix} = (I - hA)^{-1} \begin{pmatrix} q_n \\ p_n \end{pmatrix}, \quad (1.9)$$

so that in this case it is still possible to get (q_{n+1}, p_{n+1}) explicitly.

Symplectic Euler methods. We consider now the following variations of the explicit Euler method applied to the harmonic oscillator:

$$\begin{pmatrix} q_{n+1} \\ p_{n+1} \end{pmatrix} = \begin{pmatrix} q_n + h\frac{p_{n+1}}{m} \\ p_n - hkq_n \end{pmatrix} = \begin{pmatrix} 1 - h^2\frac{k}{m} & \frac{h}{m} \\ -hk & 1 \end{pmatrix} \begin{pmatrix} q_n \\ p_n \end{pmatrix}, \quad (1.10)$$

which is referred to as the *Symplectic Euler-VT method*, and

$$\begin{pmatrix} q_{n+1} \\ p_{n+1} \end{pmatrix} = \begin{pmatrix} q_n + h\frac{p_n}{m} \\ p_n - hkq_{n+1} \end{pmatrix} = \begin{pmatrix} 1 & \frac{h}{m} \\ -hk & 1 - h^2\frac{k}{m} \end{pmatrix} \begin{pmatrix} q_n \\ p_n \end{pmatrix}, \quad (1.11)$$

called the *Symplectic Euler-TV method.* Notice that the only difference with (1.7) is that for the computation of q_{n+1} in (1.10) we use the already calculated momentum p_{n+1} instead of p_n, whereas in (1.11) the already evaluated coordinate q_{n+1} is used (instead of q_n) to compute p_{n+1}.

All the schemes considered render first-order approximations to the exact solution provided by the matrix M_h in (1.5), in the sense that $\psi_h(x_0) = \varphi_h(x_0) + \mathcal{O}(h^2)$ as $h \to 0$, but there are important differences. Thus, whereas the maps (1.10) and (1.11) are area preserving, as the exact flow (since the determinant of the corresponding matrix is one), this is not the case with (1.7) and (1.9).

To further illustrate this point, let D_0 be a certain domain in $\mathbb{R}^2$ with area $S(D_0)$, and apply one step of the previous methods to any single point in D_0. Then we will end up with another domain, $D_1 = \psi_h(D_0)$, whose area is given by

$$S(D_1) = \int_{D_1} dq_1 dp_1 = \int_{D_0} \left| \frac{\partial(q_1, p_1)}{\partial(q_0, p_0)} \right| dq_0 dp_0 = \psi_h' \, S(D_0),$$

since the Jacobian $\psi_h' \equiv \left| \frac{\partial(q_1,p_1)}{\partial(q_0,p_0)} \right|$ is in fact independent of x_0. It is then clear that after n steps one has for the previous methods

$$\begin{aligned} S(D_n) &= \left(1 + h^2\frac{k}{m}\right)^n S(D_0) && \text{(Explicit Euler);} \\ S(D_n) &= \left(1 + h^2\frac{k}{m}\right)^{-n} S(D_0) && \text{(Implicit Euler);} \\ S(D_n) &= S(D_0) && \text{(Symplectic Euler schemes).} \end{aligned}$$

How does this feature manifest itself in practice? Let us analyze the situation with the following simple domain D_0.

Example 1.1. Given the harmonic oscillator (1.1) with $k = m = 1$, take initial conditions on the sector centered at $(q, p) = (3/2, 0)$ of radius $r = 1/2$ and $\theta \in [-5\pi/6, 5\pi/6]$, so that the domain is denoted as $D_0 = B_{1/2,5\pi/6}(3/2, 0)$. Apply six steps of the explicit and implicit Euler and both symplectic Euler methods with step size $h = \pi/6$ and show the exact and numerical solutions at times $t = 0$, $t = \pi/2$ and $t = \pi$.

Solution. We take a set of initial conditions (q_0, p_0) parametrized as

$$q_0 = \frac{3}{2} + \frac{1}{2}\cos\theta, \qquad p_0 = \frac{1}{2}\sin\theta,$$

for a set of values of θ, say $\theta = -\frac{5\pi}{6} + \frac{5\pi}{3}\frac{i}{N}$, $i = 0, 1, 2, \ldots, N$, and a set of points connecting the first and last point with $(3/2, 0)$. Then we compute the exact and numerical solution for each (q_0, p_0).

The results are shown in Figure 1.1, where the exact solution is given by solid lines: (left) explicit Euler (regions with light shading) and implicit Euler (regions with dark shading); (right) symplectic Euler-VT (1.10) (regions with light shading) and symplectic Euler-TV (1.11) (regions with dark shading). The flow of the field is also shown for convenience (dotted lines). Notice how the area of the transformed domains D_1, D_2 grows for the explicit Euler scheme and decreases for the implicit Euler scheme, in accordance with the previous result, whereas for the symplectic Euler methods, the initial region is only deformed but the area is constant. □

In addition to the area preservation property, there is another distinctive feature of schemes (1.10) and (1.11). This is related to the concept of *backward error analysis*. Loosely speaking, the idea is that the approximation furnished by a numerical integrator can be considered as (almost) the exact solution at discrete times of a modified differential equation. This connection allows one to explore the long-time behavior of the numerical scheme by analyzing the corresponding modified equation. In particular, the analysis of the qualitative properties and performance of a numerical method can be carried out by studying the exact solution of the associated modified differential equation. For the symplectic Euler methods, it turns out that the modified equations can also be derived from (modified) Hamiltonian functions.

Example 1.2. Apply the explicit and implicit Euler methods and both symplectic Euler methods to the harmonic oscillator (1.2) with $k = m = 1$ and initial conditions $(q_0, p_0) = (5/2, 0)$. Take $h = \pi/10$ and integrate in the time interval $t \in [0, 2\pi]$. Compare with the exact solution (1.5) and also with the exact solution of the following differential systems (taking $h = \pi/10$):

$$(i): \left\{ \begin{array}{rcl} \dot{q} & = & p + \frac{h}{2}q \\ \dot{p} & = & -q + \frac{h}{2}p \end{array} \right., \qquad (ii): \left\{ \begin{array}{rcl} \dot{q} & = & p - \frac{h}{2}q \\ \dot{p} & = & -q + \frac{h}{2}p \end{array} \right.,$$

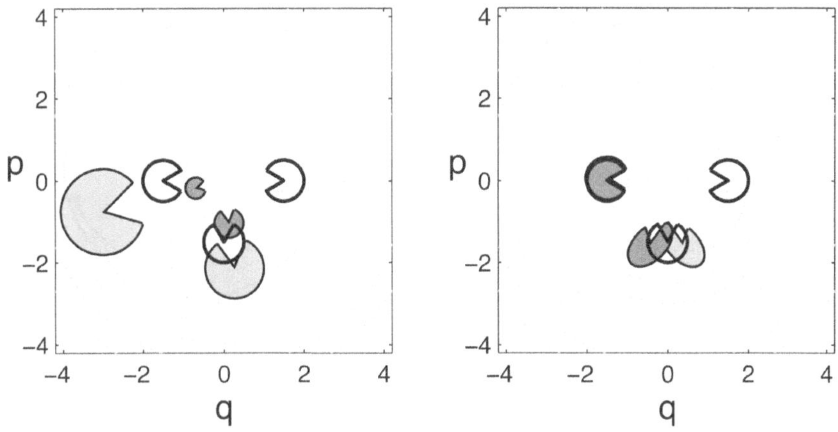

FIGURE 1.1: Evolution of the region $B_{1/2,5\pi/6}(3/2,0)$ through the exact solutions (solid lines) and approximate flows associated to the harmonic oscillator ($k = m = 1$) and several numerical schemes (Example 1.1). Left: explicit Euler (regions with light shading) and implicit Euler (regions with dark shading). Right: symplectic Euler-VT (1.10) (regions with light shading) and symplectic Euler-TV (1.11) (regions with dark shading). The flow of the field is given by dotted lines.

$$(iii): \left\{ \begin{array}{rcl} \dot{q} & = & p - \dfrac{h}{2}q \\ \dot{p} & = & -q - \dfrac{h}{2}p \end{array} \right. , \qquad (iv): \left\{ \begin{array}{rcl} \dot{q} & = & p + \dfrac{h}{2}q \\ \dot{p} & = & -q - \dfrac{h}{2}p \end{array} \right. .$$

Solution. Figure 1.2 shows in addition to the exact solution of the harmonic oscillator (solid line) the results obtained with: (left) the explicit Euler (squares), symplectic Euler-VT (1.10) (circles), exact solution of (i) (thin line) and exact solution of (ii) (dashed line); (right) the implicit Euler (squares), symplectic Euler-TV (1.11) (circles), exact solution of (iii) (thin line) and exact solution of (iv) (dashed line).

Notice that the exact solutions of (i) and (iii) follow very closely the numerical solution obtained by the explicit and the implicit Euler methods, respectively, whereas the solutions of (ii) and (iv) are closed curves which interpolate the results furnished by the symplectic Euler methods (with an error in the phases). □

It is worth remarking that the divergence of the vector field of equation (i), $f(x) = (p + hq/2, -q + hp/2)^T$, verifies $\nabla \cdot f(x) = h/2 + h/2 = h > 0$ and the vector field of equation (iii) verifies $\nabla \cdot f(x) = -h < 0$. In consequence, there is a growth (respectively, a decrease) in the area of the phase space covered by the flow of equation (i) (respect., (ii)). On the other hand, for the

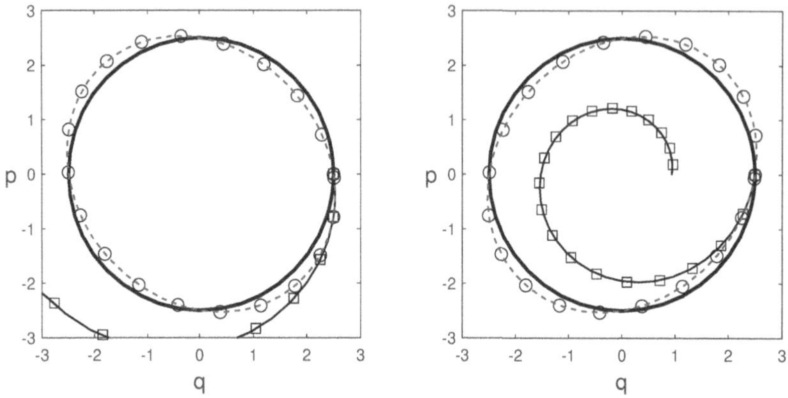

FIGURE 1.2: Solution of the equations of motion of the harmonic oscillator with initial conditions $(q_0, p_0) = (2.5, 0)$ with time step $h = \pi/10$ along the time interval $t \in [0, 2\pi]$ (Example 1.2). Left: explicit Euler (squares), symplectic Euler (1.10) (circles), the exact solution (1.5) (solid line), the exact solution of (*i*) (thin line), and the exact solution of (*ii*) (dashed line). Right: implicit Euler (squares), symplectic Euler (1.11) (circles), the exact solution (1.5) (solid line), the exact solution of (*iii*) (thin line), and the exact solution of (*iv*) (dashed line).

vector fields of equations (*ii*) and (*iv*), one has $\nabla \cdot f(x) = 0$, and then the area is preserved. Moreover, it is straightforward to verify that (*ii*) corresponds indeed to a Hamiltonian system with Hamiltonian $H = \frac{1}{2}(p^2 - hpq + q^2)$, whereas for the numerical approximation (1.10) one has

$$p_{n+1}^2 - hp_{n+1}q_{n+1} + q_{n+1}^2 = p_n^2 - hp_nq_n + q_n^2,$$

for all n. In other words, the quantity $\tilde{E} = \frac{1}{2}(p_n^2 - hp_nq_n + q_n^2)$ is an invariant of the numerical flow if the step size h is kept fixed along the integration. Similarly, (*iv*) corresponds to a Hamiltonian system with Hamiltonian $H = \frac{1}{2}(p^2 + hpq + q^2)$, whereas for the numerical approximation (1.11) one has

$$p_{n+1}^2 + hp_{n+1}q_{n+1} + q_{n+1}^2 = p_n^2 + hp_nq_n + q_n^2,$$

for all n. Thus, although the total energy E is not preserved by the numerical scheme, $\tilde{E} = E + \mathcal{O}(h)$ is a conserved quantity in both cases. □

1.1.3 Simple mathematical pendulum

The mathematical pendulum constitutes a classical example of a nonlinear Hamiltonian system. Denoting by θ the angle from the vertical suspension point, ℓ the constant length of the pendulum, m the mass of the particle, and g the acceleration of gravity, the equation of motion reads

$$m\,\ell\,\frac{d^2\theta}{dt^2} = -mg\sin\theta, \qquad \theta(0) = \theta_0, \quad \dot{\theta}(0) = \dot{\theta}_0.$$

Introducing the new variables $(q, p) = (\theta, \dot{\theta})$, the previous equation can be written as

$$\frac{d}{dt}\begin{pmatrix} q \\ p \end{pmatrix} = \begin{pmatrix} p \\ -k^2 \sin q \end{pmatrix} \tag{1.12}$$

with $k = \sqrt{g/\ell}$. In this case the Hamiltonian reads

$$H = \frac{1}{2}p^2 + k^2(1 - \cos q).$$

It is straightforward to verify that the corresponding flow φ_t is also area preserving [170]. The symplectic Euler-VT scheme for this problem is the map

$$q_{n+1} = q_n + hp_{n+1}, \qquad p_{n+1} = p_n - hk^2 \sin q_n, \tag{1.13}$$

whereas the symplectic Euler-TV scheme corresponds to

$$q_{n+1} = q_n + hp_n, \qquad p_{n+1} = p_n - hk^2 \sin q_{n+1}. \tag{1.14}$$

Next, we analyze its behavior in comparison with the explicit and implicit Euler schemes.

Example 1.3. Consider the equations of the pendulum (1.12) with $k = 1$ and the same initial conditions (q_0, p_0) as in Example 1.1. Apply 6 steps of length $h = \pi/6$ for the explicit and implicit Euler methods and the symplectic Euler methods and plot the exact and numerical solutions at $t = 0$, $t = \pi/2$ and $t = \pi$.

Solution. With a trivial adaptation to this case of the codes used to generate Figure 1.1, it is possible to obtain the results shown in Figure 1.3, where the same notation has been used. In this case the period varies for each closed trajectory. Notice that for the explicit Euler method, one has

$$\left|\frac{\partial(q_{n+1}, p_{n+1})}{\partial(q_n, p_n)}\right| = \left|1 + h^2 \cos q_n\right|,$$

and thus the map expands or contracts the area depending on the regions of the phase space considered. □

Example 1.4. Solve the pendulum equations (1.12) with $k = 1$ and initial conditions $(q_0, p_0) = (5/2, 0)$ using the explicit and implicit Euler schemes and the symplectic Euler methods for $t \in [0, T]$ with $T = 10.5$ and time step $h = T/30$. Compare the results obtained with the exact solution and the exact solution of the following modified differential equations (with $h = T/30$):

$$(i): \left\{\begin{array}{lcl} \dot{q} & = & p + \frac{h}{2}\sin q \\ \dot{p} & = & -\sin q + \frac{h}{2} p \cos q \end{array}\right., \qquad (ii): \left\{\begin{array}{lcl} \dot{q} & = & p - \frac{h}{2}\sin q \\ \dot{p} & = & -\sin q + \frac{h}{2} p \cos q \end{array}\right.,$$

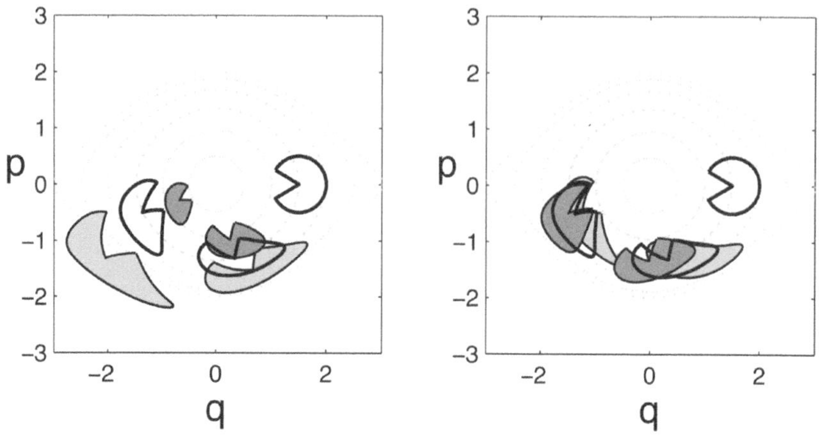

FIGURE 1.3: Same as Figure 1.1 for the pendulum problem (1.12), Example 1.3.

$$(iii): \left\{\begin{array}{rcl} \dot{q} & = & p - \frac{h}{2}\sin q \\ \dot{p} & = & -\sin q - \frac{h}{2} p\cos q \end{array}\right., \qquad (iv): \left\{\begin{array}{rcl} \dot{q} & = & p + \frac{h}{2}\sin q \\ \dot{p} & = & -\sin q - \frac{h}{2} p\cos q \end{array}\right..$$

Solution. Figure 1.4 shows, in addition to the exact solution[1] (solid line): (left) the results obtained with the explicit Euler (squares), symplectic Euler-VT (1.13) (circles), exact solution of (*i*) (thin line), and exact solution of (*ii*) (dashed line); (right) the results obtained with the implicit Euler (squares), symplectic Euler-TV (1.14) (circles), exact solution of (*iii*) (thin line) and exact solution of (*iv*) (dashed line).

Notice that, similarly to the harmonic oscillator, the exact solutions of (*i*) and (*iii*) nearly overlap the interpolating curves to the numerical solution obtained by the explicit and the implicit Euler methods, respectively, whereas the solutions of (*ii*) and (*iv*) are closed curves which are nearly the interpolation curves connecting the points from the symplectic Euler methods. We will return to this issue in Chapter 5.

Now the divergence of the vector field of equation (*i*) is $\nabla \cdot f(x) = h\cos q$, whereas for the vector field of (*iii*) reads $\nabla \cdot f(x) = -h\cos(q)$. Again, there is an increase and/or a decrease in the area of the region of the phase space. On the other hand, for the vector fields of equations (*ii*) and (*iv*), one has $\nabla \cdot f(x) = 0$, and then the area is preserved. Moreover, (*ii*) corresponds to a Hamiltonian system with $H = \frac{1}{2}p^2 - \frac{h}{2}p\sin q + (1-\cos q)$, whereas for the numerical approximation (1.13), one has

$$\frac{1}{2}p_n^2 - \frac{h}{2}p_n\sin q_n + (1-\cos q_n) = \frac{1}{2}p_0^2 - \frac{h}{2}p_0\sin q_0 + (1-\cos q_0) + \mathcal{O}(h^2),$$

[1] We refer here as *exact solution* as the solution computed numerically to high accuracy with the function `ode45` from MATLAB.

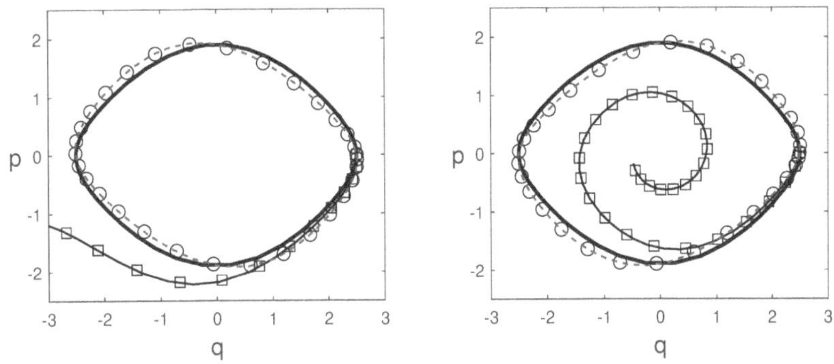

FIGURE 1.4: Same as Figure 1.2 for the pendulum problem (1.12), Example 1.4.

for all n, with the term $\mathcal{O}(h^2)$ bounded for sufficiently small h. In other words, there is a quantity close to $H = \frac{1}{2}p^2 - \frac{h}{2}p\sin q + (1-\cos q)$ that is an invariant of the numerical flow if the step size h is kept fixed along the integration. Similarly, (iv) corresponds to a Hamiltonian system with Hamiltonian $H = \frac{1}{2}p^2 + \frac{h}{2}p\sin q + (1-\cos q)$. Again, although the total energy E is not preserved by the numerical scheme, there is a conserved quantity close to this value. □

1.2 Classical paradigm of numerical integration

Obviously, there is not much point in designing numerical integrators for the harmonic oscillator and the simple pendulum, since for both models the exact solution is available in closed form; that is, one has analytic formulae expressing $q(t)$ and $p(t)$ as functions of time. This, however, is the exception rather than the general rule. Given a generic initial value problem

$$\frac{dx}{dt} = f(t,x), \qquad x(0) = x_0 \in D \subset \mathbb{R}^d \tag{1.15}$$

with f satisfying the usual requirements to guarantee existence and uniqueness of solution, it is most uncommon to find such closed form expressions for its unique solution. This being the case, one is then forced to seek approximations by means of numerical techniques. As we said at the beginning of the chapter, with the widespread use of numerical simulations on digital computers, new methods for integrating differential equations were created and thoroughly analyzed, giving rise to several sophisticated general-purpose algorithms integrated in commercial numerical libraries. This allowed scientists and practitioners alike to routinely solve problems, either "general" or "stiff," in a reliable and efficient way. All these developments have been summarized in a number of monographs over the years, particularly in [171], and form what Sanz-Serna calls "the classical paradigm" in the numerical solution of ordinary differential equations [286].

The main constituents of this classical paradigm, according to [286], are the following:

- *The goal* is to find as cheaply as possible and with a given accuracy the values $x(t_i)$ at some prescribed points where output is desired.
- *The tool* to achieve that goal is one package (or several packages) to deal with general or stiff equations, in which the subroutine evaluating the corresponding function f is inserted by the user.
- *The theoretical framework* to design and understand the methods used in the package includes the analysis of consistency, local error bounds, error constants, stability and error propagation.

Let us briefly summarize each of the main aspects of this approach. The idea of step-by-step numerical methods consists of dividing the integration interval $[t_0, t_{\rm f}]$ where the approximation is to be constructed by the mesh points $t_n = t_0 + nh$, $n = 0, 1, \ldots, N$, $h = (t_{\rm f} - t_0)/N$, with $t_N = t_{\rm f}$ and N a positive integer. Then one determines approximate values x_n for $x(t_n)$, the value of the analytical solution at t_n. These values are calculated in succession, for $n = 1, 2, \ldots, N$. The approximation x_n depends, of course, on the *step size* h used, which, in turn, can be made to vary along the integration.

We will consider *one-step* methods: x_{n+1} is expressed in terms of only the previous value x_n. As stated before, the simplest one-step integrator is the Euler method

$$x_{n+1} = x_n + h\, f(t_n, x_n). \tag{1.16}$$

It computes approximations x_{n+1} to the values $x(t_{n+1})$ of the solution using one explicit evaluation of f at the already computed value x_n.

More generally, a one-step method is expressed as

$$x_{n+1} = x_n + h\, \Phi(t_n, x_n; h), \qquad n = 0, 1, \ldots, N-1, \qquad x(t_0) = x_0 \tag{1.17}$$

for a given function Φ. Each one-step numerical method induces a one-parameter family ψ_h of maps $\psi_h : \mathbb{R} \times \mathbb{R}^d \longrightarrow \mathbb{R}^d$ in such a way that $\psi_h(t_0, x_0)$ is the numerical solution after one step of size h starting from the initial condition x_0. For the Euler method applied to the autonomous equation $\dot{x} = f(x)$ one has $\psi_h^E(x) = x + hf(x)$.

The *global error* of method (1.17) is defined as $e_n = x(t_n) - x_n$, the difference between the exact and the numerical solution at the grid point t_n, whereas the *truncation error* (also called *local error*) is given by

$$T_n = \frac{x(t_{n+1}) - x(t_n)}{h} - \Phi(t_n, x(t_n); h). \tag{1.18}$$

Notice that hT_n is nothing but the difference of both sides of (1.17) when the exact solution is inserted in the formula. It is a trivial exercise to verify that for the explicit Euler method, one has $|T_n| \leq Kh$ for $0 < h \leq h_0$, with K independent of h [306]. As a matter of fact, for each method (1.17), one may determine its corresponding truncation error, and this in turn can be used to get bounds on the magnitude of the global error.

The method is said to be *consistent* if $\Phi(t, x; 0) \equiv f(t, x)$ [208]. If the one-step scheme (1.17) is consistent, with some (mild) additional conditions on Φ and f, it can be shown that

$$\lim_{n\to\infty} x_n = x(t) \quad \text{as} \quad t_n \to t \in [t_0, t_{\mathrm{f}}] \quad \text{when} \quad h \to 0 \quad \text{and} \quad n \to \infty,$$

in which case the method is *convergent* [306]. The numerical method is said to be of *order of accuracy* r if there are constants K and h_0 such that

$$|T_n| \leq K\, h^r \qquad \text{for } 0 < h \leq h_0$$

or alternatively if $|x(t_{n+1}) - x(t_n) - h\Phi(t_n, x(t_n); h)| = \mathcal{O}(h^{r+1})$. In other words, when the exact solution is plugged into the numerical scheme, the corresponding expression is of order $\mathcal{O}(h^{r+1})$. Thus, the Euler method is of order 1, whereas Heun's method

$$x_{n+1} = x_n + \frac{h}{4}\left(f(t_n, x_n) + 3f\big(t_n + 2h/3, x_n + (2h/3) f(t_n, x_n)\big)\right) \tag{1.19}$$

is of order 2. For a scheme of order r, the global error $e_n = \mathcal{O}(h^r)$ uniformly in bounded time-intervals under the above mild conditions. Equivalently, if φ_t denotes the exact flow, the integrator is of order r if for all $x \in \mathbb{R}^d$ and all smooth f it is true that

$$\psi_h(t_n, x_n) = \varphi_h(t_n, x_n) + \mathcal{O}(h^{r+1})$$

as $h \to 0$. The extension of these concepts to variable step lengths h_n is straightforward.

One may distinguish between *explicit* and *implicit* one-step methods. Expression (1.17) corresponds indeed to an explicit method. If, on the other hand, the scheme reads

$$x_{n+1} = x_n + h\,\Phi(t_{n+1}, t_n, x_{n+1}, x_n; h),$$

then it is implicit. The simplest example corresponds, of course, to the first-order implicit Euler method

$$x_{n+1} = x_n + h\, f(t_{n+1}, x_{n+1}) \equiv \psi_h^I(x_n), \tag{1.20}$$

whereas the formula

$$x_{n+1} = x_n + \frac{h}{2}\left(f(t_n, x_n) + f(t_{n+1}, x_{n+1})\right) \equiv \psi_h^T(x_n), \tag{1.21}$$

known as the *trapezoidal rule*, provides a second-order approximation [208]. Another example of a second-order method is the *implicit midpoint rule*

$$x_{n+1} = x_n + hf\left(t_n + \frac{h}{2}, \frac{x_{n+1} + x_n}{2}\right) \equiv \psi_h^M(x_n). \tag{1.22}$$

More elaborate and efficient general-purpose algorithms, using several evaluations of f per step, have been proposed over the years for the numerical treatment of equation (1.15). Among them, the Runge–Kutta (RK) class of methods is possibly the most frequently used, in particular the classical explicit fourth-order scheme. We will analyze this class of integrators in more detail in Chapter 2.

1.2.1 Adjoint method, symmetric method

We will see that *time-symmetric* (or just *symmetric*) methods have several appealing properties. In close connection with symmetry is the concept of the *adjoint* of a method. The flow φ_t of the autonomous system $\dot{x} = f(x)$ verifies $\varphi_{-t}^{-1} = \varphi_t$, but this property is not shared by the corresponding map ψ_h of many numerical methods.

In general, if $\psi_h(x)$ represents a numerical method of order at least one, i.e., $\psi_h(x) = x + h\, f(x) + \mathcal{O}(h^2)$, then also $\psi_{-h}^{-1}(x) = x + h\, f(x) + \mathcal{O}(h^2)$, so that

$$\psi_h^* \equiv \psi_{-h}^{-1}$$

is a numerical method of order at least one. It is called the *adjoint method* of ψ_h [293]. In words, *stepping forward with the given method ψ_h is the same as stepping backward with the inverse of its adjoint ψ_h^**. Whenever an integrator satisfies

$$\psi_h^* = \psi_h,$$

it is called a *symmetric method*. Thus, a symmetric method verifies $\psi_h = \psi_{-h}^{-1}$. According to this definition, the scheme $x_{n+1} = \psi_h(x_n)$ is symmetric if and only if exchanging $h \leftrightarrow -h$ and $x_n \leftrightarrow x_{n+1}$ we achieve the same expression, i.e., $\psi_{-h}(x_{n+1}) = x_n$. It turns out that, using the adjoint, it is straightforward to construct symmetric methods: given an arbitrary method ψ_h of order $r \geq 1$, then the compositions

$$\psi_{h/2} \circ \psi_{h/2}^* \quad \text{and} \quad \psi_{h/2}^* \circ \psi_{h/2} \tag{1.23}$$

are symmetric methods of order $r \geq 2$ [293]. Moreover, it can be shown that symmetric methods are necessarily of even order.

Example 1.5. Given the autonomous equation $\dot{x} = f(x)$, we next construct the adjoint method associated with the explicit Euler method $x \mapsto y = \psi_h^E(x) = x + hf(x)$. It is clear then that $x = y - hf(x)$, or (exchanging $x \leftrightarrow y$) $y = (\psi_h^E)^{-1}(x) = x - hf(y)$, and thus

$$y = (\psi_h^E)^*(x) = (\psi_{-h}^E)^{-1}(x) = x + hf(y)$$

which corresponds to the map associated with the implicit Euler method ψ_h^I. In consequence, $\psi_h^I = (\psi_h^E)^*$.

On the other hand, the second-order compositions (1.23) formed with the explicit Euler method can be obtained as follows. Taking $y = \psi_{h/2}^E(x) = x + (h/2)f(x)$, then

$$z = \left((\psi_{h/2}^E)^* \circ \psi_{h/2}^E\right)(x) = (\psi_{h/2}^E)^*(y) = y + \frac{h}{2}f(z) = x + \frac{h}{2}(f(x) + f(z)),$$

which corresponds to the map for the trapezoidal rule ψ_h^T.

If we take instead $y = (\psi_{h/2}^E)^*(x) = x + h/2f(y)$, then

$$z = \left(\psi_{h/2}^E \circ (\psi_{h/2}^E)^*\right)(x) = \psi_{h/2}^E(y) = y + \frac{h}{2}f(y) = x + hf(y) = x + hf\left(\frac{x+z}{2}\right)$$

(where we have eliminated f from the equations $y = x + h/2f(y)$ and $z = y + \frac{h}{2}f(y)$), i.e. $y = (x+z)/2$, which corresponds to the map for the implicit midpoint rule ψ_h^M, equation (1.22). □

1.2.2 Stability

Another important concept associated with the numerical treatment of ordinary differential equations is *stability*. Intuitively, a method is stable if the corresponding numerical solution is bounded when the exact solution does not tend to infinity. In other words, the method does not magnify the differences $x_n - x(t_n)$ as t_n increases, irrespective of the time step at which the error is produced. There are several ways to express this stability condition. One of the strongest forms is A-stability [187], determined by the way the method behaves when it is applied to the simple scalar linear equation (or Dahlquist test problem)

$$\frac{dx}{dt} = -\lambda x, \tag{1.24}$$

with $\lambda > 0$, and initial condition $x(t_0) = x_0$. The exact solution reads $x(t) = x_0\,\mathrm{e}^{-\lambda t}$ and approaches zero as t increases. It is said that the numerical method is *A-stable* if, when applied to (1.24), it produces a bounded solution irrespective of λ and the step size h. If boundedness occurs only when h is small, then the method is conditionally A-stable. Otherwise, the method is unstable.

For equation (1.24), the explicit Euler method produces $x_{n+1} = (1-\lambda h)x_n$. Taking into account the initial condition, then

$$x_n = x_0(1 - \lambda h)^n,$$

so that the numerical solution is bounded as n increases only as long as $|1 - \lambda h| \leq 1$. This happens when the step size is chosen so as to satisfy the condition $h \leq 2/\lambda$, and thus the Euler method is conditionally A-stable. If h

exceeds $2/\lambda$, the numerical solution will oscillate with increasing magnitude with fixed h as $n \to \infty$, instead of converging to zero. On the other hand, for the implicit Euler method (1.20), one finds

$$x_{n+1} = \frac{1}{1 + \lambda h} x_n,$$

so that $x_n = x_0(1 + \lambda h)^{-n}$. Since the numerical solution goes to zero as n increases irrespective of the value of h, this method is A-stable. For other higher-order methods the analysis is, of course, more involved.

1.2.3 Stiffness

Although several rigorous definitions of stiffness can be found in the literature [174, 208] (sometimes not without controversy [302]), perhaps it is more appropriate at this point to provide an intuitive notion without entering into much detail. Thus, according to Iserles [193], one may say that the differential equation $\dot{x} = f(t, x)$ with $x(t_0) = x_0$ is *stiff* if the numerical solution obtained by some method requires (at least in a portion of the integration interval) a significant reduction in the step size to avoid instability. In practice, generic/classical explicit methods do not work for stiff problems. To illustrate the phenomenon of stiffness, consider a linear system of differential equations with constant coefficients

$$\dot{x} = A\,x, \qquad x(t_0) = x_0,$$

where A is a $d \times d$ matrix whose eigenvalues λ_j, $j = 1, \ldots, d$ are all real and negative. Then, it is clear that the exact solution verifies

$$\lim_{t \to +\infty} \|x(t)\| = 0,$$

whereas the same behavior holds for the numerical solution obtained by the explicit Euler method with a fixed value of h only if $h|\lambda_j| < 2$, $j = 1, \ldots, d$. If, in addition, the largest eigenvalue is much bigger than the smallest one – for example, if $|\lambda_1|$ is much greater than all the others – then λ_1 will determine how small the step size h must be, even though its influence on the actual solution may be minimal for moderate values of t.

For nonlinear problems, it is not an easy task to give a precise definition of stiffness. System $\dot{x} = f(t, x)$ exhibits stiffness if the eigenvalues of the Jacobian matrix $\partial f_i / \partial x_j$ behave in a similar fashion, although in this case the eigenvalues are no longer constant but depend on the solution and therefore vary with t [208].

Very often, stiffness is present when there are vastly different time scales in the problem, for instance, in equations modeling chemical reactions with different rates of evolution. In this case, the reaction rates correspond to the eigenvalues of the Jacobian matrix, and the ratio of the largest to the smallest eigenvalues frequently exceeds 10^6. This is the case, in particular, of the Robertson chemical reaction system [174], treated in more detail in Chapter 4 (*Example 4.10*). In such problems, of course, it is essential to use a numerical method without restrictions on the step size concerning stability. For a comprehensive treatment of this issue, the reader is referred to [125, 174].

1.3 Towards a new paradigm: Geometric numerical integration

The classical approach has allowed the construction of whole families of general purpose explicit/implicit, one-step/multistep methods of high orders of accuracy, which have been implemented and incorporated into excellent packages, so that the user only has to plug in a specific routine to evaluate the function f of his/her particular problem, no matter its distinctive features (beyond the distinction general/stiff). In spite of that, several distinguished contributors in the field of numerical analysis of differential equations already pointed out the importance of incorporating into the algorithm whatever special structure the exact equation may possess [176, 209]. This fact has been strikingly displayed by the examples examined in section 1.1.2: both for the harmonic oscillator and the simple pendulum, the first-order numerical methods preserving areas in the phase plane offered a much better qualitative and quantitative description than the explicit Euler method.

These examples belong to the more general class of mechanical systems whose equation of motion is obtained from Newton's second law. If one has a force F depending only on the position that acts on a particle of mass m and $y(t)$ designates this position, then

$$m\frac{d^2y}{dt^2} = F(y). \tag{1.25}$$

As is well known, if the force can be derived from a potential, $F = -\nabla V(y)$, then the energy of the system

$$E = \frac{1}{2}m\dot{y}^2 + V(y) \tag{1.26}$$

is constant along the evolution, so that it seems reasonable to design a numerical scheme providing approximations that come very close to keeping this energy preserved. One possible way to do that is first rewriting the second-order equation (1.25) as a first-order system by introducing the velocity $v = \dot{y}$:

$$\dot{y} = v \qquad\qquad \dot{v} = \frac{1}{m}F(y)$$

and then applying the trapezoidal rule (1.21) to both equations,

$$y_{n+1} = y_n + \frac{h}{2}(v_{n+1} + v_n), \qquad v_{n+1} = v_n + \frac{h}{2m}\big(F(y_{n+1}) + F(y_n)\big). \quad (1.27)$$

This results in an implicit scheme that preserves the energy exactly for the harmonic oscillator. An explicit method can be obtained simply by replacing v_{n+1} in the first equation of (1.27) by the approximation rendered by Euler's method, $v_{n+1} = v_n + hF(y_n)/m$ [187], so that the resulting scheme reads

$$\begin{aligned} y_{n+1} &= y_n + hv_n + \frac{h^2}{2m}F(y_n) \\ v_{n+1} &= v_n + \frac{h}{2m}\big(F(y_{n+1}) + F(y_n)\big). \end{aligned} \quad (1.28)$$

This is the so-called *velocity Verlet* method for solving (1.25). It requires one force evaluation per step, but it is only second-order accurate and conditionally stable for linear forces and small time steps. In spite of this, it does a good job in approximating the energy over long-time intervals. As a matter of fact, for many years this has been the method of choice in molecular dynamics [217]. There, in order to carry out simulations, Newton's second law has to be integrated for the motion of atoms in molecules, and the number of atoms could be as high as 100000, with six differential equations per atom. Under such circumstances, rather than accurately determine the position of each atom (something clearly unfeasible), the aim is to obtain information on macroscopic quantities like average energies, conformational distributions, etc., and it is in this setting where the Verlet method clearly supersedes other classical schemes.

The reason for the success of the Verlet method in molecular dynamics does not reside certainly in its outstanding accuracy or stability, and so one has to look beyond the classical consistency/stability approach.

There are other problems where the approximation rendered by general-purpose schemes based on the traditional ideas of consistency, stability, or order is qualitatively meaningless. As in [194], suppose one has to solve numerically the matrix differential equation

$$\dot{Y} = [A, Y] \equiv AY - YA, \qquad Y(t_0) = Y_0 \in \mathbb{R}^{d\times d}, \quad (1.29)$$

with A a skew-symmetric $d \times d$ matrix. The solution is given by

$$Y(t) = \mathrm{e}^{tA}\, Y_0\, \mathrm{e}^{-tA} = \mathrm{e}^{tA}\, Y_0\, \mathrm{e}^{tA^T}. \quad (1.30)$$

Since $Y(t)$ and $Y(0)$ are orthogonally similar, they have the same eigenvalues, whereas the numerical approximations Y_n obtained by a numerical scheme of the form (1.17) will not, in general, share this property.

More generally, consider equation (1.29), but now with an explicitly time-dependent skew-symmetric $d \times d$ matrix $A(t)$. If Y_0 is a symmetric matrix, then the exact solution can be factorized as $Y(t) = Q(t)Y_0Q^T(t)$, with $Q(t)$ an orthogonal matrix satisfying the equation

$$\dot{Q} = A(t)Q, \qquad Q(0) = I \tag{1.31}$$

and $Y(t)$ and $Y(0)$ have again the same eigenvalues. As in the previous case, when a numerical scheme of the form (1.17) is applied to (1.31), in general, the approximations Q_n will no longer be orthogonal matrices, and therefore Y_n and Y_0 will not be orthogonally similar. As a result, the *isospectral* character of the system (1.29) is lost in the numerical description, i.e., the eigenvalues of Y_n and Y_0 will no longer be the same. This is already evident for the explicit Euler scheme and Heun's method (1.19) and is generally true for the class of explicit Runge–Kutta methods [194]. What one would really like is to have a specific numerical integrator for equation (1.29) such that the corresponding numerical approximations still preserve the main qualitative feature of the exact solution, namely its isospectral character, and, furthermore, in a computationally efficient way [82].

It has been in trying to address this type of problem that the field of *geometric numerical integration* has emerged since the late 1980s. Here, rather than taking primarily into account such prerequisites as consistency and stability, the aim is to reproduce the qualitative features of the solution of the differential equation which is being discretized, in particular its geometric properties. The motivation for developing such structure-preserving algorithms arises independently in areas of research as diverse as celestial mechanics, molecular dynamics, control theory, particle accelerator physics, and numerical analysis [170, 194, 218, 243, 244]. Although diverse, the systems appearing in these areas have one important common feature: they all preserve some underlying geometric structure which influences the qualitative nature of the phenomena they produce. In the field of geometric numerical integration, these properties are built into the numerical method, which gives the method an improved qualitative behavior but also allows for a significantly more accurate long-time integration than with general-purpose methods [196].

A very important consideration is that, while the classical paradigm is aimed at creating general-purpose software, in geometric numerical integration the interest often lies in integrators tailored to the problem or class of problems at hand.

1.4 Symplectic integration

The first and perhaps most familiar examples of geometric integrators are symplectic integration algorithms in classical Hamiltonian dynamics, as these systems are ubiquitous in applications. In particular, they appear in classical and quantum mechanics, optics, condensed matter physics, molecular dynamics, and celestial mechanics.

Much attention will be devoted within this book to symplectic integrators specifically designed for the numerical treatment of Hamiltonian systems. For this reason we believe it is convenient to summarize some of the main properties of Hamiltonian dynamical systems. Only this way will we be able to identify their most salient features and eventually to incorporate them into the numerical integrators. More comprehensive treatments can be found in the classical references [2, 17, 159, 271], among others.

1.4.1 Crash course on Hamiltonian dynamics: Hamiltonian systems

Let us consider in general a system with d degrees of freedom and phase space variables $x = (q,p)^T = (q_1, \ldots, q_d, p_1, \ldots, p_d)^T$, where (q_i, p_i), $i = 1, \ldots, d$ is the usual pair of canonical conjugate coordinate and momentum, respectively. The coordinates q and momenta p define in the $2d$-dimensional phase space a point (q,p) representing the state of the system. Here we follow the usual notation of classical mechanics books by introducing first the coordinates and then the momenta [271, 316, 326]. Given the Hamiltonian function $H(q,p)$ defined on $D \subset \mathbb{R}^d \times \mathbb{R}^d$, the equations of motion are formulated as

$$\dot{q}_i = \frac{\partial H}{\partial p_i}, \qquad \dot{p}_i = -\frac{\partial H}{\partial q_i}, \qquad i = 1, \ldots, d. \tag{1.32}$$

This Hamiltonian system can be written in a more compact notation as

$$\dot{x} = J \nabla_x H(x), \tag{1.33}$$

where $x = (q,p)^T \in \mathbb{R}^{2d}$, J is the basic canonical matrix

$$J = \begin{pmatrix} O_d & I_d \\ -I_d & O_d \end{pmatrix}, \tag{1.34}$$

I_d is the $d \times d$ identity matrix, O_d is the corresponding zero matrix, and

$$\nabla_x = \left(\frac{\partial}{\partial x_1}, \ldots, \frac{\partial}{\partial x_{2d}} \right)^T.$$

Both the simple harmonic oscillator and the mathematical pendulum of section 1.1.2 are 1-degree-of-freedom Hamiltonian systems. The matrix J verifies the following properties: (i) $J^2 = -I_{2d}$ or $J^{-1} = -J$; (ii) $\det(J) = 1$; (iii) $J^T = -J$; (iv) $J^T J = I_{2d}$.

By introducing the Poisson bracket of two scalar functions $F(q,p)$ and $G(q,p)$ of phase space variables [316] as

$$\{F, G\} \equiv \sum_{i=1}^{d} \left(\frac{\partial F}{\partial q_i} \frac{\partial G}{\partial p_i} - \frac{\partial F}{\partial p_i} \frac{\partial G}{\partial q_i} \right),$$

we have

$$\{F,G\} = \sum_{i,j=1}^{2d} \frac{\partial F}{\partial x_i} J_{ij} \frac{\partial G}{\partial x_j} \equiv (\nabla F)^T J\,(\nabla G). \tag{1.35}$$

In particular,

$$\{q_i,q_j\} = \{p_i,p_j\} = 0, \qquad \{q_i,p_j\} = \delta_{ij}, \quad \text{with} \quad \delta_{ij} = \begin{cases} 1 & i=j \\ 0 & i \neq j. \end{cases} \tag{1.36}$$

Equivalently, we can write $\{x_i,x_j\} = J_{ij}$ in terms of $x = (q,p)^T$ and the canonical matrix (1.34). The Poisson bracket has several algebraic properties of interest. It is linear,

$$\{aF + bG, H\} = a\{F,H\} + b\{G,H\},$$

for all functions F, G, H and constants a, b. It is antisymmetric,

$$\{F,G\} = -\{G,F\},$$

and verifies the *Jacobi identity*:

$$\{F,\{G,H\}\} + \{G,\{H,F\}\} + \{H,\{F,G\}\} = 0. \tag{1.37}$$

In particular, the set of all functions of the variables q,p,t (which forms a linear space) with the Lie product of any two functions defined as the Poisson bracket forms a Lie algebra. This is sometimes called the Lie algebra of Hamiltonian functions (see Appendix A.2).

Notice that the equations of motion (1.32) can be written in terms of the Poisson bracket simply as

$$\dot{q}_i = \{q_i,H\}, \qquad \dot{p}_i = \{p_i,H\}, \qquad i = 1,\ldots,d,$$

or $\dot{x}_i = \{x_i,H\}$. As a matter of fact, for any function $F(q,p,t)$ it is true that

$$\frac{dF}{dt} = \sum_{i=1}^{d} \left(\frac{\partial F}{\partial q_i}\dot{q}_i + \frac{\partial F}{\partial p_i}\dot{p}_i \right) + \frac{\partial F}{\partial t} = \{F,H\} + \frac{\partial F}{\partial t}. \tag{1.38}$$

In consequence, if the Hamiltonian function $H(q,p)$ does not depend explicitly on time, then

$$\frac{dH}{dt} = 0,$$

so that it is constant along solutions. In systems of the form $H(q,p) = T(p) + V(q)$ this corresponds to the principle of conservation of mechanical energy: kinetic energy $T(p)$ plus potential energy $V(q)$.

Moreover, the flow φ_t of a Hamiltonian system also preserves volume in phase space according to Liouville's theorem [16]: for any region D in the phase space, one has

$$\text{volume of } \varphi_t(D) = \text{ volume of } D.$$

This, in fact, is a consequence of the most distinctive feature of a Hamiltonian system: its flow φ_t is a *symplectic* transformation, which means that its Jacobian matrix $\varphi_t'(x)$ verifies

$$\varphi_t'(x)^T J \, \varphi_t'(x) = J \qquad \text{for } t \geq 0, \tag{1.39}$$

or, in virtue of the properties of the canonical matrix J,

$$\varphi_t'(x) J \, \varphi_t'(x)^T = J.$$

Notice that in general the matrix $\varphi_t'(x)$ depends on x and t. However, the particular combination $\varphi_t'(x)^T J \, \varphi_t'(x)$ is x and t independent. Therefore, such a map surely has very special properties. In fact, the symplecticity of the flow is related to the existence in phase space of a non-degenerate closed differential two-form, so that the phase space is a symplectic manifold [2, 16]. In connection with this two-form there are the so-called Poincaré integral invariants, which are also preserved by the evolution. If the kth Poincaré integral invariant is integrated over an arbitrary domain of dimension $2k$ ($1 \leq k \leq d$), one obtains an invariant which is proportional to the sum of oriented areas of the projections onto the spaces $(q_{i_1}, \ldots, q_{i_k}, p_{i_1}, \ldots, p_{i_k})$, with $1 \leq i_m \leq d$ [316]. In particular, when $k = d$ we recover Liouville's theorem on the preservation of phase space volume. If $d = 1$ this is equivalent to area preservation.

The symplectic character of the Hamiltonian flow thus places very stringent conditions on the global geometry of the corresponding dynamics. In consequence, it makes sense to consider, when carrying out simulations of Hamiltonian systems, numerical schemes that do respect these restrictions.

1.4.2 Symplectic integrators of first and second-order

Symplectic Euler methods

All the characteristic properties of Hamiltonian systems above enumerated have motivated the search for numerical integrators that preserve them, and more specifically their symplectic character, since all traditional methods lead to maps that are not symplectic (even when in some cases the property of energy conservation is built into them). This failure was already identified in 1956 by de Vogelaere in his pioneering paper [123][2], when he stated that the worst effect of the errors introduced by the method of integration "*is probably to destroy the contact transformation property*" of the exact motion of protons in particle accelerators. In consequence, he devoted himself to the task of designing "*a method of integration which, if there was no round-off error, would give a solution with the contact transformation property.*" Here "contact transformation" has to be understood as "symplectic transformation."

[2]See also [301] for the context of the work and the preprint itself, typeset in LaTeX.

For a general Hamiltonian $H(q,p)$, de Vogelaere proposed the following scheme for numerically solving equations (1.32):

$$q_{n+1} = q_n + h\,\nabla_p H(q_{n+1}, p_n), \qquad p_{n+1} = p_n - h\,\nabla_q H(q_{n+1}, p_n). \quad (1.40)$$

We notice at once that the q-variable is treated by the implicit Euler method and the p-variable by the explicit Euler method. Of course, the treatment of both variables can be interchanged, thus resulting in the method

$$q_{n+1} = q_n + h\,\nabla_p H(q_n, p_{n+1}), \qquad p_{n+1} = p_n - h\,\nabla_q H(q_n, p_{n+1}). \quad (1.41)$$

The proof that both schemes are symplectic is straightforward. We consider only method (1.40) and $d = 1$ for simplicity. Differentiating with respect to q_n, p_n yields

$$\begin{aligned}
\frac{\partial q_{n+1}}{\partial q_n} &= 1 + hH_{pq}\frac{\partial q_{n+1}}{\partial q_n}; \qquad \frac{\partial q_{n+1}}{\partial p_n} = h\left(H_{pq}\frac{\partial q_{n+1}}{\partial p_n} + H_{pp}\right)\\
\frac{\partial p_{n+1}}{\partial q_n} &= -hH_{qq}\frac{\partial q_{n+1}}{\partial q_n}; \qquad \frac{\partial p_{n+1}}{\partial p_n} = 1 - h\left(H_{qq}\frac{\partial q_{n+1}}{\partial p_n} + H_{qp}\right),
\end{aligned}$$

where H_{qq}, H_{qp}, H_{pp} denote second partial derivatives evaluated at (q_{n+1}, p_n). Working out these relations, we can compute the Jacobian matrix

$$\mathcal{N} \equiv \begin{pmatrix} \dfrac{\partial q_{n+1}}{\partial q_n} & \dfrac{\partial q_{n+1}}{\partial p_n} \\ \dfrac{\partial p_{n+1}}{\partial q_n} & \dfrac{\partial p_{n+1}}{\partial p_n} \end{pmatrix} \quad (1.42)$$

and check out that, indeed, $\mathcal{N}^T J \mathcal{N} = J$.

Integrators (1.40) and (1.41) are both of order 1 and can be appropriately called *symplectic Euler methods*. In fact, the adjoint of (1.40) is precisely method (1.41). These schemes are implicit for general Hamiltonian systems, but if H has the separable form $H(q,p) = T(p) + V(q)$, then they are explicit. Specifically, method (1.40) reads

$$q_{n+1} = q_n + h\nabla_p T(p_n), \qquad p_{n+1} = p_n - h\nabla_q V(q_{n+1}), \quad (1.43)$$

whereas scheme (1.41) is given by

$$p_{n+1} = p_n - h\nabla_q V(q_n), \qquad q_{n+1} = q_n + h\nabla_p T(p_{n+1}). \quad (1.44)$$

We note in passing that the symplectic Euler schemes (1.10) and (1.13) are particular examples of method (1.44), whereas (1.11) and (1.14) belong to the class (1.43).

Störmer–Verlet scheme

Taking ψ_h as the map corresponding to the symplectic Euler method (1.41), i.e.,

$$\psi_h: \begin{aligned} q_{n+1} &= q_n + h\,\nabla_p H(q_n, p_{n+1}),\\ p_{n+1} &= p_n - h\,\nabla_q H(q_n, p_{n+1}), \end{aligned}$$

and composing $\psi_{h/2} \circ \psi^*_{h/2}$ as in (1.23), yields the scheme

$$\begin{aligned} q_{n+1/2} &= q_n + \frac{h}{2} \nabla_p H(q_{n+1/2}, p_n) \\ p_{n+1} &= p_n - \frac{h}{2} \big(\nabla_q H(q_{n+1/2}, p_n) + \nabla_q H(q_{n+1/2}, p_{n+1}) \big) \\ q_{n+1} &= q_{n+1/2} + \frac{h}{2} \nabla_p H(q_{n+1/2}, p_{n+1}), \end{aligned} \tag{1.45}$$

whereas the composition $\psi^*_{h/2} \circ \psi_{h/2}$ gives

$$\begin{aligned} p_{n+1/2} &= p_n - \frac{h}{2} \nabla_q H(q_n, p_{n+1/2}) \\ q_{n+1} &= q_n + \frac{h}{2} \big(\nabla_p H(q_n, p_{n+1/2}) + \nabla_p H(q_{n+1}, p_{n+1/2}) \big) \\ p_{n+1} &= p_{n+1/2} - \frac{h}{2} \nabla_q H(q_{n+1}, p_{n+1/2}). \end{aligned} \tag{1.46}$$

Both schemes are known as Störmer–Verlet/leapfrog methods, depending on the context in which they are used. They are of order 2 and symplectic since they are obtained as the composition of symplectic maps. Reference [169] contains a thorough review of the different variants of the Störmer–Verlet method, with interesting historical remarks, a complete analysis of its preservation properties, and up to four different proofs of symplecticity.

Methods (1.45) and (1.46) are explicit for separable Hamiltonians. In particular, if

$$H(q, p) = \frac{1}{2} p^T M^{-1} p + V(q), \tag{1.47}$$

where M is a positive definite mass matrix, then method (1.45) reduces to

$$\begin{aligned} q_{n+1/2} &= q_n + \frac{h}{2} M^{-1} p_n \\ p_{n+1} &= p_n - h \nabla_q V(q_{n+1/2}) \\ q_{n+1} &= q_{n+1/2} + \frac{h}{2} M^{-1} p_{n+1}, \end{aligned} \tag{1.48}$$

whereas (1.46) leads to

$$\begin{aligned} p_{n+1/2} &= p_n - \frac{h}{2} \nabla_q V(q_n) \\ q_{n+1} &= q_n + h M^{-1} p_{n+1/2} \\ p_{n+1} &= p_{n+1/2} - \frac{h}{2} \nabla_q V(q_{n+1}). \end{aligned} \tag{1.49}$$

A helpful interpretation of scheme (1.48) goes by the saying that there is first a "drift" (uniform motion on half the subinterval $[t_n, t_{n+1}]$), then a "kick" is applied, and finally there is another uniform motion on the second half. In the second variant (1.49) the sequence is reversed: "kick" + "drift" + "kick." Notice also that with the substitution $f(q_n) = -M^{-1}\nabla_q V(q_n)$, $p_n = Mv_n$ in (1.49), we recover the velocity Verlet method (1.28) (with the trivial changes of f by F and q by y).

In the particular case of the second-order differential equation $\ddot{q} = f(q)$, where f does not depend on $\dot{q}$, method (1.49) reduces, after elimination of p, to the familiar scheme

$$q_{n+1} - 2q_n + q_{n-1} = h^2 f(q_n). \tag{1.50}$$

More in general, for the partitioned system

$$\dot{q} = g(q, v), \qquad \dot{v} = f(q, v), \tag{1.51}$$

the previous integrators adopt the form

$$\begin{aligned} q_{n+1/2} &= q_n + \frac{h}{2} g(q_{n+1/2}, v_n) \\ v_{n+1} &= v_n + \frac{h}{2}\left(f(q_{n+1/2}, v_n) + f(q_{n+1/2}, v_{n+1}\right) \\ q_{n+1} &= q_{n+1/2} + \frac{h}{2} g(q_{n+1/2}, v_{n+1}) \end{aligned} \tag{1.52}$$

and

$$\begin{aligned} v_{n+1/2} &= v_n + \frac{h}{2} f(q_n, v_{n+1/2}) \\ q_{n+1} &= q_n + \frac{h}{2}\left(g(q_n, v_{n+1/2}) + g(q_{n+1}, v_{n+1/2})\right) \\ v_{n+1} &= v_{n+1/2} + \frac{h}{2} f(q_{n+1}, v_{n+1/2}) \end{aligned} \tag{1.53}$$

called, respectively, the *position* and *velocity* Verlet methods.

Geometric properties of the Störmer–Verlet method

In addition to symplecticity when applied to Hamiltonian systems, the Störmer–Verlet method (in all of its variants) preserves many other geometric properties of the exact flow associated with the differential system. As a trivial consequence of its symplectic character, it is clear that it also preserves volume in phase space.

Since the Störmer–Verlet method is the composition of a first-order method and its adjoint, then it is symmetric with respect to changing the direction of time: if we denote by $S_h^{[2]}$ the numerical flow defined by method (1.52), then $(S_h^{[2]})^{-1} = S_{-h}^{[2]}$.

This time-symmetry implies *reversibility* of the numerical solution for certain differential equations. For instance, if $g(q, v) = v$ in (1.51) and f only depends on q, i.e., we have the first-order system

$$\dot{q} = v, \qquad \dot{v} = f(q),$$

then inverting the direction of the initial velocity does not change the trajectory, only the direction of motion along this trajectory. In other words, for the exact flow it is true that

$$\varphi_t(q, v) = (\bar{q}, \bar{v}) \quad \text{implies} \quad \varphi_t(\bar{q}, -\bar{v}) = (q, -v).$$

This is a generalization of what happens for the flow of the simple harmonic oscillator (1.5). Alternatively, φ_t is reversible with respect to the reflection $\rho : (q, v) \longmapsto (q, -v)$. It turns out that the same is true for the Störmer–Verlet numerical flow, i.e.,

$$S_h^{[2]}(q, v) = (\bar{q}, \bar{v}) \quad \text{implies} \quad S_h^{[2]}(\bar{q}, -\bar{v}) = (q, -v)$$

for all q, v and h. More generally, $S_h^{[2]}$ is ρ-reversible for any map ρ of the form $\rho(q, v) = (\rho_1(q), \rho_2(v))$, so that $\rho \circ S_h^{[2]} = (S_h^{[2]})^{-1} \circ \rho$ [169].

A function $I(x)$ is a *first integral* (or constant of motion) of the differential equation $\dot{x} = f(x)$ if I, computed along every solution, does not vary with time, i.e., $I(x(t))$ is constant or

$$\dot{I} = \nabla I(x)\dot{x} = \nabla I(x) f(x) = 0 \qquad \text{for all } \ x.$$

As we know, the total energy in a mechanical system (1.25) is a constant of motion. More generally, in a Hamiltonian system, since $\dot{I} = \{I, H\}$ (equation (1.38)), then a necessary and sufficient condition for $I(x)$ to be a first integral of the system is that the Poisson bracket $\{H, I\} \equiv 0$ is identically zero.

Linear first integrals such as the linear momentum are all preserved by the Störmer–Verlet method. This is also true for quadratic first integrals of the form $I(q, p) = p^T(Cq + c)$ for Hamiltonian systems, where C is a constant square matrix and c is a constant vector [169]. In particular, the angular momentum in N-body problems is preserved if the acting forces only depend on the distances of the particles. In general, the energy $H(q, p)$ is *not* conserved by the scheme, although the energy error is of order $\mathcal{O}(h^2)$ over exponentially large time intervals. As a matter of fact, no integration method can preserve energy *and* symplecticity in general, as established by the classical theorem of Ge and Marsden [156]. More specifically, assume ψ_t is the numerical flow corresponding to a symplectic integrator preserving the energy for the autonomous Hamiltonian H, and that $\varphi_t\big|_{H=c}$ denotes the exact flow restricted to the surface $H = c$. Then, there exists a function $\tau = \tau(c, t)$ defined on a neighborhood of 0 such that $\psi_{\tau(c,t)}\big|_{H=c} = \varphi_t\big|_{H=c}$. In other words, the numerical flow coincides with the exact flow up to a reparameterization of time [142].

All these favorable properties, in addition to its simple formulation, help to understand why the Störmer–Verlet/leapfrog method is probably the most used geometric integrator and has been so even before the notion of geometric numerical integration arose, especially in molecular dynamics [217, 295], condensed matter simulations [99], sampling with the hybrid Monte Carlo method [265], etc. There is one important observation worth noticing, however: *the step size h has to be constant* for the method to have these advantageous geometric properties. Otherwise, if the step size is changed along the integration process, they no longer hold. We will return to this point in Chapter 5 in more detail.

Implicit midpoint and trapezoidal rules

For the general system $\dot{x} = f(x)$ the midpoint rule

$$x_{n+1} = \psi_h^M(x_n) = x_n + hf\left(\frac{x_{n+1} + x_n}{2}\right) \tag{1.54}$$

and the trapezoidal rule

$$x_{n+1} = \psi_h^T(x_n) = x_n + \frac{h}{2}\left(f(x_{n+1}) + f(x_n)\right) \tag{1.55}$$

are both symmetric (the formulae are left unaltered after exchanging $x_n \leftrightarrow x_{n+1}$ and $h \leftrightarrow -h$) and of order 2. Moreover, in *Example 1.5* we showed that

$$\psi_h^T = \psi_{h/2}^I \circ \psi_{h/2}^E, \qquad \psi_h^M = \psi_{h/2}^E \circ \psi_{h/2}^I,$$

where ψ_h^E and ψ_h^I denote, respectively, the explicit and implicit Euler methods. In consequence,

$$\psi_h^M = \pi_h^{-1} \circ \psi_h^T \circ \pi_h, \quad \text{with} \quad \pi_h = \psi_{h/2}^I.$$

In the terminology of dynamical systems, the trapezoidal and midpoint rules are said to be *conjugate* by the ($\mathcal{O}(h)$-near to identity) map π_h (the implicit Euler method), which can be regarded as a change of coordinates. Many dynamical properties of interest (Lyapunov exponents, periodic orbits, phase space averages, etc.) are indeed invariant under changes of coordinates, and so conjugate methods provide the same characterization of these properties (although, of course, trajectories corresponding to the same initial condition are different).

Since ψ_h^T is a symmetric method that admits an expansion in powers of h, there exists a (first-order) method ψ_h such that $\psi_h^T = \psi_{h/2} \circ \psi_{h/2}$ [170, p. 154], so that we can also write

$$\begin{aligned}\psi_h^M &= (\psi_{h/2}^I)^{-1} \circ \psi_{h/2} \circ \psi_{h/2} \circ \psi_{h/2} \circ \psi_{h/2}^{-1} \circ \psi_{h/2}^I \\ &= \hat{\pi}_h^{-1} \circ \psi_h^T \circ \hat{\pi}_h\end{aligned}$$

where $\hat{\pi}_h = \psi_{h/2}^{-1} \circ \psi_{h/2}^I$ is a $\mathcal{O}(h^2)$ near to the identity transformation, since both schemes are consistent. In consequence, for every numerical trajectory of the midpoint rule there exists another trajectory of the trapezoidal rule which is $\mathcal{O}(h^2)$-close on compact sets.

For Hamiltonian systems, in accordance with (1.33), the midpoint rule particularizes to

$$x_{n+1} = x_n + hJ\,\nabla H\big((x_{n+1} + x_n)/2\big), \tag{1.56}$$

which turns out to be symplectic, as shown in [170]. Moreover, it is reversible with respect to any linear-reversing symmetry and preserves *any* quadratic first integral of the system. In particular, it is energy preserving for quadratic Hamiltonians. In addition, it is linearly stable for all time steps.

In accordance with the previous result, the trapezoidal method applied to a Hamiltonian system is conjugate to a symplectic method, and the change of coordinates is of the form $\pi_h(x) = x + \mathcal{O}(h^2)$. Thus, a trajectory of the non-symplectic trapezoidal rule is very similar to a trajectory of the symplectic midpoint rule [293].

1.5 Illustration: the Kepler problem

The so-called Kepler problem describes the motion of two bodies which attract each other according to the universal gravitational law. It is an example of an integrable Hamiltonian system; its long-term behavior is well understood, and the exact solution can be obtained without much difficulty (yet in an implicit form). For this reason it is an appropriate candidate to validate and test the efficiency of numerical integrators. As a matter of fact, we will use it several times within this book, either as a test bench for the methods or as a part of more involved systems (such as the motion of the outer Solar System). Here we introduce the system and illustrate a number of interesting features shown by the previous schemes.

As is well known, the motion of two bodies attracting each other through the gravitational law can be described, using relative coordinates, by the differential equation

$$\ddot{q} = -\mu\,\frac{q}{r^3}, \qquad \text{with} \qquad r = \|q\| = \sqrt{q^T q}, \qquad \mu = GM, \tag{1.57}$$

the symbol G stands for the gravitational constant, and M is the sum of the masses of the two bodies. This equation can be obtained from the Hamiltonian function

$$H(q,p) = T(p) + V(q) = \frac{1}{2}p^T p - \mu\,\frac{1}{r} \tag{1.58}$$

by means of (1.32). For negative energies, the trajectory is bounded and periodic with period

$$t_K = \frac{2\pi\mu}{\sqrt{-8H_0^3}},$$

where $H_0 = H(q(0), p(0))$ is the energy (which is a first integral of the motion).

The exact flow of the system, $(q(t), p(t)) = \varphi_t(q(0), p(0)) \equiv \varphi_t(q_0, p_0)$ is given by [122, p. 165]

$$q(t) = f\, q_0 + g\, p_0, \qquad p(t) = f_p\, q_0 + g_p\, p_0,$$

where

$$\begin{aligned} f &:= 1 + \frac{(\cos x - 1)a}{r_0}, & g &:= t + \frac{\sin x - x}{w}, \\ f_p &:= -\frac{aw \sin x}{r_0(1 - \sigma \cos x + \psi \sin x)}, & g_p &:= 1 + \frac{\cos x - 1}{1 - \sigma \cos x + \psi \sin x}. \end{aligned} \tag{1.59}$$

Here x is given by the implicit equation

$$w\, t = x - \sigma \sin x + \psi(1 - \cos x) \tag{1.60}$$

and

$$\psi = \frac{u}{wa^2}, \quad \sigma = 1 - \frac{r_0}{a}, \quad w = \sqrt{\frac{\mu}{a^3}}, \quad a = -\frac{\mu}{2E}, \quad E = \frac{1}{2} p_0^T p_0 - \mu \frac{1}{r_0},$$

$$u = q_0^T p_0, \quad r_0 = \|q_0\|.$$

The implicit equation (1.60) can be solved by applying, for instance, the Newton–Raphson method, with the iteration

$$x^{[j]} = x^{[j-1]} - \left(\frac{x^{[j-1]} - \sigma \sin(x^{[j-1]}) + \psi(1 - \cos(x^{[j-1]})) - w\, t}{1 - \sigma \cos(x^{[j-1]}) + \psi \sin(x^{[j-1]})} \right).$$

It can be started with $x^{[0]} = wta/r_0$ for a fast convergence. The algorithm has been implemented as the MATLAB function `phiKepler.m`, to be used as follows:

```
[q p] = phiKepler(q0,p0,t,mu).
```

The exact solution obtained in this way can be used for comparison and testing of different numerical schemes, just by measuring the error in coordinates and momenta along the time integration.

Since we are dealing with a central force, the motion takes place in a plane, and so we may analyze it as a two-dimensional problem. Taking $\mu = 1$ and initial conditions

$$q_1(0) = 1 - e, \quad q_2(0) = 0, \quad p_1(0) = 0, \quad p_2(0) = \sqrt{\frac{1+e}{1-e}}, \tag{1.61}$$

if $0 \leq e < 1$ the total energy is $H = H_0 = -1/2$, the solution is periodic with period 2π and the trajectory is an ellipse of eccentricity e. The initial conditions correspond, in fact, to the pericenter, and the semi-major axis of the ellipse is 1.

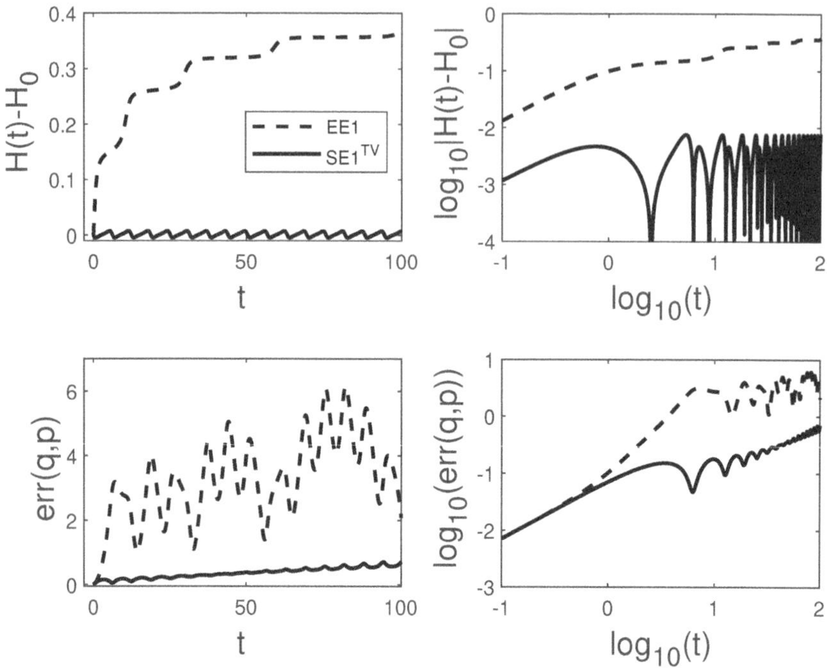

FIGURE 1.5: Errors in the numerical integration of the Kepler problem with initial conditions given by (1.61) with $e = 0.2$ using the explicit Euler method (EE1) and the symplectic Euler-TV ($\mathrm{SE1}^{\mathrm{TV}}$). Top figures show the error in energy, and bottom figures show the 2-norm error in position and momenta. The right figures show the same results in a double logarithmic scale.

1.5.1 Error growth

As a first test we take eccentricity $e = 1/5$ and apply the explicit Euler method (EE1) and the symplectic Euler-TV scheme (1.43) ($\mathrm{SE1}^{\mathrm{TV}}$) in the interval $t \in [0, 100]$ with step size $h = \frac{1}{20}$. Figure 1.5 shows the error in energy (top) and in phase space (bottom) at each step. The error in phase space is computed as the Euclidean norm of the difference between the point computed by the numerical scheme and the exact solution $(q(t), p(t))$. Since the errors achieved by the two methods differ by several orders of magnitude, we also show the same results in a log-log diagram (right). Notice that the error in energy just oscillates for the symplectic Euler method without any secular component, whereas there is an error growth in energy for the explicit Euler scheme. With respect to the error in positions and momenta, a linear error growth for the symplectic method and a faster error growth for the non-symplectic one can be observed. Similar results can be obtained by applying the implicit Euler method and the symplectic Euler-VT scheme (1.44).

Is this behavior typical of symplectic and non-symplectic integrators? To get some insight, we repeat the same experiment but this time applying second-order methods. Specifically, we consider the (non-symplectic) Heun's method (1.19) (H2), the (symplectic) Störmer–Verlet method (1.48) (S2), the (non-symplectic) trapezoidal method (T2) (although it is conjugate to the symplectic midpoint method) and finally the symmetric and symplectic scheme (S2m) defined by

$$\begin{aligned} q_{n+1/2} &= q_n + \frac{h}{2} M^{-1} p_n \\ p_{n+1} &= p_n - h\, \nabla_q V(q_{n+1/2}) + \frac{h^3}{24} \nabla_q \Big(\nabla_q V(q_{n+1/2})^T \nabla_q V(q_{n+1/2}) \Big) \\ q_{n+1} &= q_{n+1/2} + \frac{h}{2} M^{-1} p_{n+1}. \end{aligned} \tag{1.62}$$

This map can be seen as the Störmer–Verlet/leapfrog method (1.48), but applied to a slightly modified potential (depending also on h). We will return to this scheme in Chapter 3. Notice that for the problem at hand $\nabla_q(\nabla_q V^T \nabla_q V) = -4q/r^6$, and so the additional computational cost due to the inclusion of this term is relatively small.

The corresponding results are shown in Figure 1.6. We notice that the behavior of the error growth produced with Heun's method is quite similar to the Euler scheme, whereas all the symplectic methods show the same pattern in the error as the Euler-TV integrator in Figure 1.5. It is worth remarking that the trapezoidal rule behaves with respect to the error as a symplectic method, since it is conjugate to symplectic.

In Figure 1.7 we show the results obtained (in double logarithmic scale) when $e = 0$, in other words, when a circular trajectory is considered. Now the error in energy drops to round off for the trapezoidal method,[3] whereas the error in phase space is only slightly smaller than that corresponding to the Störmer–Verlet method S2. Surprisingly, although the energy error achieved by scheme (1.62) is only slightly smaller than that of S2, the error in phase space behaves much better.

1.5.2 Efficiency plots

When accurate results are desired, high-order methods are usually preferable to low-order schemes. The former are in general more costly per time step, but this extra cost can be compensated if larger steps can be used. Generally speaking, one can say that the most efficient scheme is the one providing a prescribed accuracy with the lowest computational effort.

Since the computational cost may depend on the implementation, the particular compiler used, and the architecture of the computer, it is customary to estimate this cost by counting how many times the most costly part of the algorithm is evaluated. In most cases this corresponds to the evaluation of the vector field $f(t,x)$. Thus, if a method requires n_f evaluations of f per step and

[3]The error corresponds, in fact, to the truncation error due to the value of the tolerance used to solve the implicit equations.

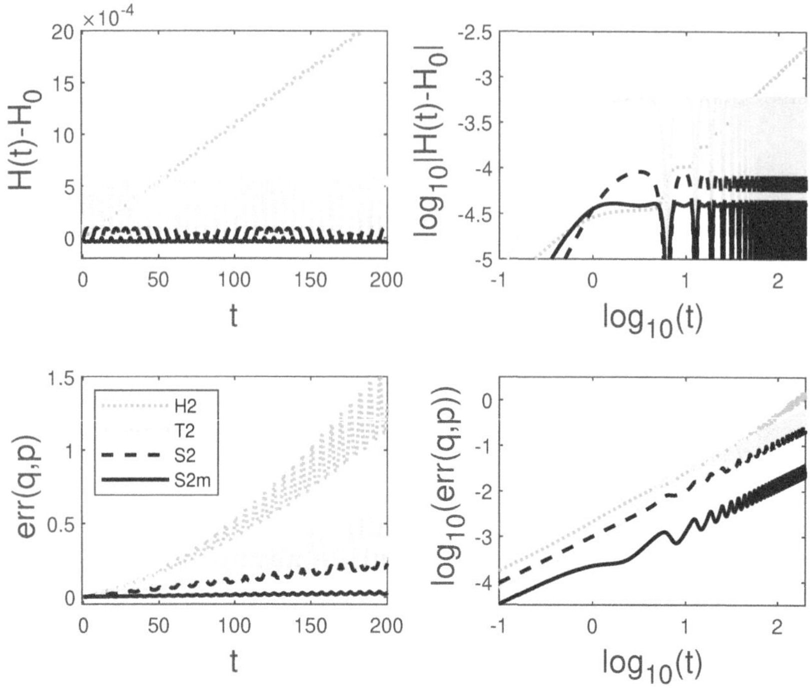

FIGURE 1.6: Errors in the numerical integration of the Kepler problem with initial conditions given by (1.61) with $e = 0.2$ obtained with the following second-order methods: Heun (H2), trapezoidal rule (T2), the symplectic Störmer–Verlet (1.48) (S2) and the version (1.62) with a modified potential (S2m).

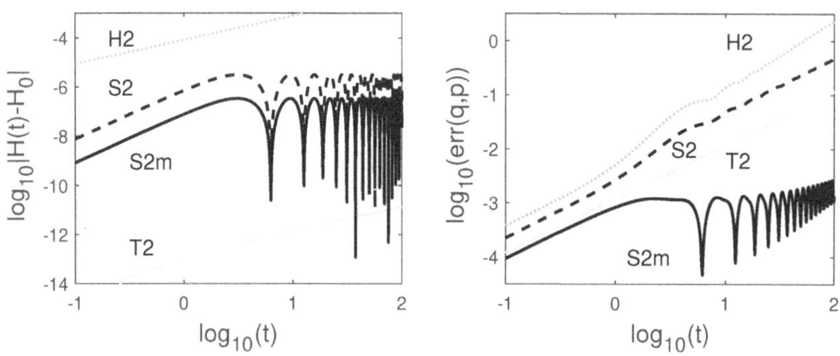

FIGURE 1.7: Same as Figure 1.6 for initial conditions with $e = 0$: Heun's method (H2); symplectic Störmer–Verlet method (S2); symplectic Störmer–Verlet method with modified potential (1.62) (S2m) and trapezoidal method (T2).

the whole integration until the final time, t_{f}, is carried out with N time steps of length h, then we say that the cost of the method is $cost = n_f N = n_f t_{\mathrm{f}}/h$. The explicit Euler method requires just one evaluation per step, while the Heun method requires two[4]. Then, to compare the Euler and Heun methods at the same computational cost, a time step twice as large must be used for Heun. One should recall, however, that this is just an estimate, since for implicit methods the number of evaluations of the vector field depends on the particular algorithm used and the number of iterations required to solve the implicit nonlinear equations, the region of the phase space, etc.

The relative performance of different integrators can be visually illustrated by plotting the error in terms of the required computational cost (estimated, as before, by counting the number of function evaluations). For simplicity, we estimate the cost of an implicit method by considering a fixed number of iterations at each step in the whole integration.

If we have a method of order r, then we may say that its error (neglecting higher-order contributions) is $\mathcal{E} = Ch^r$, with C a constant independent of h, so that we can write it as a function of the computational cost, namely

$$\mathcal{E} = Ch^r = C\left(\frac{n_f t_{\mathrm{f}}}{cost}\right)^r = D\ cost^{-r}. \tag{1.63}$$

In consequence, if we plot $\log(\mathcal{E})$ versus $\log(cost)$ we will obtain, in the limit $h \to 0$, i.e., $N \to \infty$, a straight line with slope $-r$, so that the order of the scheme can be deduced at once from this efficiency diagram.

Let us obtain the corresponding efficiency plot of the previous integrators when they are applied to the Kepler problem with initial conditions (1.61) and eccentricities $e = 1/5$ and $e = 0$. The time interval considered is $t \in [0, 100]$ and the time steps are $h = 10/2^i$, $i = 1, \ldots, 10$. Figure 1.8 shows the maximum error in energy and position using the explicit Euler method (EE1), the Heun method (H2), the symplectic Störmer–Verlet method (1.48) (S2) and the modified version (1.62) (S2m).

We notice at once that the symplectic methods are not only qualitatively but also quantitatively superior. In particular, scheme (1.62) shows slightly smaller errors in energy, but they considerably diminish in the determination of the positions and momenta. This is clearly manifest for the circular trajectory. As we will analyze later in the book, the reason is that the second-order method (1.62) is in fact conjugate (under an appropriate change of coordinates) to a fourth-order scheme: this fourth-order behavior manifests itself in the error in positions but not in the error in energy.

Numerical integrations with large time steps require a small number of evaluations of f and so this corresponds to the left part of the diagrams. With sufficiently large steps, the numerical solution (both for symplectic and non-symplectic methods) is unbounded, but the symplectic schemes lead to

[4] In this count we neglect that the Heun method has to store the two evaluations of the vector field and needs to compute more additions and multiplications.

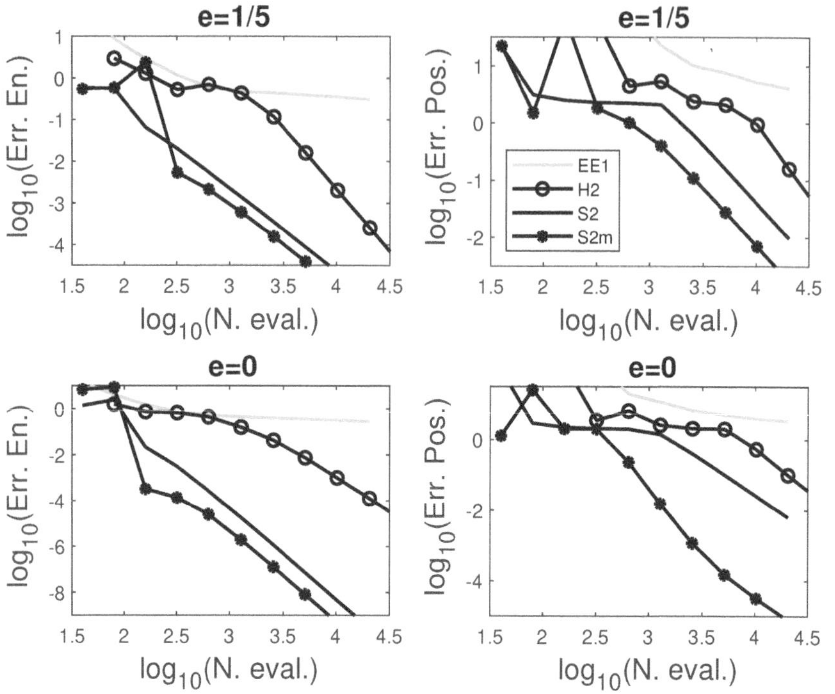

FIGURE 1.8: (Left) Maximum error in energy versus number of force evaluations and (right) maximum error in coordinates and momenta versus number of force evaluations along the integration on the interval $t \in [0, 100]$ for initial conditions given by (1.61) with $e = 1/5$ (left diagrams) and $e = 0$ (right diagrams) for the explicit Euler (EE1), the Heun's method (H2), the symplectic Störmer–Verlet method (1.48) (S2), and its modified version (1.62) (S2m).

bounded numerical solutions for larger values of h. In addition, the superiority of the symplectic integrators is evident when longer time integrations are considered.

1.6 What is to be treated in this book (and what is not)?

In this chapter we have provided a glimpse of some of the issues involved in geometric numerical integration, introducing some elementary geometric integrators and illustrating them in comparison with other schemes of the same order of accuracy in several particularly simple autonomous Hamiltonian systems. It is clear, however, that there are many other topics in the field that deserve further attention, even if this book is only intended as a basic introduction to the subject.

Thus, although we have seen that the Störmer–Verlet methods provide a qualitatively correct description of systems possessing geometric structures, in the sense that they preserve several symmetries and first integrals, they only render second-order approximations to the exact solution. This, of course, does not represent any particular concern for molecular dynamics but *is* an issue in other problems where high accuracy is required, as is the case for instance in the computation of ephemerides in the Solar System. A natural question then arises: is it possible to construct higher-order symplectic (and, in general, geometric) integrators? And if the answer is affirmative, how can we do that? This is the topic addressed in Chapter 3, where high-order integrators are constructed by *composition* of basic low-order methods, and the idea of *splitting* is analyzed in detail. These schemes have found applications in two areas of recent interest, namely in sampling statistical distributions and in quantum simulations of quantum systems, as we will see in Chapter 7.

Before that, in Chapter 2 we will investigate high-order classical integration methods (mainly within the class of Runge–Kutta integrators) from the structure preservation point of view, showing that an important subclass of schemes are indeed symplectic.

Besides splitting and composition methods, there are, of course, other types of geometric integrators worth considering. Thus, in Chapter 4 we review symplectic integrators constructed directly from the theory of generating functions and the important category of variational integrators, which mimic the continuous Hamiltonian system by verifying a discrete analogue of Hamilton's variational principle by construction. There we also review numerical methods specifically designed for dynamical systems that preserve volume and consider the more restricted, but relevant in physical applications, case of differential equations defined in matrix Lie groups. Finally, we will consider schemes obtained by linear combinations of splitting methods and their preservation properties.

In addition to the construction of new numerical algorithms, an important aspect of geometric integration is the explanation of the relationship between the preservation of the geometric properties of a numerical method and the observed favorable error propagation in long-time integration. This is the subject of Chapter 5, where the idea of backward error analysis is elaborated in more detail.

In Chapter 6 we analyze the applicability of splitting and composition methods to certain classes of partial differential equations (PDEs), namely the Schrödinger equation and other evolution equations. Here the approach is first to apply a space discretization, in which spatial derivatives are discretized (either by finite differences or pseudo-spectral methods), thus leading to a system of coupled ordinary differential equations, typically of large dimension, that can be integrated by geometric integrators (if the semi-discretization process does not destroy the desired geometric property of the continuous equation). In the process, we will present and analyze a special class of splitting methods formulated in terms of *complex* coefficients, which are useful in particular for the time integration of semi-discretized diffusion equations. Finally, in Chapter 7 we collect some compelling applications where geometric integrators are being used nowadays.

Many other aspects of geometric numerical integration are not covered in this book. There are at least two different reasons for this. On the one hand, our aim here is simply to provide an introduction to the subject, and as such we have decided to focus on a set of problems that (we believe) are at the core of geometric numerical integration, hoping in this way to provide the reader the necessary tools for tackling other, perhaps less elementary, problems. On the other hand, there are excellent review papers and monographs available in the market which adequately describe these more advanced topics, and so we refer the reader to them for an exhaustive treatment.

Among the issues not treated here, we mention the following:

- *Multistep methods in geometric integration.* Multistep methods are at the core of widely used codes for the numerical integration of non-stiff differential equations (Adams methods) and also for stiff problems (BDF or backward differentiation formulae) [171, 174, 193]. It is thus quite natural to examine how they behave in geometric integration and, in particular, their applicability in long-time integrations of Hamiltonian systems. This issue is treated at length in [170] and references therein, as well as the more general class of *general linear methods* [79, 171]. Here, by contrast, we will restrict ourselves to one-step methods.

- *Projection methods.* They essentially consist of combining a step of an arbitrary method and then enforcing (by hand) the property one wishes to preserve at the discrete level. First integrals like energy can be preserved in this way at the end of a step (or several steps), although this procedure may destroy other properties of the method (such as symplecticity) and may not give a good long-time behavior if applied inappropriately. As usual, for a detailed treatment of projection methods, the reader is referred to [170].

- *Energy-preserving methods.* Autonomous Hamiltonian systems preserve, in addition to the symplectic structure, the total energy. Since no integration method can preserve both properties for general Hamiltonian functions [156], much interest has been devoted to the construction of the so-called energy-momentum methods, i.e., schemes that preserve by design the energy and momentum (both linear and angular) in geometric mechanics. A list of references on this subject can be found in [233]. Also of interest for the preservation of energy and other first integrals is the class of discrete-gradient systems, analyzed in [97, 98, 242, 278].

- *Variable time steps.* Whereas in general-purpose software packages for the integration of ordinary differential equations, it is a standard practice to incorporate a device for adapting the time step during the integration interval in accordance with some previously accorded tolerance, this is not the case for geometric integrators. As a matter of fact, a simple choice of h as a function of the point x generally compromises the favorable geometric properties the integrator may have and affect, its long-time behavior. The usual practice is then, rather than adapting the time step, to adapt the system so that it can be integrated with a constant time step and still have the obvious advantages of variable step methods. Several techniques exist depending on the particular problem considered (see, e.g., [34, 62, 100, 172, 188, 218, 252, 299, 300]).

- *Constrained mechanical systems.* An ordinary differential equation with a constraint forms what is called a *differential-algebraic equation* (DAE), $\dot{x} = f(x, \lambda)$, $g(x) = 0$ (here λ is typically a Lagrange multiplier), for which a myriad of numerical methods have been proposed [174]. When the DAE has an extra structure to be preserved by discretization, as occurs in mechanical systems, specific methods have been proposed and widely used in applications. A particularly convenient integrator is the RATTLE algorithm: it is time-symmetric, symplectic, and convergent of order 2 for general Hamiltonians, and thus it can be used as a low order scheme to construct higher-order approximations by composition [170, 218, 279].

- *Poisson systems and rigid body dynamics.* Poisson systems constitute a generalization of Hamiltonian systems obtained when the canonical matrix J in (1.33) is replaced with a smooth matrix-valued function $S(x)$, so that the system reads $\dot{x} = S(x)\nabla_x H(x)$. Now the Poisson bracket (1.35) generalizes to $\{F, G\} = \nabla F^T S \, \nabla G$ and $S(x)$ is such that this generalized bracket is bilinear, skew-symmetric and satisfies the Jacobi identity. The function $H(x)$ (still called Hamiltonian) is conserved by the corresponding flow. A function $C(x)$ is called a Casimir function of the Poisson system if $\nabla C(x)^T S(x) = 0$ for all x, and is also a first integral. Systems of this form appear in many different applications [170, 234, 267], so it is quite natural to look for numerical methods that preserve at the discrete level the main characteristics of the flow of a Poisson system. These are the so-called *Poisson integrators* [170, 334]. A highly relevant example in this setting is the dynamics of a rigid body [170, 218].

- *Highly oscillatory problems.* The Störmer–Verlet method (1.49) applied to the simple harmonic oscillator $\ddot{y} + \omega^2 y = 0$ is only stable with a time step h such that $h\omega < 2$. This means that the error due to the scheme grows without bound unless h satisfies the previous inequality. In general, for an equation $\ddot{q} = -\nabla V(q)$, $q \in \mathbb{R}^d$, this restriction still applies, where ω now represents the largest eigenfrequency of the Hessian matrix $\nabla^2 V(q)$ along the numerical solution. In any case, such a restriction represents a severe bottleneck in the performance of the method, especially in applications where the potential $V(q)$ contains terms acting on different timescales ("fast" and "slow" forces). In this situation, the solution is highly oscillatory on the slow timescale, and several alternatives have been proposed to integrate the system more efficiently: the "mollified impulse method" [153, 290], heterogeneous multiscale methods [85, 131, 137], exponential integrators [115, 185], stroboscopic averaging methods [80], etc. A very useful tool for the analysis in this setting is provided by modulated Fourier expansions [116, 117, 170]. Nowadays, the numerical analysis of highly oscillatory problems remains a very active area of research (see, e.g., [86, 119, 136, 154, 192]).

- *Dynamics of geometric integrators.* Geometric numerical integrators are designed in such a way that they inherit the structural properties possessed by the vector field defining the differential equation, with the goal of providing a faithful description of the continuous dynamical system (its *phase portrait*). This includes preservation of equilibrium points, periodic and quasi-periodic orbits, and more generally all the invariant sets. These questions are analyzed in detail in [170, 305] and references therein. A brief overview can be found in [244]. It has been shown, in particular, that symplectic integrators do preserve the invariant tori in phase space guaranteed to exist in virtue of the Kolmogorov–Arnold–Moser (KAM) theorem [29, 170, 297].

- *Hamiltonian partial differential equations.* In addition to the aforementioned approach of discretizing a PDE with a certain structure first in space and then applying a geometric integrator to the resulting system of ODEs, there are particular classes of PDEs for which a Hamiltonian structure can be associated. These include, in particular, nonlinear wave equations and two-dimensional rotating shallow-water equations [218]. It is possible then to apply a symplectic discretization by first carrying out a spatial truncation that reduces the PDE to a system of Hamiltonian ODEs and then using an appropriate symplectic integrator. Other popular geometric methods for PDEs are the so-called multisymplectic integrators [75, 118, 152, 218, 245, 281]. In any case, all these techniques are essentially restricted to smooth solutions of the PDE under consideration.

1.7 Exercises

1. Compute the exact solution at $t = \pi$ of the differential equations

$$(i): \left\{ \begin{array}{rcl} \dot{q} & = & p + \frac{h}{2} q \\ \dot{p} & = & -q + \frac{h}{2} p \end{array} \right. , \qquad (ii): \left\{ \begin{array}{rcl} \dot{q} & = & \left(1 - \frac{h^2}{3}\right) p + \frac{h}{2} q \\ \dot{p} & = & -\left(1 - \frac{h^2}{3}\right) q + \frac{h}{2} p, \end{array} \right.$$

for $h = \frac{\pi}{100}$ and $h = \frac{\pi}{1000}$, and compare with the numerical solution by the explicit Euler method applied to the harmonic oscillator at the final time. What do you conclude?

2. Apply the trapezoidal, midpoint and Heun methods to (1.24), writing all methods in an explicit form, $x_{n+1} = \phi_h(\lambda)x_n$. Compare the results obtained with the exact solution $x_{n+1} = \mathrm{e}^{-\lambda h}x_n$.

3. Verify that the implicit scheme (1.27), when applied to equation (1.25) with $F(y) = -y$ expressed as a first-order system, preserves exactly the energy (1.26).

4. Verify that the solution of the matrix equation (1.29) is given by (1.30).

5. Consider the linear system of equations $\dot{x} = A(t)x$ given by

$$\frac{d}{dt}\begin{pmatrix} x_1 \\ x_2 \end{pmatrix} = \begin{pmatrix} -21+\alpha t & 20 \\ 20 & -21 \end{pmatrix}\begin{pmatrix} x_1 \\ x_2 \end{pmatrix}, \qquad \begin{pmatrix} x_1(0) \\ x_2(0) \end{pmatrix} = \begin{pmatrix} 1 \\ 0 \end{pmatrix}. \tag{1.64}$$

For $\alpha = 0$ the eigenvalues are $\lambda_1 = -1$ and $\lambda_2 = -41$, and the solution reads

$$x_1(t) = \frac{1}{2}(\mathrm{e}^{-t} + \mathrm{e}^{-41t}), \qquad x_2(t) = \frac{1}{2}(\mathrm{e}^{-t} - \mathrm{e}^{-41t}).$$

(i) Take $\alpha = 1$ and compute numerically the solution at $t = 1$ with the explicit Euler method and with the exponential explicit Euler method given by the scheme: $x_{n+1} = \mathrm{e}^{hA(t_n)}x_n$ for $h = 1, \frac{1}{2}, \frac{1}{4}$. Compare the results obtained with the exact solution: $x_1(1) = 0.239616$, $x_2(1) = 0.233847$.

(ii) Compute the solution given by the exponential midpoint method with $h = 1$, $x_1 = \mathrm{e}^{hA(1/2)}x_0$. What do you conclude?

6. Verify that the Jacobian matrix $\mathcal{N}$ given by (1.42) verifies $\mathcal{N}^T J \mathcal{N} = J$. Verify that the transformation defined by (1.41) is symplectic.

7. Verify that the Jacobi identity is indeed satisfied for any functions $F(q,p)$, $G(q,p)$ and $H(q,p)$ of the canonical variables q, p.

8. Given the functions $T(p)$ and $V(q)$ in one dimension, compute the following Poisson brackets in terms of the derivatives of $T(p)$ and $V(q)$:

$$(i)\ \{V(q),T(p)\}, \quad (ii)\ \{V(q),\{V(q),T(p)\}\}, \quad (iii)\ \{T(p),\{V(q),T(p)\}\}.$$

Repeat the computation for any $d > 1$.

9. Given the Hamiltonian of the one-dimensional pendulum with $T(p) = \frac{1}{2}p^2$ and $V(q) = 1 - \cos q$, compute the following Poisson brackets:

(i) $\{V(q), T(p)\}$,

(ii) $\{V(q), \{V(q), T(p)\}\}$,

(iii) $\{V(q), \{V(q), \{V(q), T(p)\}\}\}$.

Repeat the computation for the quadratic function $T(p) = \frac{1}{2}p^T M^{-1} p$, where M is a positive definite matrix, and a general function $V(q)$, and give the results in terms of derivatives of $V(q)$.

10. Verify that the midpoint rule (1.56) is a second-order symplectic integrator.

11. Given the domain $D_0 = B_{1/2}(3/2, 0)$ of area $S(D_0) = \pi/4$, apply 2 steps of length $h = \pi/6$ of the explicit Euler method for the pendulum with $k = 1$ and compute $S(D_2)$.

    Hint:

$$\left|\frac{\partial(q_2, p_2)}{\partial(q_1, p_1)}\right| \left|\frac{\partial(q_1, p_1)}{\partial(q_0, p_0)}\right| = (1 + h^2 \cos q_1)(1 + h^2 \cos q_0);$$

next, write q_1 in terms of q_0, p_0 and then use polar coordinates.

Chapter 2

Classical integrators and preservation of properties

Numerical methods for solving initial value problems defined by ordinary differential equations can be classified into two classes: those which use one starting value x_n at each step and additional, auxiliary function evaluations to construct x_{n+1} (*one-step* methods), such as those collected in Chapter 1, and those which are based on several previous values $x_{n-1}, x_{n-2}, \ldots$ and/or function evaluations at these points to construct x_{n+1} (*multistep* methods).

In both cases, the resulting schemes can achieve a high order of accuracy and excellent results in many situations. Thus, for nonstiff initial value problems, explicit Runge–Kutta methods up to order 8 are widely used, as well as multistep methods up to order 12. When very stringent accuracies are required, then high-order extrapolation methods or high-order Taylor series expansions of the solution are useful alternatives. It is thus quite natural to analyze the behavior of these standard integrators with respect to the preservation of whatever geometric properties the exact solution may possess. In particular, if the differential equation is derived from a Hamiltonian function, it seems appropriate to examine whether the numerical approximation provided by the schemes preserves the symplectic character of the system. In this chapter we review the main features of different standard one-step integrators from the point of view of geometric numerical integration.

2.1 Taylor series methods

Perhaps the simplest (and oldest) one-step methods are those based on the Taylor series expansion of the exact solution. The idea is just to approximate the solution of the initial value problem

$$\dot{x} = f(t, x), \qquad x(t_0) = x_0, \qquad x \in \mathbb{R}^d \tag{2.1}$$

at t_{n+1} by the rth-degree Taylor polynomial of $x(t)$ evaluated at t_n [208]. For the time being, we allow the time step to vary from step to step along the integration process so that $t_{n+1} = t_n + h_n$ and the rth-order approximation

DOI: 10.1201/9781003527626-2

reads

$$x_{n+1} = x_n + \frac{dx}{dt}(t_n, x_n)h_n + \cdots + \frac{1}{r!}\frac{d^r x}{dt^r}(t_n, x_n)h_n^r,$$

or equivalently

$$x_{n+1} = x_n + f(t_n, x_n)h_n + \cdots + \frac{1}{r!}\frac{d^{r-1} f(t_n, x_n)}{dt^{r-1}}h_n^r. \tag{2.2}$$

The explicit Euler scheme is recovered, of course, by putting $r = 1$. In general, the truncation error is $T_n = \mathcal{O}(h^r)$ so that the accuracy of the method can be increased arbitrarily by taking larger values of r. These methods have an obvious weakness: they fail if any one of the necessary derivatives of f does not exist at the point where they have to be computed. On the other hand, converting this simple idea into a practical integration method requires evaluating efficiently the total derivatives $f^{(j)}(t_n, x_n) \equiv d^j f/dt^j(t_n, x_n)$ appearing in (2.2). A possible approach consists in differentiating recursively the function f, but the complexity of the procedure grows in a manner that this is not affordable in practical situations. Thus, for instance,

$$\begin{aligned}
\frac{d^2x}{dt^2} &= \frac{df}{dt}(t,x) = f_t(t,x) + f_x(t,x)\dot{x} = f_t(t,x) + f_x(t,x)f(t,x) \equiv f_t + f_x f, \\
\frac{d^3x}{dt^3} &= f_{tt} + f_{tx}f + (f_{tx} + f_{xx}f)f + f_x(f_t + f_x f),
\end{aligned}$$

and so on. Since the complexity of the analytical expressions for the functions $f^{(k)}$ increases a great deal with k, this has been one of the main reasons why Taylor methods have been traditionally neglected for most practical purposes. However, with the advent of automatic differentiation techniques, it has been possible to construct recurrence formulas for the Taylor coefficients quite efficiently. The idea is to decompose the function f into a sequence of arithmetic operations and use standard unary or binary functions, together with the chain rule [28].

Although the Taylor method is only conditionally stable, for large r the stability domain is reasonably large [27] and thus it can be used even with moderately stiff equations. On the other hand, within this framework it is quite straightforward to implement variable step size and even variable order techniques so that both h and r can be adjusted along the integration interval. In contrast with other methods, here there are no rejected steps in the process, since the step size is chosen once the series are generated to obtain the required precision.

Taylor series methods have proved their usefulness in obtaining approximate solutions with a very high accuracy for ODE systems of low dimension. In particular, they have made it possible to compute periodic orbits in dynamical systems with several hundred digits of accuracy, as well as solving directly variational equations in the study of indicators for detecting chaos [28]. To carry out such a task, special packages implementing Taylor methods have been designed and used for integrating systems of low dimension

[1, 198]. They allow one to automatically select the step size and the order along the integration process according to the accuracy requirements. On the other hand, for systems of large dimension such as those resulting from a semi-discretization of a partial differential equation, the computational cost of this class of methods largely increases and their efficiency suffers accordingly.

Another important application of Taylor series methods refers to the construction of the so-called validated or interval methods for ordinary differential equations [319]. When these schemes return a solution, then the problem is guaranteed to have a unique solution, and an enclosure of the true solution is obtained with a validated error bound. This can be used in computer-assisted proofs in dynamical systems.

From the previous considerations, it is clear that all the geometric properties the differential equation may have are preserved by these schemes up to the order of the truncation error, i.e., the degree r of the polynomial approximating the solution. If r is large enough, the error in the preservation of invariants will be below the round-off error.

Example 2.1. A third-order scheme based on the Taylor series for the pendulum (1.12) with $k = 1$ reads as follows for the step $(q_n, p_n) \mapsto (q_{n+1}, p_{n+1})$. Recall that $x = (q,p)^T$, $f = (p, -\sin(q))^T$, $f_t = (0,0)^T$, so that

$$f_x f = \begin{pmatrix} -\sin q \\ -p\cos q \end{pmatrix}, \quad f_{xx} f f = \begin{pmatrix} 0 \\ p^2 \sin q \end{pmatrix}, \quad f_x f_x f = \begin{pmatrix} -p\cos q \\ \frac{1}{2}\sin 2q \end{pmatrix}$$

and then

$$\begin{aligned} \begin{pmatrix} q_{n+1} \\ p_{n+1} \end{pmatrix} &= \begin{pmatrix} q_n \\ p_n \end{pmatrix} - h \begin{pmatrix} -p_n \\ \sin q_n \end{pmatrix} - \frac{h^2}{2} \begin{pmatrix} \sin q_n \\ p_n \cos q_n \end{pmatrix} + \\ & \frac{h^3}{6} \begin{pmatrix} -p_n \cos q_n \\ (p_n^2 + \cos q_n)\sin q_n \end{pmatrix}. \end{aligned}$$

□

2.2 Runge–Kutta methods

2.2.1 Introduction

For simplicity in the presentation, we take a constant step size h in what follows. The general idea of Runge–Kutta (RK) methods is to produce a formula for $\Phi(t_n, x_n; h)$ in the one-step integrator

$$x_{n+1} = x_n + h\,\Phi(t_n, x_n; h) \tag{2.3}$$

such that (2.3) agrees with the Taylor expansion (2.2) up to terms of $\mathcal{O}(h^r)$, $r > 1$, *without computing any derivative of* f, but only reevaluating f at intermediate points between (t_n, x_n) and (t_{n+1}, x_{n+1}) [181]. We illustrate this idea with a simple example. Consider

$$\Phi(t_n, x_n; h) = b_1 f(t_n, x_n) + b_2 f\big(t_n + c_2 h, x_n + a_{21} h f(t_n, x_n)\big), \tag{2.4}$$

where b_1, b_2, c_2, a_{21} are constants to be determined. Expanding Φ in powers of h, we get

$$\begin{aligned}\Phi(t_n, x_n; h) \;&=\; b_1 f + b_2\Big(f + c_2 h f_t + a_{21} h f_x f + \frac{1}{2}(c_2 h)^2 f_{tt} \\ &\quad + c_2 a_{21} h^2 f_{tx} f + \frac{1}{2}(a_{21} h)^2 f_{xx} f^2 + \mathcal{O}(h^3)\Big),\end{aligned}$$

where the arguments (t_n, x_n) have been dropped for simplicity. Comparing the constant terms with (2.2), we find

$$b_1 + b_2 = 1,$$

whereas for the coefficients in h we have

$$b_2\, c_2 = \frac{1}{2}, \qquad b_2\, a_{21} = \frac{1}{2}.$$

Solving these equations results in

$$a_{21} = c_2, \qquad b_1 = 1 - \frac{1}{2c_2}, \qquad b_2 = \frac{1}{2c_2}, \qquad c_2 \neq 0,$$

so that there is a one-parameter family of second-order methods parametrized by $c_2 \neq 0$. The truncation error (1.18) for this family reads

$$\begin{aligned}T_n = h^2\Big(&(\frac{1}{6} - \frac{c_2}{4})(f_{tt} + f_{xx} f^2) + (\frac{1}{3} - \frac{c_2}{2}) f_{tx} f \\ &+ \frac{1}{6}(f_x f_t + f_x^2 f)\Big) + \mathcal{O}(h^3).\end{aligned}$$

We see then that it is not possible to cancel the term in h^2 for an arbitrary f; in other words, it is not possible to achieve order 3 with this family of schemes. Three particular choices of c_2 lead to the following well-known methods:

(i) $c_2 = \frac{1}{2}$. In this case the scheme reads

$$x_{n+1} = x_n + h f\left(t_n + \frac{1}{2} h, x_n + \frac{1}{2} h f(t_n, x_n)\right) \tag{2.5}$$

and is called the *modified Euler method*.

(ii) $c_2 = 1$, so that

$$x_{n+1} = x_n + \frac{1}{2} h\big(f(t_n, x_n) + f(t_n + h, x_n + h f(t_n, x_n))\big). \tag{2.6}$$

The resulting scheme is known as the *improved Euler method* [306].

(iii) $c_2 = \frac{2}{3}$, in which case $b_1 = 1/4$, $b_2 = 3/4$, $a_{21} = 2/3$, resulting in Heun's method (1.19).

Notice that Φ in (2.4) may be considered as a weighted average of values of f computed at different points. This same strategy can be pursued to achieve higher-order approximations, although the analysis turns out to be much more complicated. Perhaps one of the most frequently used methods within this class is the *classical fourth-order Runge–Kutta method* [208]. Here Φ is a weighted average of four evaluations of the function f, with the arguments in each evaluation depending on the preceding one. Specifically, the scheme reads

$$x_{n+1} = x_n + \frac{1}{6}h(k_1 + 2k_2 + 2k_3 + k_4), \tag{2.7}$$

with

$$\begin{aligned} k_1 &= f(t_n, x_n) \\ k_2 &= f\big(t_n + \frac{1}{2}h, x_n + \frac{1}{2}hk_1\big) \\ k_3 &= f\big(t_n + \frac{1}{2}h, x_n + \frac{1}{2}hk_2\big) \\ k_4 &= f(t_n + h, x_n + hk_3). \end{aligned} \tag{2.8}$$

More generally, given an integer s (the *number of stages*) and real coefficients a_{ij}, b_i, c_i ($i = 1, \ldots, s$, $j = 1, \ldots, i-1$), the method

$$\begin{aligned} k_1 &= f(t_n, x_n) \\ k_2 &= f(t_n + c_2 h, x_n + ha_{21}k_1) \\ k_3 &= f(t_n + c_3 h, x_n + h(a_{31}k_1 + a_{32}k_2)) \\ \cdots &\quad \cdots \\ k_s &= f(t_n + c_s h, x_n + h(a_{s1}k_1 + \cdots + a_{s,s-1}k_{s-1})) \\ x_{n+1} &= x_n + h(b_1 k_1 + \cdots + b_s k_s) \end{aligned} \tag{2.9}$$

is called an *s-stage explicit Runge–Kutta method* [79, 171]. Typically (but not always), $c_i = \sum_{j=1}^{i-1} a_{ij}$, so that all points where f is evaluated are first-order approximations to the solution.

2.2.2 General formulation

It is possible to consider even more general (implicit) methods for the first-order ordinary differential system (2.1) within this family simply by allowing coefficients $a_{i,j} \neq 0$ for $j = 1, \ldots, s$ in expression (2.9). In other words, the general class of *s-stage Runge–Kutta methods* is characterized by the real numbers a_{ij}, b_i $(i,j=1,\ldots,s)$ and $c_i=\sum_{j=1}^{s} a_{ij}$, as

$$\begin{aligned} k_i &= f\Big(t_n + c_i h, x_n + h\sum_{j=1}^{s} a_{ij}k_j\Big), \qquad i=1,\ldots,s \\ x_{n+1} &= x_n + h\sum_{i=1}^{s} b_i\, k_i. \end{aligned} \tag{2.10}$$

For simplicity, the associated coefficients are usually displayed with the so-called *Butcher tableau* [79, 171] as follows:

$$\begin{array}{c|ccc} c_1 & a_{11} & \ldots & a_{1s} \\ \vdots & \vdots & & \vdots \\ c_s & a_{s1} & \ldots & a_{ss} \\ \hline & b_1 & \ldots & b_s \end{array} \tag{2.11}$$

If the method is explicit ($a_{ij}=0$, $j\geq i$), the zero a_{ij} coefficients (in the upper triangular part of the tableau) are omitted for clarity. Notice that the computation of one step with an explicit RK method thus requires s evaluations of the function f. With the notation (2.11), the method of Heun (1.19) and the classical fourth-order Runge–Kutta method (2.7)–(2.8) can be expressed as

$$\begin{array}{c|cc} 0 & & \\ \frac{2}{3} & \frac{2}{3} & \\ \hline & \frac{1}{4} & \frac{3}{4} \end{array} \qquad\qquad \begin{array}{c|cccc} 0 & & & & \\ \frac{1}{2} & \frac{1}{2} & & & \\ \frac{1}{2} & 0 & \frac{1}{2} & & \\ 1 & 0 & 0 & 1 & \\ \hline & \frac{1}{6} & \frac{2}{6} & \frac{2}{6} & \frac{1}{6} \end{array} \tag{2.12}$$

respectively. If some $a_{ij}\neq 0$, $j\geq i$, the scheme is implicit and requires numerically solving a system of nonlinear equations at every step. More specifically, if we express method (2.10) in the alternative but equivalent form

$$\begin{aligned} Y_i &= x_n + h\sum_{j=1}^{s} a_{ij} f(t_n + c_j h, Y_j), \qquad i=1,\ldots,s \\ x_{n+1} &= x_n + h\sum_{i=1}^{s} b_i\, f(t_n + c_i h, Y_i), \end{aligned} \tag{2.13}$$

then the Y_i are determined by solving a system of sd nonlinear equations of the form

$$y = X_n + h\, G(h,y), \tag{2.14}$$

where $y=(Y_1,\ldots,Y_s)^T$, $X_n=(x_n,\ldots,x_n)^T\in\mathbb{R}^{sd}$ and G is a function which depends on the method. A standard procedure to get $Y_1,\ldots,Y_s$ from (2.14) is applying fixed point iteration:

$$y^{[j]} = X_n + h\, G(h, y^{[j-1]}), \qquad j=1,2,\ldots. \tag{2.15}$$

When h is sufficiently small, the iteration starts with $y^{[0]} = x_n$ and stops once $\|y^{[j]} - y^{[j-1]}\|$ is smaller than a prefixed tolerance. Of course, more sophisticated techniques can be used [171, 293]. Finally, x_{n+1} is computed with the last formula of (2.13).

As usual, a Runge–Kutta method is said to be of order r if for all sufficiently regular problems (2.1) the local error satisfies

$$x_1 - x(t_0 + h) = \mathcal{O}(h^{r+1}) \quad \text{as} \quad h \to 0$$

for the first step. The order can be checked by computing the Taylor series expansion of $x(t_0 + h)$ and x_1 around $h = 0$. By equating coefficients of successive powers of h in both expansions, it results in the following *order conditions*:

$$\sum_{i=1}^{s} b_i = 1 \qquad \text{(for order 1)}$$

$$\sum_{i=1}^{s} b_i c_i = 1/2 \qquad \text{(for order 2)}$$

$$\sum_{i=1}^{s} b_i c_i^2 = 1/3, \qquad \sum_{i,j=1}^{s} b_i a_{ij} c_j = 1/6 \qquad \text{(for order 3)}$$

and four more equations for order 4 [171]. The analysis becomes increasingly difficult for higher orders and thus special techniques have been developed to systematically obtain the order conditions for an arbitrary order. In this context, the so-called B-series constitute an invaluable tool to derive the order conditions a method has to satisfy, as well as to analyze the accuracy and qualitative properties of the numerical integrator [78, 77, 173, 264].

2.2.3 Collocation methods

Many Runge–Kutta methods can be derived by applying the idea of *collocation*. The starting point is to choose the number of stages s and abscissae c_i, $i = 1, \ldots, s$. Next one determines a collocation polynomial $u(t)$ of degree $\leq s$ such that

$$\begin{aligned} u(t_0) &= x_0 \\ \dot{u}(t_0 + c_i h) &= f(t_0 + c_i h, u(t_0 + c_i h)), \qquad i = 1, \ldots, s \end{aligned}$$

and then define the numerical solution as $x_1 = u(t_0 + h)$ [293]. The internal stages k_i in (2.10) are then the values of the collocation polynomial at the intermediate points $t_0 + c_i h$:

$$\dot{u}(t_0 + c_i h) = k_i, \qquad i = 1, \ldots, s.$$

For $s = 1$ the polynomial is of the form $u(t) = x_0 + (t - t_0)k_1$, with

$$k_1 = f(t_0 + c_1 h, x_0 + c_1 h k_1).$$

With $c_1 = 0$, $c_1 = 1$ we recover the explicit and implicit Euler methods, respectively, whereas $c_1 = 1/2$ leads to the midpoint rule

$$x_{n+1} = x_n + h\, f\big(t_0 + h/2, (x_n + x_{n+1})/2\big). \tag{2.16}$$

The three of them are thus collocation methods.

The weights b_i of the RK scheme obtained by collocation coincide with the weights of the interpolatory quadrature rule in the interval $[0, 1]$ based on the abscissae c_i [293]. In other words, the b_i, $i = 1, \ldots, s$, are the unique choice such that the quadrature formula

$$\int_0^1 p(t)dt \approx \sum_{i=1}^{s} b_i\, p(c_i) \tag{2.17}$$

gives no error if the integrand $p(t)$ is a polynomial of degree $\leq s-1$. The order of any s-stage RK method obtained by collocation is $\geq s$. Order $r = s+\sigma$, $\sigma = 1, 2, \ldots$ can be achieved if the quadrature (2.17) integrates exactly polynomials of degree $\leq s + \sigma - 1$. The maximum order corresponds to $\sigma = s$, which is obtained when the abscissae c_i are the zeros of the s-th shifted Legendre polynomial

$$\frac{d^s}{dx^s}\big(x^s(x-1)^s\big).$$

The corresponding RK method is called the *Gauss s-stage method* and is of order $2s$. In accordance with the preceding comments, no other RK method achieves order $2s$ with s stages [293].

For $s = 1$ (order 2) one has $a_{11} = 1/2$, $b_1 = 1$ and the method reads

$$\begin{aligned} Y_1 &= x_0 + \frac{1}{2} h f(t_0 + h/2, Y_1) \\ x_1 &= x_0 + h f(t_0 + h/2, Y_1) \end{aligned}$$

whence, by eliminating f from both equations, leads to $Y_1 = (x_0 + x_1)/2$, so that the implicit midpoint rule (2.16) is recovered. The case $s = 2$ (order 4) is also widely used in applications. The corresponding tableaux are

$$s = 1: \quad \begin{array}{c|c} \frac{1}{2} & \frac{1}{2} \\ \hline & 1 \end{array} \qquad\qquad s = 2: \quad \begin{array}{c|cc} \frac{1}{2} - \frac{\sqrt{3}}{6} & \frac{1}{4} & \frac{1}{4} - \frac{\sqrt{3}}{6} \\ \frac{1}{2} + \frac{\sqrt{3}}{6} & \frac{1}{4} + \frac{\sqrt{3}}{6} & \frac{1}{4} \\ \hline & \frac{1}{2} & \frac{1}{2} \end{array} \tag{2.18}$$

respectively. Other choices of the quadrature rule result in different collocation methods (e.g., Radau and Lobatto methods, etc. [170]).

2.3 Partitioned Runge–Kutta methods

Sometimes it is preferable to integrate some components of $\dot{x} = f(t,x)$ with a numerical method and the remaining components with a different method. Suppose we write $x = (q,p)^T$ (with q and p possibly of different dimension) and partition accordingly the right-hand side function as $f = (f_1, f_2)^T$ [293]. If, for the time being, we only consider the autonomous case, we then have

$$\dot{q} = f_1(q,p), \qquad \dot{p} = f_2(q,p). \tag{2.19}$$

Now let us assume that the q components are integrated by an RK method and the p components are integrated with a different RK formula. Then the overall scheme is called a *Partitioned Runge–Kutta* (PRK) method and is defined by

$$\begin{aligned} k_i &= f_1\Big(q_0 + h\sum_{j=1}^{s} a_{ij}k_j, p_0 + h\sum_{j=1}^{s} \hat{a}_{ij}\ell_j\Big) \\ \ell_i &= f_2\Big(q_0 + h\sum_{j=1}^{s} a_{ij}k_j, p_0 + h\sum_{j=1}^{s} \hat{a}_{ij}\ell_j\Big) \\ q_1 &= q_0 + h\sum_{i=1}^{s} b_i k_i, \qquad p_1 = p_0 + h\sum_{i=1}^{s} \hat{b}_i \ell_i. \end{aligned} \tag{2.20}$$

If the PRK method (2.20) is of order r, then the RK method with coefficients (a_{ij}, b_j) and the RK method with coefficients $(\hat{a}_{ij}, \hat{b}_j)$ are both of order r. The converse is not true, however: if both RK schemes are of order r, then the combined PRK method may be of order $< r$ [291].

Example 2.2. If equations (2.19) are derived from a Hamiltonian function $H(q,p)$ as

$$f_1(q,p) = \nabla_p H(q,p), \qquad f_2(q,p) = -\nabla_q H(q,p), \tag{2.21}$$

and we take $s = 1$, $b_1 = 1$, $a_{11} = 1$ (the implicit Euler method), and $\hat{b}_1 = 1$, $\hat{a}_{11} = 0$ (the explicit Euler method) we get

$$k_1 = f_1(q_0 + hk_1, p_0), \qquad \ell_1 = f_2(q_0 + hk_1, p_0)$$

and therefore

$$\begin{aligned} q_1 &= q_0 + h\,k_1 = q_0 + hf_1(q_1, p_0) = q_0 + h\,\nabla_p H(q_1, p_0) \\ p_1 &= p_0 + h\,\ell_1 = p_0 + hf_2(q_1, p_0) = p_0 - h\,\nabla_q H(q_1, p_0), \end{aligned}$$

i.e., we recover the symplectic Euler method (1.40). On the other hand, the Störmer–Verlet method of the form (1.46) applied to (2.19) with (2.21) *is* a PRK method with coefficients given by the following tableaux:

$$\begin{array}{c|cc} 0 & & \\ 1 & 1/2 & 1/2 \\ \hline & 1/2 & 1/2 \end{array} \qquad\qquad \begin{array}{c|cc} 1/2 & 1/2 & \\ 1/2 & 1/2 & \\ \hline & 1/2 & 1/2 \end{array} \tag{2.22}$$

In general, when PRK methods are applied to separable Hamiltonian systems, one tableau is used for the q variables and the other for the p variables [293]. Other examples of PRK methods used in geometric numerical integration are the Lobatto IIIA-IIIB methods, of order $2s-2$ [170, 171]. □

2.4 Runge–Kutta–Nyström methods

In the particular case of systems of the form

$$\dot{q} = v, \qquad\qquad \dot{v} = g(t,q), \tag{2.23}$$

or equivalently

$$\ddot{q} = g(t,q), \tag{2.24}$$

the k_i variables can be eliminated in (2.20) so that we can write

$$\begin{aligned} \ell_i &= g\Big(t_0 + c_i h, q_0 + hc_i v_0 + h^2 \sum_{j=1}^{s} \tilde{a}_{ij}\ell_j\Big), \\ q_1 &= q_0 + hv_0 + h^2 \sum_{i=1}^{s} \tilde{b}_i \ell_i, \qquad v_1 = v_0 + h\sum_{i=1}^{s} \hat{b}_i \ell_i, \end{aligned} \tag{2.25}$$

where $c_i = \sum_{j=1}^{s} a_{ij}$ and the coefficients $\tilde{b}_i$, $\tilde{a}_{ij}$ are related with those of method (2.20) as follows. Denoting by A, $\hat{A}$ and $\tilde{A}$ the matrices whose elements are a_{ij}, $\hat{a}_{ij}$ and $\tilde{a}_{ij}$, respectively, one has

$$\tilde{A} = A\hat{A}, \qquad \tilde{b} = \hat{A}^T b.$$

The resulting scheme is called a *Runge–Kutta–Nyström* (RKN) method [293]. As we have seen in Chapter 1, one important application of this class of schemes is the numerical simulations arising in Newtonian mechanics, with the force being proportional to the acceleration (second derivative of position).

Example 2.3. In the autonomous case, $\ddot{q} = g(q)$, the velocity Verlet method in the form (1.28) is a RKN method. It can be recovered from (2.25) (when $g = F/m$) by taking $\tilde{a}_{11} = \tilde{a}_{12} = \tilde{a}_{22} = 0$, $\tilde{a}_{21} = 1/2$, $\tilde{b}_1 = 1/2$, $\tilde{b}_2 = 0$, $\hat{b}_1 = \hat{b}_2 = 1/2$ and $c_1 = 0$, $c_2 = 1$. □

Equation (2.24) written as the first-order differential system (2.23), has a vector field with a very particular structure, which is reflected in this class of methods. Thus, the number of order conditions is reduced for methods of order ≥ 4, and this results in schemes with less stages. Thus, in particular, the fourth-order method with tableau [171]

$$\begin{array}{c|c} c & \tilde{A} \\ \hline & \bar{b} \\ \hline & \hat{b} \end{array} \quad \equiv \quad \begin{array}{c|ccc} 0 & & & \\ 1/2 & 1/8 & & \\ 1 & 0 & 1/2 & \\ \hline & 1/6 & 1/3 & 0 \\ \hline & 1/6 & 4/6 & 1/6 \end{array} \tag{2.26}$$

has only three stages.

2.5 Preservation of invariants in Runge–Kutta schemes

Let us now examine how Runge–Kutta methods behave with respect to the preservation of first integrals, and in particular the conditions the coefficients of a given RK, PRK or RKN method must satisfy to ensure this property.

We recall that if the function $I(x)$ is such that $\nabla I(x) f(x) = 0$ for the differential equation $\dot{x} = f(x)$, then it is preserved along the motion: $I(x(t)) = I(x_0)$ for all t. It turns out that B-series (and its generalizations) also constitute a very powerful tool in this analysis [264].

It can be easily shown that all explicit and implicit Runge–Kutta methods conserve linear invariants, whereas partitioned Runge–Kutta methods conserve linear invariants if $b_i = \hat{b}_i$ for all i [170].

On the other hand, if the coefficients of a RK method satisfy the relations

$$b_i a_{ij} + b_j a_{ji} - b_i b_j = 0 \quad \text{for all} \quad i, j = 1, \ldots, s \tag{2.27}$$

then the scheme preserves quadratic invariants of the form

$$I(x) = x^T C x + d^T x,$$

with C a symmetric matrix and d a vector. As a matter of fact, the only collocation methods that verify (2.27) are the Gauss methods. What is more remarkable is that every RK method satisfying (2.27), when applied to the integration of the Hamiltonian system (1.33), is also a symplectic method. This important result was discovered independently by Lasagni [211], Sanz-Serna [287] and Suris [307] in 1988. Thus, the Gauss collocation methods of section 2.2.3 are symplectic. That condition (2.27) is also necessary for symplecticity can be shown in particular by using B-series.

The relations (2.27) guarantee that the resulting RK method preserves two different properties: quadratic invariants and symplecticity. As a matter of fact, both properties are related: the preservation of symplecticity by the Hamiltonian flow may be considered as an integral of the flow corresponding to the associated variational system [66]. In addition, by imposing (2.27) on the coefficients, the order conditions of an RK method are no longer independent [292]. Thus, for instance, if (2.27) holds, then the second-order condition in (2.16) is a consequence of the first one and thus symplectic RK methods of order ≥ 1 are automatically of order ≥ 2 [291].

Although symplectic Runge–Kutta methods preserve linear and quadratic integrals of motion, no other integrals can be preserved. If the integral of motion is, say, cubic, then the RK scheme preserves either the symplectic structure or the integral, but not both.

Notice that if we put $i = j$ in (2.27), we get $2a_{ii} - b_i = 0$ for $b_i \neq 0$, which implies that $a_{ii} \neq 0$. Therefore, *symplectic Runge–Kutta methods are necessarily implicit.*

The conditions for a partitioned Runge–Kutta method to be symplectic when applied to autonomous Hamiltonian systems are [262, 293]

$$\begin{aligned} b_i \hat{a}_{ij} + \hat{b}_j a_{ji} - b_i \hat{b}_j &= 0 \qquad \text{for} \quad i,j = 1,\dots,s \\ b_i - \hat{b}_i &= 0 \qquad \text{for} \quad i = 1,\dots,s, \end{aligned} \tag{2.28}$$

whereas if the system is explicitly time-dependent, then one must impose in addition [291]

$$c_i = \hat{c}_i, \qquad i = 1,\dots,s.$$

Again, symplectic PRK methods are implicit when applied to general Hamiltonian systems (1.32), but if H is separable, $H(q,p) = T(p) + V(q)$, then explicit schemes do exist (although they are equivalent to splitting methods, as we will make clear in Chapter 3).

Symplecticity conditions (2.28) also imply preservation of invariants of the form

$$I(q,p) = q^T C p + d_1^T q + d_2^T p,$$

where now C is a constant square matrix and d_1, d_2 are constant vectors [218]. First integrals of this type arise from the symmetry of H under linear canonical transformations.

Finally, if the coefficients of a Runge–Kutta–Nyström method satisfy

$$\begin{aligned} \tilde{b}_i &= \hat{b}_i(1 - c_i) \qquad \text{for} \quad i = 1,\dots,s \\ \hat{b}_i(\tilde{b}_j - \tilde{a}_{ij}) &= \hat{b}_j(\tilde{b}_i - \tilde{a}_{ji}) \qquad \text{for} \quad i,j = 1,\dots,s \end{aligned} \tag{2.29}$$

then the method is symplectic when applied to Hamiltonian systems of the form [293]

$$H(q,p) = \frac{1}{2} p^T M^{-1} p + V(q).$$

One can easily check that the coefficients of method (2.26) do not satisfy these symplectic conditions.

2.6 Numerical examples

For illustrating the performance of the different types of numerical integrators reviewed in this chapter, we introduce two systems widely used as a test bench in addition to the mathematical pendulum and the gravitational two-body (Kepler) problem of Chapter 1.

(i) Lorenz equations. These quadratic ordinary differential equations represent three modes (one in velocity and two in temperature) of the Oberbeck–Boussinesq equations for fluid convection in a two-dimensional layer heated from below and read

$$\begin{aligned} \dot{x} &= \sigma(y-x), \\ \dot{y} &= rx - y - xz, \\ \dot{z} &= -bz + xy, \end{aligned} \tag{2.30}$$

where $\sigma, r, \beta > 0$ are real parameters. As is well known, for certain values of these parameters, the system possesses a chaotic attractor [166].

(ii) ABC flow. The ABC flow (after Arnold, Beltrami and Childress) is a widely studied example of a volume-preserving three-dimensional flow. Its phase space is the 3-torus and its equations are

$$\begin{aligned} \dot{x} &= B\cos y + C\sin z, \\ \dot{y} &= C\cos z + A\sin x, \\ \dot{z} &= A\cos x + B\sin y. \end{aligned} \tag{2.31}$$

Also in this case the solutions depend in a very sensitive way on the initial conditions [170].

To test the performance of the different methods, the following numerical integrations are carried out:

- The pendulum (1.12) with $k = 1$ and initial conditions $q(0) = 2.5$, $p(0) = 0$ along the time span $t \in [0, 20]$. We measure the maximum error in energy (which is a first integral) along the integration for each numerical method.
- The Kepler problem (1.57) with $\mu = 1$ and initial conditions (1.61) with $e = 0.2$ along the time span $t \in [0, 20]$. Again, we measure the maximum error in energy along the integration.
- The Lorenz equations (2.30) with $\sigma = 10, r = 28, b = 8/3$ (for which a chaotic attractor exists) and initial conditions $x(0) = 1, y(0) = 2, z(0) = 3$ in the time interval $t \in [0, 5]$. We now compute $\|\mathbf{x}_{ex}(5) - \mathbf{x}_{ap}(5)\|$, where $\mathbf{x}_{ex}$ is the exact (vector) solution computed numerically to high accuracy and $\mathbf{x}_{ap}$ is the solution provided by a given numerical method.

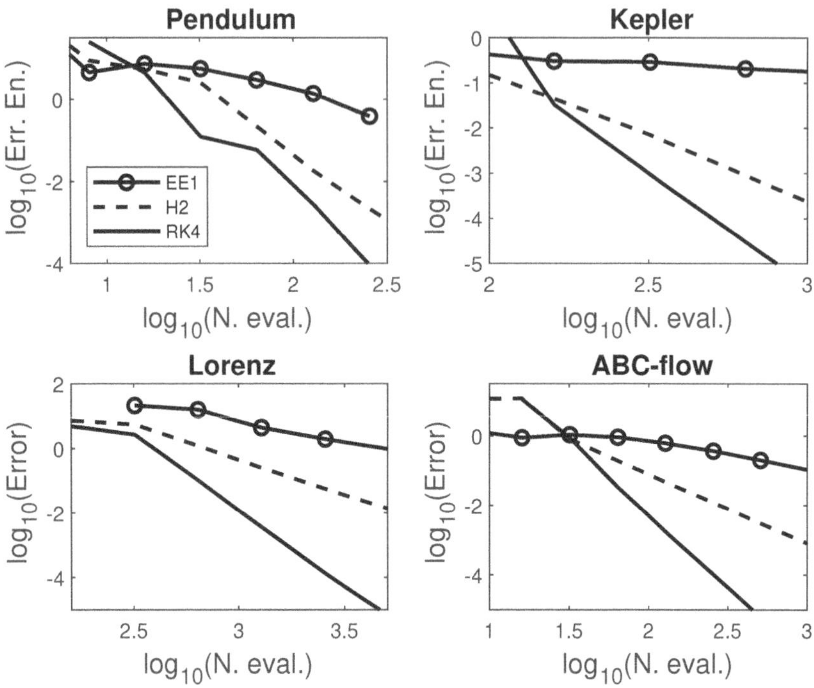

FIGURE 2.1: (Top) Maximum error in energy versus number of force evaluations and (bottom) error in coordinates at the final time versus number of evaluations of the vector field for the first-order explicit Euler method (EE1), the two-stage second-order Heun method (H2), and the standard four-stage fourth-order RK method (RK4).

- The ABC flow equations (2.31) with $A = 1, B = 2, C = 3$ and initial conditions $x(0) = 1, y(0) = 2, z(0) = 3$ for $t \in [0, 5]$. Here we also compute $\|\mathbf{x}_{ex}(5) - \mathbf{x}_{ap}(5)\|$.

We first compare the performance of three Runge–Kutta methods: the first-order explicit Euler method (EE1), the two-stage second-order Heun's method (H2), and the classical fourth-order scheme (2.7)–(2.8) (RK4). Figure 2.1 collects the efficiency diagrams corresponding to each of the aforementioned systems. As predicted, the first-order method shows a very poor performance, the second-order scheme provides a clear improvement, but it is the fourth-order method that is the most efficient when accuracy is an issue.

Figure 2.2 shows the error in energy and position (in phase space) as a function of time for $t \in [0, 300]$ and $e = 0.2$ for the Kepler problem achieved by some of the previous fourth-order methods, in particular: RK4, the classical RK method (2.7); RKN4, the three-stage RKN method with coefficients given in (2.26); and RKGL4, the two-stage implicit RK method with coefficients

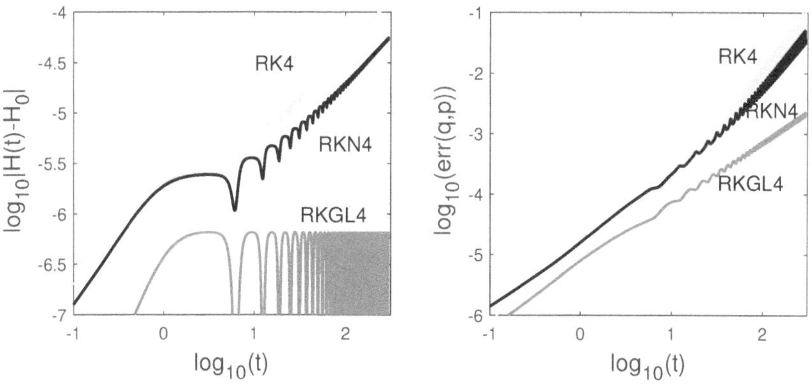

FIGURE 2.2: Errors in the numerical integration of the Kepler problem with initial conditions given by (1.61) with $e = 0.2$ and time step $h = \frac{1}{10}$ using the following fourth-order methods: (RK4) the four-stage RK method (2.7); (RKN4) the three-stage RKN method with coefficients given in (2.26); and (RKGL4) the two-stage implicit RK method with coefficients given in (2.18).

given in (2.18), and all of them using the same time step $h = \frac{1}{10}$. We observe that the implicit RKGL4 method shows no secular error growth in energy and only linear error growth in positions due to its symplectic character.

Finally, we collect in Figure 2.3 the efficiency plots corresponding to these fourth-order methods applied to the test systems enumerated above. To estimate the cost of implicit methods, we take two as the number of iterations per step required, for a total of six evaluations of the vector field per step. The RKN method is used only for the pendulum and Kepler problems since they can be written as second-order differential equations. For these two problems, it shows a clear improvement with respect to the RK4 method. The implicit RKGL4 method is, in most cases, the most efficient one, and this relative improvement grows when solving Hamiltonian problems over longer time intervals, but requires numerically solving implicit equations whose convergence could fail in some cases. Efficient implementations have been proposed, however [12].

2.7 Exercises

1. Show that the classical explicit Runge–Kutta method (2.7) is indeed a method of at most order 4. (*Hint*: apply the scheme to equation $\dot{x} = t^\alpha$, $\alpha = 0, 1, 2, 3, 4$).

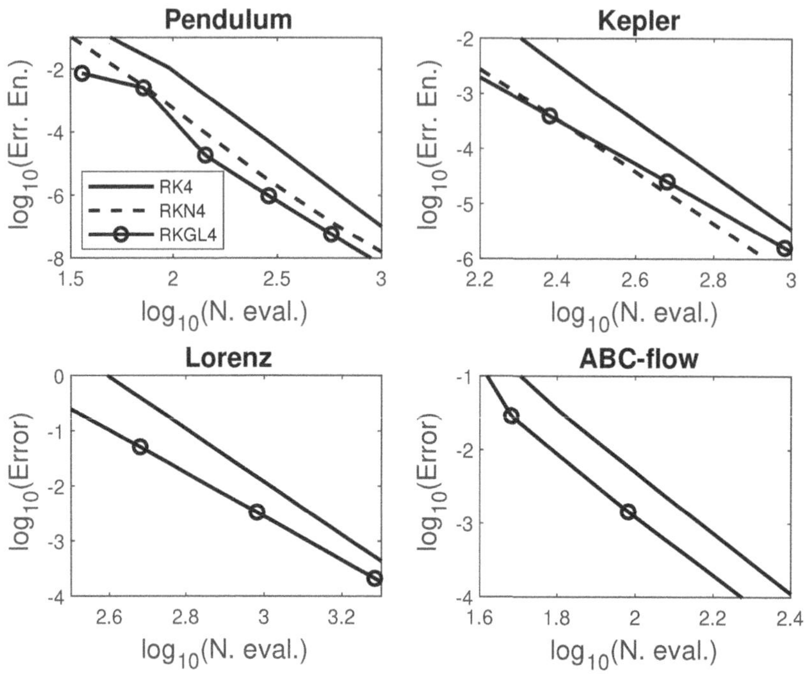

FIGURE 2.3: Efficiency plots with respect to energy (top) and coordinates (bottom) obtained with the fourth-order methods considered in Figure 2.2 (the RKN method is only used for solving second-order differential equations, top figures).

2. Verify that the classical explicit Runge–Kutta method (2.7) satisfies the order conditions (2.16).

3. Verify explicitly that the Störmer–Verlet method (1.46) applied to equations (2.19) with (2.21) is a partitioned Runge–Kutta method with coefficients given by (2.22).

4. (Hairer–Lubich–Wanner, [170]). Suppose that b_i, a_{ij} and b_i, $\hat{a}_{ij}$ are the coefficients of two Runge–Kutta methods. An additive Runge–Kutta method for equation $\dot{x} = f^{[1]}(x) + f^{[2]}(x)$ reads

$$
\begin{aligned}
k_i &= f^{[1]}\big(x_0 + h\sum_{j=1}^{s} a_{ij}k_j\big) + f^{[2]}\big(x_0 + h\sum_{j=1}^{s} \hat{a}_{ij}k_j\big) \\
x_1 &= x_0 + h\sum_{i=1}^{s} b_i k_i.
\end{aligned}
$$

Show that this scheme is a partitioned Runge–Kutta method applied to the equation

$$\dot{x} = f^{[1]}(x) + f^{[2]}(y), \qquad \dot{y} = f^{[1]}(x) + f^{[2]}(y)$$

with $x(0) = y(0) = x_0$.

5. Given the one-dimensional Hamiltonian

$$H = \frac{1}{2}p^2 + V(q), \tag{2.32}$$

show that the second-order Taylor method applied to the associated Hamilton equations verifies

$$\det \begin{pmatrix} \dfrac{\partial q_{n+1}}{\partial q_n} & \dfrac{\partial q_{n+1}}{\partial p_n} \\ \dfrac{\partial p_{n+1}}{\partial q_n} & \dfrac{\partial p_{n+1}}{\partial p_n} \end{pmatrix} = 1 + \frac{h^3}{2} V_{qqq}(q_n)p_n + \frac{h^4}{4} V_{qq}(q_n)^2.$$

Write also the fourth-order Taylor method in terms of $V(q)$ and its derivatives, and apply it to the pendulum problem.

6. Consider the d-dimensional Hamiltonian

$$H = \frac{1}{2}p^T M^{-1} p + V(q), \tag{2.33}$$

with M a symmetric positive definite matrix. Write the second-order Taylor method applied to the associated Hamilton equations and apply it to the Kepler problem.

7. Consider the matrix differential equation

$$\dot{X} = AX, \qquad X(0) = I$$

where X, A are $d \times d$ matrices. The explicit and implicit RK methods can be written in this case as polynomial and rational approximations to the exact solution, $X(t) = \mathrm{e}^{tA}$. *(i)* Write the general two-stage second-order RK method (2.4) and the fourth-order method (2.7) applied to this problem. Compare the results with the second- and fourth-order Taylor approximations to the exponential. *(ii)* Write the implicit second- and fourth-order Gauss methods given in (2.18) as an explicit method for this problem. Compare the results with the second- and fourth-order Padé approximations to the exponential, i.e.,

$$P_2(z) = \frac{1 + z/2}{1 - z/2}, \qquad P_4(z) = \frac{1 + z/2 + z^2/12}{1 - z/2 + z^2/12},$$

where $P_n(z) = \mathrm{e}^z + \mathcal{O}(z^{n+1})$.

8. Consider the van der Pol equation

$$\ddot{y} + a(1 - y^2)\dot{y} + y = 0 \tag{2.34}$$

with $y(0) = 1$, $\dot{y}(0) = 0$ and $a = \frac{1}{4}$. Write the equation as a first-order system of equations and compute numerically the exact solution at $T = 5$. Obtain the efficiency plot of the fourth-order methods used in Figure 2.3.

Chapter 3

Splitting and composition methods

The geometric numerical integrators introduced in Chapter 1, such as symplectic Euler schemes and Störmer–Verlet methods, provide, in spite of their low order of accuracy, a fairly good qualitative description of dynamical systems possessing distinctive geometric features. Although this low order does not constitute a hindrance for, say, molecular dynamics applications, there are other important areas of applications where a higher degree of precision is very welcome in addition to the preservation of qualitative properties. A familiar example is the long-term numerical integration of the Solar System, both forward (to analyze e.g., the existence of chaos [212, 309]) and backward in time (to study the insolation quantities of the Earth [214]).

In this chapter we review two popular techniques to raise the order of geometric integrators and achieve the desired goal of constructing very accurate high-order structure-preserving algorithms: *composition* and *splitting*.

In a *composition method*, its associated mapping ψ_h is a composition of several simpler maps for the problem at hand. The situation is the following: sometimes one has a method ϕ_h (the basic method) with good geometric properties and the idea is to construct a new method of the form

$$\psi_h = \phi_{\gamma_s h} \circ \phi_{\gamma_{s-1} h} \circ \cdots \circ \phi_{\gamma_1 h}, \tag{3.1}$$

where the γ_i are suitable constants chosen in such a way that ψ_h is of higher order than ϕ_h. In this way one increases the order of accuracy while preserving the desirable properties the basic method ϕ_h has, as long as the geometric property is preserved by composition. This is true, in particular, if ϕ_h is symplectic or volume preserving. In addition, if ϕ_h is time-symmetric and the coefficients in (3.1) verify $\gamma_i = \gamma_{s+1-i}$ for $i = 1, 2, \ldots$, then the composition ψ_h is also time-symmetric. Simple examples of symmetric composition methods are the Störmer–Verlet schemes (1.48) and (1.49) of Chapter 1.

With respect to *splitting*, suppose that the (time-independent) vector field in the differential equation

$$\dot{x} = f(x), \qquad x(0) = x_0 \in \mathbb{R}^d \tag{3.2}$$

can be decomposed as

$$f = f_1 + \cdots + f_m, \qquad m \geq 2, \tag{3.3}$$

DOI: 10.1201/9781003527626-3

so that each initial value problem $\dot{x} = f_j(x)$, $x(0) = x_0$ is easier to solve than (3.2). Most commonly, they can be integrated exactly in closed form. Then, by combining the solutions of each subproblem in an appropriate way, it is possible to get accurate approximations to the solution of (3.2). This, in a nutshell, is the underlying idea of *splitting methods*.

We provide in this chapter a short review of splitting and composition methods, since they constitute a very important class of geometric integrators. We first relate them with some of the low-order algorithms we have introduced in Chapter 1. Then we examine the conditions they have to satisfy to achieve a higher order of approximation, the simplifications that take place when the equation (3.2) possesses some special structure, the idea of processing a given integrator, and how the methods can be adapted to deal with non-autonomous problems. Finally, we collect some recommended schemes and illustrate them with numerical examples.

A more detailed treatment of splitting and composition methods can be found in the review papers [243] and [52].

3.1 Lie–Trotter and Strang splitting

One of the simplest examples corresponds to a linear differential equation defined in $\mathbb{R}^d$ and $m = 2$, namely

$$\dot{x} = f_1(x) + f_2(x) = F_1\, x + F_2\, x, \qquad x(0) = x_0, \tag{3.4}$$

where F_1 and F_2 are $d \times d$ matrices and $x \in \mathbb{R}^d$, and whose solution reads

$$x(t) = \mathrm{e}^{t(F_1+F_2)} x_0.$$

Closely associated with (3.4) is the matrix differential equation

$$\frac{dX}{dt} = (F_1 + F_2)X, \qquad X(0) = I, \tag{3.5}$$

in the sense that the solution of (3.4) can be written as $x(t) = X(t)\, x_0$, where $X(t)$ is the solution of (3.5),

$$X(t) = \mathrm{e}^{t(F_1+F_2)}.$$

It often happens that evaluating the action of $X(t)$ on x_0 is difficult and/or computationally expensive. In that case, we can get an approximate solution at the final time $t = t_{\mathrm{f}}$ by subdividing the interval $[0, t_{\mathrm{f}}]$ into N steps of length h, with $Nh = t_{\mathrm{f}}$, and computing the sequence

$$x_{n+1} = \mathrm{e}^{hF_2}\, \mathrm{e}^{hF_1}\, x_n, \qquad n \geq 0, \tag{3.6}$$

so that $x_{n+1} \approx x(t_{n+1})$ with $t_{n+1} = (n+1)h$). This is the so-called *Lie–Trotter scheme.* When the matrices commute, the sequence produces the exact solution. Otherwise, a direct calculation shows that

$$e^{h(F_1+F_2)} - e^{hF_2}e^{hF_1} = \frac{1}{2}h^2[F_1, F_2] + \mathcal{O}(h^3) \qquad \text{as} \qquad h \to 0,$$

and so the previous approximation is only of first order of accuracy. Another version of the method is possible, of course, by reversing the order of F_1 and F_2, namely

$$x_{n+1} = e^{hF_1}\, e^{hF_2}\, x_n, \qquad n \geq 0 \tag{3.7}$$

has the same order of accuracy and properties as (3.6). If we have $m > 2$ matrices in (3.5), the Lie–Trotter scheme reads

$$x_{n+1} = e^{hF_m} \cdots e^{hF_2}\, e^{hF_1}\, x_n$$

(or any other permutation of the matrices F_j).

If we instead consider a symmetrized version of (3.6), then

$$e^{h(F_1+F_2)} - e^{\frac{1}{2}hF_1}e^{hF_2}e^{\frac{1}{2}hF_1} = Ch^3 + \mathcal{O}(h^4), \tag{3.8}$$

where the constant C can be obtained either by comparing Taylor series or by applying the Baker–Campbell–Hausdorff (BCH) formula (see Appendix A.4) as $C = \frac{1}{24}([F_1, [F_1, F_2]] + 2[F_2, [F_1, F_2]])$. Therefore, the sequence

$$x_{n+1} = e^{\frac{1}{2}hF_1}e^{hF_2}e^{\frac{1}{2}hF_1}\, x_n, \qquad n \geq 0 \tag{3.9}$$

produces a second-order approximation for the solution of (3.4). This corresponds to the *Strang* (splitting) *scheme.* If the roles of F_1 and F_2 are interchanged, one has another version of the Strang splitting scheme.

Simple generalizations to the case $m > 2$ include, in particular, the product

$$e^{\frac{1}{2}hF_1}\, e^{\frac{1}{2}hF_2} \cdots e^{hF_m} \cdots e^{\frac{1}{2}hF_2}\, e^{\frac{1}{2}hF_1}.$$

Higher-order splitting methods could in principle be constructed by including more exponentials with their corresponding coefficients in a time step, namely

$$\Psi(h) = e^{a_{s+1}hF_1}\, e^{b_s hF_2}\, e^{a_s hF_1} \cdots e^{b_1 hF_2}\, e^{a_1 hF_1}. \tag{3.10}$$

The number s (called the *number of stages*) as well as the coefficients a_j, b_j are chosen so that

$$\Psi(h) = e^{h(F_1+F_2)} + \mathcal{O}(h^{r+1}) \qquad \text{as} \qquad h \to 0$$

for a given order r. In (3.10), the first and last exponentials correspond to F_1. This format is convenient for implementation, since the last exponential in one step can be concatenated with the first at the next step, thus reducing by one the number of evaluations. This corresponds to the well known FSAL

(First Same As Last) property. The situation when one has an exponential of F_2 as the first and last term is recovered by taking $a_1 = a_{s+1} = 0$.

Example 3.1. Consider again the simple harmonic oscillator (1.1), with $k = m = 1$ for simplicity. The equations of motion (1.2) can be written as

$$\begin{aligned}\dot{x} \equiv \begin{pmatrix} \dot{q} \\ \dot{p} \end{pmatrix} &= \underbrace{\begin{pmatrix} p \\ 0 \end{pmatrix}}_{f_1} + \underbrace{\begin{pmatrix} 0 \\ -q \end{pmatrix}}_{f_2} = \Bigg[\underbrace{\begin{pmatrix} 0 & 1 \\ 0 & 0 \end{pmatrix}}_{F_1} + \underbrace{\begin{pmatrix} 0 & 0 \\ -1 & 0 \end{pmatrix}}_{F_2}\Bigg]\begin{pmatrix} q \\ p \end{pmatrix} \\ &= (F_1 + F_2)\, x, \end{aligned} \tag{3.11}$$

with solution (1.5) for $t = h$

$$x(h) = \mathrm{e}^{h(F_1+F_2)} x_0 = \begin{pmatrix} \cos h & \sin h \\ -\sin h & \cos h \end{pmatrix} \begin{pmatrix} q_0 \\ p_0 \end{pmatrix}, \tag{3.12}$$

whereas

$$\mathrm{e}^{hF_1} = I + hF_1 = \begin{pmatrix} 1 & h \\ 0 & 1 \end{pmatrix}, \qquad \mathrm{e}^{hF_2} = I + hF_2 = \begin{pmatrix} 1 & 0 \\ -h & 1 \end{pmatrix}.$$

Then, the Lie–Trotter scheme (3.7)

$$x(h) \simeq x_1 = \mathrm{e}^{hF_1}\mathrm{e}^{hF_2} x_0 = \begin{pmatrix} 1-h^2 & h \\ -h & 1 \end{pmatrix} \begin{pmatrix} q_0 \\ p_0 \end{pmatrix} \tag{3.13}$$

reproduces the numerical solution obtained by the symplectic Euler-VT method (1.10), whereas the sequence (3.6) leads to the symplectic Euler-TV scheme:

$$x(h) \simeq x_1 = \mathrm{e}^{hF_2}\mathrm{e}^{hF_1} x_0 = \begin{pmatrix} 1 & h \\ -h & 1-h^2 \end{pmatrix} \begin{pmatrix} q_0 \\ p_0 \end{pmatrix}.$$

Proceeding analogously, we can show that the Strang splitting (3.9) and $x_{n+1} = \mathrm{e}^{\frac{h}{2}F_2}\mathrm{e}^{hF_1}\mathrm{e}^{\frac{h}{2}F_2} x_n$ lead to both variants (1.52) and (1.53) of the Störmer–Verlet method. Then, it is clear that the symplectic Euler and the Störmer–Verlet schemes *are* particular examples of splitting methods. □

The Lie–Trotter scheme can be easily generalized to any system (3.2) when the solution is no longer given by exponentials, as in the linear case. If $m = 2$, it is equivalent to

Algorithm 1 (Lie–Trotter). Starting from $x_0 = x(0)$, for $n \geq 0$:

- solve $\dot{y}_1 = f_1(y_1)$, $y_1(t_n) = x_n$, in $[t_n, t_{n+1}]$;
- set $y_{n+\frac{1}{2}} = y_1(t_{n+1})$;
- solve $\dot{y}_2 = f_2(y_2)$, $y_2(t_n) = y_{n+\frac{1}{2}}$, in $[t_n, t_{n+1}]$;

- finally, set $x_{n+1} = y_2(t_{n+1})$.

Alternatively, Algorithm 1 can be formally expressed as

$$x_{n+1} = \chi_h(x_n) \equiv \varphi_h^{[2]}\big(\varphi_h^{[1]}(x_n)\big) = \big(\varphi_h^{[2]} \circ \varphi_h^{[1]}\big)(x_n), \tag{3.14}$$

where we have denoted by $\varphi_t^{[j]}(x_0)$ the solutions of the subproblems $\dot{y} = f_j(y)$, $y(0) = x_0$. Analogously, the Strang splitting (3.9) is generalized as

Algorithm 2 (Strang). From $x_0 = x(0)$, for $n \geq 0$:

- solve $\dot{y}_1 = f_1(y_1),\ \ y_1(t_n) = x_n$, in $[t_n, t_{n+\frac{1}{2}}]$, with $t_{n+\frac{1}{2}} = (n + \frac{1}{2})h$;
- set $y_{n+\frac{1}{2}} = y_1(t_{n+\frac{1}{2}})$;
- solve $\dot{y}_2 = f_2(y_2),\ \ y_2(t_n) = y_{n+\frac{1}{2}}$, in $[t_n, t_{n+1}]$;
- set $\hat{y}_{n+\frac{1}{2}} = y_2(t_{n+1})$;
- solve $\dot{y}_1 = f_1(y_1),\ \ y_1(t_{n+\frac{1}{2}}) = \hat{y}_{n+\frac{1}{2}}$, in $[t_{n+\frac{1}{2}}, t_{n+1}]$;
- finally, set $x_{n+1} = y_1(t_{n+1})$

or, in short,

$$x_{n+1} = S_h^{[2]}(x_n) \equiv \big(\varphi_{h/2}^{[1]} \circ \varphi_h^{[2]} \circ \varphi_{h/2}^{[1]}\big)(x_n). \tag{3.15}$$

Even in this case, when the Lie–Trotter method and the Strang splitting are applied to nonlinear ODEs evolving in $\mathbb{R}^d$, they can also be formally expressed as products of exponentials of differential operators. The key point is to reformulate the original nonlinear problem into a linear one with the Lie formalism. Although a more detailed treatment is carried out in Appendix A.1, we next introduce the important concepts of *Lie operator/derivative* and *Lie transformation* in the simplest setting of a scalar autonomous differential equation to get some insight into the general procedure.

An interlude: Lie derivatives and Lie transformations

Given the initial value problem

$$\dot{x} = f(x), \qquad x(0) = x_0 \in \mathbb{R},$$

its solution $x(t)$ satisfies

$$\left.\frac{dx(t)}{dt}\right|_{x=x_0} = f(x_0) \equiv \left.\left(f(y)\ \frac{d}{dy}(y)\right)\right|_{y=x_0}$$

or

$$\left.\frac{dx(t)}{dt}\right|_{x=x_0} = f(x_0)\frac{d}{dx_0}(x_0) \equiv L_{f(x_0)}(x_0), \tag{3.16}$$

in terms of the *Lie operator* or *Lie derivative* in one dimension, $L_{f(y)} \equiv f(y)\frac{d}{dy}$. This is a linear operator acting on functions, so that, formally, x_0 in (3.16) is a variable. Analogously, for the second derivative, we have

$$\left.\frac{d^2x(t)}{dt^2}\right|_{x=x_0} = f(y)\,\frac{d}{dy}\left(f(y)\frac{d}{dy}\right)(y)\bigg|_{y=x_0} = L^2_{f(x_0)}(x_0),$$

and similar expressions for higher-order derivatives. This allows us to write the exact solution at $t = h$ as

$$x(h) = \varphi_h(x_0) = \sum_{n\geq 0}\frac{h^n}{n!}\left.\frac{d^nx(t)}{dt^n}\right|_{x=x_0} = \sum_{n\geq 0}\frac{h^n}{n!}L^n_{f(x_0)}(x_0) = \mathrm{e}^{hL_{f(x_0)}}[\mathrm{Id}](x_0),$$

where we have introduced the identity map $\mathrm{Id}(x) = x$. The object $\mathrm{e}^{hL_{f(x_0)}}$ is the *Lie transformation* associated to the solution flow and has some remarkable properties. In particular,

$$\mathrm{e}^{hL_{f(x_0)}}\big([\mathrm{Id}](x_0)\big)^n = \big(\mathrm{e}^{hL_{f(x_0)}}[\mathrm{Id}](x_0)\big)^n$$

for any $n \in \mathbb{N}$, so that, for any analytic function $g(x)$,

$$\left(\mathrm{e}^{tL_f}g\right)(x_0) \equiv \mathrm{e}^{tL_{f(x_0)}}[g](x_0) = g(\mathrm{e}^{tL_{f(x_0)}}(x_0)) = g(x(t)) = g(\varphi_t(x_0)).$$

Lie–Trotter and Strang in terms of Lie transformations

From the previous considerations, the maps (3.14) and (3.15) can be formulated in terms of Lie transformations as follows. If $F_1 \equiv L_{f_1}$ and $F_2 \equiv L_{f_2}$ denote the Lie derivatives associated with f_1 and f_2, respectively, it holds that

$$g(\varphi_t^{[1]}(x)) = \left(\mathrm{e}^{tF_1}g\right)(x), \qquad g(\varphi_t^{[2]}(x)) = \left(\mathrm{e}^{tF_2}g\right)(x)$$

for any analytic function $g : \mathbb{R}^d \to \mathbb{R}$. Then, for the first-order approximation $\chi_h = \varphi_h^{[2]} \circ \varphi_h^{[1]}$ furnished by Algorithm 1, one has $g\big(\chi_h(x)\big) = \big(\Psi(h)g\big)(x)$, where $\Psi(h)$ is a series of linear differential operators defined as

$$\Psi(h) = \mathrm{e}^{hF_1}\,\mathrm{e}^{hF_2}. \tag{3.17}$$

Notice that the exponentials of Lie derivatives in (3.17) appear in the reverse order with respect to the maps in the integrator [170, p. 88]. The same procedure can be carried out with the Strang splitting, resulting in the product

$$\Psi(h) = \mathrm{e}^{\frac{h}{2}F_1}\,\mathrm{e}^{hF_2}\,\mathrm{e}^{\frac{h}{2}F_1} \tag{3.18}$$

and in fact to other splitting methods applied to nonlinear differential equations. Thus, in the end we can express them as products of exponentials of linear operators acting on a certain function space [52]. This observation is very useful when analyzing the order conditions for a method to be of a given order. In particular, one has the same order conditions for linear and nonlinear ODEs.

Lie–Trotter and Strang as symplectic integrators

Suppose now we have a Hamiltonian $H(q,p)$ that can be decomposed as

$$H(q,p) = H_1(q,p) + H_2(q,p), \tag{3.19}$$

so that H_1 and H_2 define themselves as Hamiltonian systems. In that case, we can apply Algorithm 1 by composing both maps and forming the first-order scheme

$$x_{n+1} = \chi_h(x_n) \equiv \left(\varphi_h^{[H_1]} \circ \varphi_h^{[H_2]}\right)(x_n), \quad n = 0,1,2,\ldots, \tag{3.20}$$

where $x_n = (q_n, p_n)^T$. Similarly, Algorithm 2 gives the second-order scheme

$$x_{n+1} = S_h^{[2]}(x_n) \equiv \left(\varphi_{h/2}^{[H_1]} \circ \varphi_h^{[H_2]} \circ \varphi_{h/2}^{[H_1]}\right)(x_n), \quad n = 0,1,2,\ldots. \tag{3.21}$$

Notice that, since both χ_h and $S_h^{[2]}$ are defined as compositions of flows of Hamiltonian systems and the composition of symplectic maps is also symplectic [16], then both (3.20) and (3.21) are *symplectic integrators* [293].

In particular, if $H(q,p)$ is the sum of kinetic and potential energies, the splitting $H(q,p) = T(p) + V(q)$ leads to the subsystems

$$\begin{pmatrix} \dot q \\ \dot p \end{pmatrix} = \begin{pmatrix} \nabla_p T(p) \\ 0 \end{pmatrix} \quad \text{and} \quad \begin{pmatrix} \dot q \\ \dot p \end{pmatrix} = \begin{pmatrix} 0 \\ -\nabla_q V(q) \end{pmatrix}, \tag{3.22}$$

with solutions

$$\varphi_t^{[T]} : \begin{pmatrix} q_0 \\ p_0 \end{pmatrix} \longmapsto \begin{pmatrix} q_0 + t\,\nabla_p T(p_0) \\ p_0 \end{pmatrix} \tag{3.23}$$

and

$$\varphi_t^{[V]} : \begin{pmatrix} q_0 \\ p_0 \end{pmatrix} \longmapsto \begin{pmatrix} q_0 \\ p_0 - t\,\nabla_q V(q_0) \end{pmatrix}. \tag{3.24}$$

Then, the first-order scheme (3.20) with $H_1 = T$, $H_2 = V$, reduces to

$$p_{n+1} = p_n - h\,\nabla_q V(q_n), \qquad q_{n+1} = q_n + h\,\nabla T_p(p_{n+1}), \tag{3.25}$$

the symplectic Euler-VT method (1.10) and (1.44). Composing the maps in the reverse order, $\varphi_h^{[V]} \circ \varphi_h^{[T]}$, or taking the adjoint of (3.25), results in

$$q_{n+1} = q_n + h\,\nabla_p T(p_n), \qquad p_{n+1} = p_n - h\,\nabla_q V(q_{n+1}), \tag{3.26}$$

i.e., the symplectic Euler-TV method of (1.11) and (1.43).

On the other hand, the symmetric version

$$S_h^{[2]} \equiv \chi_{h/2}^* \circ \chi_{h/2} = \varphi_{h/2}^{[V]} \circ \varphi_h^{[T]} \circ \varphi_{h/2}^{[V]} \tag{3.27}$$

results in the Störmer–Verlet method (1.46). The role of $\varphi_h^{[T]}$ and $\varphi_h^{[V]}$ can be reversed, thus leading to the scheme

$$\tilde S_h^{[2]} \equiv \chi_{h/2} \circ \chi_{h/2}^* = \varphi_{h/2}^{[T]} \circ \varphi_h^{[V]} \circ \varphi_{h/2}^{[T]}, \tag{3.28}$$

a particular case of method (1.45).

3.2 Raising the order by composition

3.2.1 Composition of Strang maps

Taking as a basic method the Strang splitting $S_h^{[2]} = \varphi_{h/2}^{[1]} \circ \varphi_h^{[2]} \circ \varphi_{h/2}^{[1]}$, the composition

$$\psi_h = S_{\gamma_s h}^{[2]} \circ S_{\gamma_{s-1} h}^{[2]} \circ \cdots \circ S_{\gamma_1 h}^{[2]} \tag{3.29}$$

is at least of order three if

$$\sum_{j=1}^{s} \gamma_j = 1 \qquad \text{and} \qquad \sum_{j=1}^{s} \gamma_j^3 = 0. \tag{3.30}$$

The smallest value of s for which equations (3.30) admit real solutions is $s = 3$. In that case, by imposing the symmetry $\gamma_1 = \gamma_3$, we get a method of order 4, known as the *triple jump* [332]:

$$S_h^{[4]} = S_{\gamma_3 h}^{[2]} \circ S_{\gamma_2 h}^{[2]} \circ S_{\gamma_1 h}^{[2]}, \qquad \text{with} \qquad \gamma_1 = \gamma_3 = \frac{1}{2 - 2^{1/3}}, \qquad \gamma_2 = 1 - 2\gamma_1. \tag{3.31}$$

In general, the recursion

$$S_h^{[2k]} = S_{\gamma_1 h}^{[2k-2]} \circ S_{(1-2\gamma_1)h}^{[2k-2]} \circ S_{\gamma_1 h}^{[2k-2]}, \qquad \text{with} \qquad \gamma_1 = \frac{1}{2 - 2^{1/(2k-1)}} \tag{3.32}$$

can be used to get methods of arbitrarily high order $2k$ ($k \geq 2$) [87, 120] starting from $S_h^{[2]}$ (notice that such methods can be written in the form (3.29)). The price to be paid is the existence of large positive and negative coefficients γ_j and the great number of elementary flows in (3.32) for high orders. The alternative formed by the five maps composition (*quintuple jump*)

$$S_h^{[2k]} = S_{\gamma_1 h}^{[2k-2]} \circ S_{\gamma_1 h}^{[2k-2]} \circ S_{(1-4\gamma_1)h}^{[2k-2]} \circ S_{\gamma_1 h}^{[2k-2]} \circ S_{\gamma_1 h}^{[2k-2]}, \qquad \gamma_1 = \frac{1}{4 - 4^{1/(2k-1)}}, \tag{3.33}$$

also giving methods $S_h^{[2k]}$ of order $2k$ of the form (3.29) is popular in applications. In general, however, other choices for the coefficients γ_j in (3.29) are more appropriate if one is interested in achieving orders ≥ 6 with a lower number of elementary flows and relatively small coefficients.

Condition (3.30) can be derived by using the approach based on linear differential operators discussed in Appendix A.2. Specifically, if the Lie transformation associated with the Strang map $S_h^{[2]}$ is written as

$e^{h/2F_1}e^{hF_2}e^{h/2F_1} = e^{Y(h)}$, where

$$
\begin{aligned}
Y(h) &:= \log(e^{h/2F_1}e^{hF_2}e^{h/2F_1}) \\
&= \sum_{n=1}^{\infty} h^{2n-1} Y_{2n-1} \\
&= h(F_1+F_2) + h^3\left(-\frac{1}{24}[F_1,[F_1,F_2]] - \frac{1}{12}[F_2,[F_1,F_2]]\right) + \cdots,
\end{aligned}
$$

then the Lie transformation $\Psi(h)$ associated with (3.29) formally satisfies

$$\Psi(h) = e^{Y(\gamma_1 h)} \cdots e^{Y(\gamma_s h)}, \tag{3.34}$$

so that $g(\psi_h(x)) = (\Psi(h)g)(x)$ for any $x \in \mathbb{R}^d$ and any smooth function $g : \mathbb{R}^d \to \mathbb{R}$. In that case, application of the Baker–Campbell–Hausdorff leads to

$$
\begin{aligned}
\Psi(h) &= \exp\left(\gamma_1 h Y_1 + \gamma_1^3 h^3 Y_3 + \cdots\right) \cdots \exp\left(\gamma_s h Y_1 + \gamma_s^3 h^3 Y_3 + \cdots\right) \\
&= \exp\left(h(\sum_{j=1}^{s}\gamma_j)(F_1+F_2)\right) + h^3\left(\sum_{j=1}^{s}\gamma_j^3\right) Y_3 + \mathcal{O}(h^4),
\end{aligned} \tag{3.35}
$$

and thus composition (3.29) is of order at least 3 if (3.30) holds. In fact, (3.35) is also true if in (3.29) the Strang map $S_h^{[2]}$ is replaced by any second order time-symmetric map.

3.2.2 Composition of Lie-Trotter maps

We can also consider an analogous composition to (3.29), but this time with the Lie–Trotter scheme $\chi_h = \varphi_h^{[1]} \circ \varphi_h^{[2]}$ as the basic method. In that case, the Lie transformation of χ_h is of the form $e^{Y(h)}$ with

$$Y(h) = h(F_1+F_2) + h^2 Y_2 + h^3 Y_3 + \cdots, \tag{3.36}$$

so that the Lie transformation $\Psi(h)$ associated to $\chi_{\gamma_s h} \circ \cdots \circ \chi_{\gamma_2 h} \circ \chi_{\gamma_1 h}$ is

$$
\begin{aligned}
\Psi(h) &= e^{Y(\gamma_1 h)} \cdots e^{Y(\gamma_s h)} \\
&= \exp\left(h\,(\sum_{j=1}^{s}\gamma_j)(F_1+F_2)\right) + h^2\left(\sum_{j=1}^{s}\gamma_j^2\right) Y_2 + \mathcal{O}(h^3).
\end{aligned}
$$

In consequence, the scheme is of order 2 if

$$\sum_{j=1}^{s}\gamma_j = 1, \qquad \sum_{j=1}^{s}\gamma_j^2 = 0, \tag{3.37}$$

but this system of equations does not admit real solutions.

The situation is different, however, if one composes χ_h with its adjoint $\chi_h^* = \varphi_h^{[2]} \circ \varphi_h^{[1]}$, that is, for the map

$$\psi_h = \chi_{\alpha_{2s}h} \circ \chi^*_{\alpha_{2s-1}h} \circ \cdots \circ \chi_{\alpha_2 h} \circ \chi^*_{\alpha_1 h}. \tag{3.38}$$

If $e^{Y(h)}$, with $Y(h)$ given by (3.36), is the Lie transformation of χ_h, then the Lie transformation of its adjoint χ_h^* is $e^{-Y(-h)}$ and the Lie transformation $\Psi(h)$ of (3.38) satisfies

$$\begin{aligned}\Psi(h) &= e^{-Y(-\alpha_1 h)} e^{Y(\alpha_2 h)} \cdots e^{-Y(-\alpha_{2s-1} h)} e^{Y(\alpha_{2s} h)} \\ &= \exp\left(h\left(\sum_{j=1}^{2s} \alpha_j\right)(F_1 + F_2)\right) + h^2 \left(\sum_{j=1}^{s} (\alpha_{2j}^2 - \alpha_{2j-1}^2) \right) Y_2 + \mathcal{O}(h^3).\end{aligned} \tag{3.39}$$

Clearly, composition (3.38) is at least of order 2 if the coefficients satisfy

$$\sum_{j=1}^{2s} \alpha_j = 1, \qquad \sum_{j=1}^{s} (\alpha_{2j}^2 - \alpha_{2j-1}^2) = 0.$$

The same argument holds when χ_h is a first order integrator other than Lie-Trotter.

The simplest situation corresponds to $s = 1$, in which case we recover the Strang splitting, $S_h^{[2]} = \chi_{h/2} \circ \chi^*_{h/2}$. In fact, the general scheme (3.38) can be rewritten as the splitting method

$$\psi_h = \varphi^{[1]}_{a_{s+1}h} \circ \varphi^{[2]}_{b_s h} \circ \varphi^{[1]}_{a_s h} \circ \cdots \circ \varphi^{[1]}_{a_2 h} \circ \varphi^{[2]}_{b_1 h} \circ \varphi^{[1]}_{a_1 h}. \tag{3.40}$$

To see this, let $\varphi_h^{[i]}$ ($i = 1, 2$) be the exact flows. Then[1], clearly, $\varphi^{[i]}_{\beta h} \circ \varphi^{[i]}_{\delta h} = \varphi^{[i]}_{(\beta+\delta)h}$. In (3.40), $a_1 = \alpha_1$, and for $j = 1, \ldots, s$,

$$a_{j+1} = \alpha_{2j} + \alpha_{2j+1}, \qquad b_j = \alpha_{2j-1} + \alpha_{2j} \tag{3.41}$$

(with $\alpha_{2s+1} = 0$). Conversely, any integrator of the form (3.40) satisfying the condition $\sum_{j=1}^{s+1} a_j = \sum_{j=1}^{s} b_j$ can be expressed in the form (3.38), as shown in [237].

Any splitting scheme (3.40) with a palindromic sequence of coefficients, that is, satisfying for all j

$$a_{s-j+2} = a_j, \qquad b_{s-j+1} = b_j$$

is time-symmetric and thus of even order. Written in the composition format (3.38), it is time-symmetric if

$$\alpha_{2s-j+1} = \alpha_j.$$

[1]This property is not satisfied, in general, if the exact flows are replaced by numerical approximations.

Another connection with symplectic partitioned Runge–Kutta methods can be established as follows. Suppose we have a Hamiltonian system of the form $H(q,p) = T(p) + V(q)$. In that case we can take $\varphi_h^{[1]} = \varphi_h^{[T]}$ and $\varphi_h^{[2]} = \varphi_h^{[V]}$, the exact flows given by (3.23) and (3.24), so that the algorithm corresponding to the splitting method (3.40) for the step $(q_0, p_0) \mapsto (q_1, p_1)$ reads

$$
\begin{aligned}
&Q_0 = q_0, \qquad\qquad\qquad\qquad P_0 = p_0 \\
&\quad \text{for} \quad i = 1, \ldots, s \\
&\qquad Q_i = Q_{i-1} + h a_i \nabla_p T(P_{i-1}) \\
&\qquad P_i = P_{i-1} - h b_i \nabla_q V(Q_i) \\
&q_1 = Q_s + h a_{s+1} \nabla_p T(P_s), \qquad p_1 = P_s.
\end{aligned}
$$

This is precisely the algorithm corresponding to an s-stage explicit symplectic partitioned Runge–Kutta method (2.20) applied to equations

$$\dot{q} = \nabla_p T(p), \qquad \dot{p} = -\nabla_q V(q).$$

Thus, symplectic partitioned RK methods applied to $H(q,p) = T(p) + V(q)$ *are* splitting methods.

3.3 Order conditions

Composition methods of the form (3.38) and splitting methods (3.40) are formulated in terms of a set of coefficients α_i, a_i, b_i so as to achieve a (presumably) higher order than the basic method χ_h or $\varphi_h^{[2]} \circ \varphi_h^{[1]}$. The important point is then how to obtain these coefficients. To do that one must first formulate and then solve the so-called *order conditions*. These are, generally speaking, systems of polynomial equations the coefficients in the map ψ_h must satisfy so that the corresponding integration method is of the prescribed order. The complexity of the polynomials obviously increases with the number of flows in the composition, and their degree also increases with the order of the method. In consequence, solving the equations is by no means a trivial task when high-order methods are desired.

3.3.1 Splitting methods

In the analysis of the order conditions for splitting methods, and without loss of generality, we may consider the linear case (3.5) so that the treatment is essentially based on matrices or, more generally, on linear operators. Thus, the integrator (3.40) corresponds in this setting to the product of exponentials

$$\Psi(h) = \mathrm{e}^{a_{s+1} h F_1} \, \mathrm{e}^{b_s h F_2} \, \mathrm{e}^{a_s h F_1} \, \cdots \, \mathrm{e}^{a_2 h F_1} \, \mathrm{e}^{b_1 h F_2} \, \mathrm{e}^{a_1 h F_1}, \tag{3.42}$$

intended to approximate $e^{h(F_1+F_2)}$.

We should recall that, in the case of systems of ODEs $\dot{x} = f_1(x) + f_2(x)$ with (non-necessarily linear) vector fields f_1 and f_2, we can use the Lie formalism and consider, instead of the map

$$\psi_h = \varphi^{[1]}_{a_{s+1}h} \circ \varphi^{[2]}_{b_s h} \circ \varphi^{[1]}_{a_s h} \circ \cdots \circ \varphi^{[1]}_{a_2 h} \circ \varphi^{[2]}_{b_1 h} \circ \varphi^{[1]}_{a_1 h},$$

the product of Lie transformations

$$\Psi(h) = e^{a_1 h F_1}\, e^{b_1 h F_2}\, e^{a_2 h F_1} \cdots e^{a_s h F_1}\, e^{b_s h F_2}\, e^{a_{s+1} h F_1} \tag{3.43}$$

formulated in terms of the Lie derivatives F_j of f_j and associated with ψ_h (in the sense that $g(\psi_h(x)) = \Psi(h)g(x)$ for arbitrary smooth $g \in C^\infty(\mathbb{R}^d, \mathbb{R})$). Again, notice that the ordering of the exponentials in (3.43) is reversed with respect to the ordering of the maps.

A standard approach to get the order conditions consists in applying formally the Baker–Cambpell–Hausdorff (BCH) formula to express $\Psi(h)$ as one exponential of a series of operators in powers of h, and finally to compare this series with $e^{h(F_1+F_2)}$. In this way, we get

$$\begin{aligned} \log(\Psi(h)) &= h(v_1 F_1 + v_2 F_2) + h^2 v_{12} F_{12} + h^3 (v_{212} F_{212} + v_{112} F_{112}) \\ &\quad + h^4 (v_{2212} F_{2212} + v_{1212} F_{1212} + v_{1112} F_{1112}) + \mathcal{O}(h^5), \end{aligned} \tag{3.44}$$

where

$$\begin{aligned} &F_{12} = [F_1, F_2], \qquad F_{212} = [F_2, F_{12}], \qquad F_{112} = [F_1, F_{12}] \\ &F_{2212} = [F_2, F_{212}], \qquad F_{1212} = [F_1, F_{212}], \qquad F_{1112} = [F_1, F_{112}] \end{aligned} \tag{3.45}$$

and $v_1, v_2, v_{12}, v_{212}, \ldots$ are polynomials of homogeneous degree in the parameters a_j, b_j. In particular,

$$\begin{aligned} v_1 &= \sum_{i=1}^{s+1} a_i, \qquad v_2 = \sum_{i=1}^{s} b_i, \qquad v_{12} = \frac{1}{2} v_1 v_2 - \sum_{1 \le i < j \le s+1} b_i a_j, \\ v_{212} &= -\frac{1}{12} v_1 v_2^2 + \frac{1}{2} \sum_{1 \le i \le j < k \le s} b_i a_j b_k, \\ v_{112} &= \frac{1}{12} v_1^2 v_2 - \frac{1}{2} \sum_{1 \le i < j \le k \le s+1} a_i b_j a_k, \end{aligned}$$

with $b_{s+1} = 0$. From (3.44), it is clear that a characterization of the order of the splitting scheme (3.40) is obtained by requiring the consistency conditions $v_1 = v_2 = 1$, that is,

$$\sum_{j=1}^{s+1} a_j = \sum_{j=1}^{s} b_j = 1 \tag{3.46}$$

TABLE 3.1: Number of independent order conditions for general splitting methods, c_r, and for RKN-type splitting methods, d_r. The number m_r corresponds to the number of order conditions for compositions of a second-order time-symmetric method.

Order r	1	2	3	4	5	6	7	8	9	10	11
c_r	2	1	2	3	6	9	18	30	56	99	186
d_r	2	1	2	2	4	5	10	14	25	39	69
m_r	1	0	1	1	2	2	4	5	8	11	17

(ensuring that the scheme (3.40) is at least of order 1) and $v_{12} = v_{122} = v_{112} = \cdots = 0$ up to the required order. The set of order conditions thus obtained will be independent in the general case if the operators F_1, F_2, F_{12}, F_{212}, $F_{112}, \ldots$ form a basis of the free Lie algebra $\mathcal{L}(F_1, F_2)$ generated by $\{F_1, F_2\}$ (see Appendix A.3 for more details).

Of course, the exact characterization of the order conditions depends on the specific basis of the free Lie algebra one takes, in the sense that a different set of polynomial functions on $(a_1, b_1, \ldots, a_s, b_s, a_{s+1})$ will be obtained. In any case, the number of independent conditions arising at each order r can be obtained just by determining the dimension of $\mathcal{L}_r(F_1, F_2)$, the linear span of all commutators containing n operators F_1, F_2. This number, denoted as c_r, is collected in Table 3.1 (see [260, 243]). Thus, the total number of independent order conditions for a splitting method of order r is $c_1+c_2+\cdots+c_r$. Notice that the dimensions involved and therefore the number of order conditions increase very fast with the order of the method. Thus, in particular, a method of order eight involves (besides the consistency conditions (3.46)) $c_2 + \cdots + c_8 = 69$ order conditions, whereas constructing a scheme of order 10 requires solving a polynomial system of 224 equations.

If the method (3.42) is time-symmetric, then, as we know, the even order conditions are automatically satisfied, and thus the total number to achieve order r is reduced to $\sum_{k=1}^{r/2} c_{2k-1}$.

3.3.2 Composition methods

The same procedure can be applied to get the order conditions for the composition

$$\psi_h = \chi_{\alpha_{2s}h} \circ \chi^*_{\alpha_{2s-1}h} \circ \cdots \circ \chi_{\alpha_2 h} \circ \chi^*_{\alpha_1 h}. \tag{3.47}$$

In that case, we have to repeatedly apply the BCH formula to the corresponding operator

$$\Psi(h) = \mathrm{e}^{-Y(-\alpha_1 h)}\mathrm{e}^{Y(\alpha_2 h)} \cdots \mathrm{e}^{-Y(-\alpha_{2s-1} h)}\mathrm{e}^{Y(\alpha_{2s} h)},$$

so that

$$\log(\Psi(h)) = hw_1Y_1 + h^2w_2Y_2 + h^3(w_3Y_3 + w_{12}[Y_1, Y_2]) \\ + h^4(w_4Y_4 + w_{13}[Y_1, Y_3] + w_{112}[Y_1, [Y_1, Y_2]]) + \mathcal{O}(h^5), \tag{3.48}$$

where $w_{j_1 \cdots j_m}$ are now polynomials of degree $n = j_1 + \cdots + j_m$ in the parameters $\alpha_1, \ldots, \alpha_{2s}$. The first such polynomials are [49]

$$w_1 = \sum_{i=1}^{2s} \alpha_i, \qquad w_2 = \sum_{i=1}^{2s} (-1)^i \alpha_i^2, \qquad w_3 = \sum_{i=1}^{2s} \alpha_i^3 \tag{3.49}$$

$$w_{12} = \sum_{j_2=1}^{2s} (-1)^{j_2} \alpha_{j_2}^2 \sum_{j_1=1}^{j_2^*} \alpha_{j_1} - w_3 - w_1 w_2, \qquad w_4 = \sum_{j=1}^{2s} (-1)^j \alpha_j^4$$

and more cumbersome expressions for w_{13}, w_{112}, etc. Here $j_2^* = j_2 - 1$ if j_2 is even, and $j_2^* = j_2$ if j is odd. The order conditions are then

$$w_1 = 1, \qquad w_{j_1 \cdots j_m} = 0 \quad \text{whenever} \quad 2 \le j_1 + \cdots + j_m \le r$$

to get a method of order r. In particular, a composition method of order 3 must satisfy $w_1 = 1$, $w_2 = w_3 = w_{12} = 0$, whereas a symmetric fourth-order method only has to verify $w_1 = 1$, $w_3 = w_{12} = 0$ since conditions $w_2 = w_4 = w_{13} = w_{112} = 0$ are already satisfied due to symmetry.

In virtue of the close relationship between the splitting method (3.40) and the composition (3.38) by means of (3.41), it is clear that both types of schemes require exactly the same number of order conditions to achieve a given order r (except for consistency). Moreover, the polynomials $v_{j_1,\ldots,j_m}$ in (3.44) can be rewritten as linear combinations of the polynomials (on the parameters α_i) in (3.48) provided that $v_1 = v_2 = w_1$ holds. A detailed analysis of the relation among different sets of order conditions in composition methods can be found in [49].

Different strategies can be applied to reduce the number of order conditions in composition methods. One of them consists of taking

$$\alpha_{2j} = \alpha_{2j-1}, \qquad j = 1, 2, \ldots \tag{3.50}$$

in (3.47), so that the resulting composition integrator is of the form (3.29),

$$\psi_h = S_{h\gamma_s}^{[2]} \circ \cdots \circ S_{h\gamma_1}^{[2]}, \tag{3.51}$$

with $\gamma_j = 2\alpha_{2j}$ and $S_h^{[2]} = \chi_{h/2} \circ \chi_{h/2}^*$. Then, the number of order conditions for a method of order r is $m_1 + m_2 + \cdots + m_r$. The first values of m_k are collected in Table 3.1.

If, in addition, the composition (3.51) is palindromic, i.e., the coefficients verify

$$\gamma_j = \gamma_{s+1-j}, \qquad j = 1, 2, \ldots, \tag{3.52}$$

then the number of order conditions to reach order $r = 2p$, i.e.,

$$m_1 + m_3 + \cdots + m_{2p-1},$$

is drastically reduced, especially when considering high orders. The resulting methods are usually referred to as *symmetric compositions of symmetric schemes* (SS) [243]. An eighth-order method within this class requires seven-order conditions, and 15 if the scheme is of order 10 [199, 303].

Whereas the previous treatment leads easily to the number of order conditions, obtaining explicit expressions for the polynomials $v_{j_1,\ldots,j_m}$ (or $w_{j_1,\ldots,j_m}$) is a much more difficult task when the order increases, even when an efficient algorithm does exist for generating the BCH formula in a basis of $\mathcal{L}(F_1, F_2)$ [94]. To ameliorate this difficulty, other alternative characterizations have been proposed, making use in particular of Lyndon words and Lyndon multi-indices that allow us to get explicit expressions for the order conditions. The interested reader is referred to [52] for a detailed discussion of this topic.

3.3.3 Negative coefficients

Splitting and composition methods of order $r \geq 3$ necessarily involve some negative coefficients. This can already be observed in the simple triple jump scheme (3.31), and in fact has been established as a general theorem in [158, 298, 311]. A simple proof can be obtained as follows [36]: given the existing relationship between the splitting method (3.40) and the composition (3.38), it is clear that any splitting scheme of order $r \geq 3$ has to verify the condition

$$w_3 = \sum_{j=1}^{2s} \alpha_j^3 = \sum_{j=1}^{s} (\alpha_{2j-1}^3 + \alpha_{2j}^3) = 0, \tag{3.53}$$

where the coefficients α_j are related to a_j, b_j through (3.41). Since, for all $x, y \in \mathbb{R}$, it is true that $x^3 + y^3 < 0$ implies $x + y < 0$, then there must exist some $j \in \{1, \ldots, s\}$ in the sum of (3.53) such that

$$\alpha_{2j-1}^3 + \alpha_{2j}^3 < 0 \qquad \text{and thus} \qquad \alpha_{2j-1} + \alpha_{2j} = b_j < 0.$$

Obviously, we can also write (by taking $\alpha_0 = 0$, $\alpha_{2s+1} = 0$)

$$w_3 = \sum_{i=0}^{2s+1} \alpha_j^3 = \sum_{i=0}^{s} (\alpha_{2j}^3 + \alpha_{2j+1}^3) = 0$$

just by grouping terms in a different way, and thus, by repeating the argument, there must exist some $j \in \{0, \ldots, s\}$ such that

$$\alpha_{2j} + \alpha_{2j+1} = a_{j+1} < 0.$$

The proof clearly shows the origin of the existence of backward time steps: the equation $w_3 = 0$ can be satisfied only if at least one a_j *and* one b_j are negative.

This constitutes, then, an inherent feature of splitting methods, and, although it does not imply any particular limitation for carrying out numerical simulations in a number of important problems (e.g., classical Hamiltonian dynamics), in other applications it is a hindrance. For instance, suppose the ordinary differential equation one has to integrate by the splitting method originates from a partial differential equation that is ill-posed for negative times, such as those involving the Laplacian operator. In that case, the corresponding flow may not be defined for negative times. We will turn to this issue in Chapter 6.

3.3.4 Stability

Instead of equation (1.24), the model problem to characterize the stability of a splitting method is the simple harmonic oscillator (1.1), $\ddot{y} + \omega^2 y = 0$, $\omega = \sqrt{k/m} > 0$ [226, 239], with the standard splitting

$$\begin{pmatrix} \dot{q} \\ \dot{p} \end{pmatrix} = \Bigg[\underbrace{\begin{pmatrix} 0 & 1/m \\ 0 & 0 \end{pmatrix}}_{F_1} + \underbrace{\begin{pmatrix} 0 & 0 \\ -k & 0 \end{pmatrix}}_{F_2}\Bigg]\begin{pmatrix} q \\ p \end{pmatrix}, \tag{3.54}$$

and an exact solution (1.5). There are at least two reasons for this choice of model. First, if a numerical method already provides unbounded numerical solutions for a given h on this system, one cannot expect good behavior for more general problems. Second, there are physically relevant problems that, once formulated in appropriate coordinates, are expressed as a system of uncoupled harmonic oscillators. Thus, a precise characterization of the stability of a splitting method on (3.54) can be useful to build accurate and stable algorithms for their numerical treatment. This is, in particular, the case of the linear system

$$\dot{q} = M^{-1}p, \qquad \dot{p} = -Nq, \tag{3.55}$$

where M and N are $d \times d$ symmetric positive definite matrices. Writing $M = LL^T$ and introducing new variables $\tilde{q} = L^T q$, then $\ddot{\tilde{q}} = -L^{-1}NL^{-T}\tilde{q}$. Since N is symmetric positive definite, then $L^{-1}NL^{-T}$ is diagonalizable with positive eigenvalues. A new change of variables reduces the system to a set of d uncoupled scalar harmonic oscillators $\ddot{y}_i = -\omega_i^2 y_i$, with ω_i^2 the eigenvalues of $L^{-1}NL^{-T}$ [72].

Application of the splitting method

$$x_{n+1} = \mathrm{e}^{ha_{s+1}F_1}\,\mathrm{e}^{hb_sF_2}\,\cdots\,\mathrm{e}^{ha_2F_1}\,\mathrm{e}^{hb_1F_2}\,\mathrm{e}^{ha_1F_1}x_n$$

to (3.54) results in the map

$$\begin{pmatrix} q_{n+1} \\ p_{n+1} \end{pmatrix} = \tilde{M}_z \begin{pmatrix} q_n \\ p_n \end{pmatrix}, \qquad \tilde{M}_z = \begin{pmatrix} K_1(z) & K_2(z) \\ K_3(z) & K_4(z) \end{pmatrix}, \tag{3.56}$$

with $z = h\omega$. Here $K_1(z)$ and $K_4(z)$ (respectively, $K_2(z)$, $K_3(z)$) are even (respectively, odd) polynomials in z,

$$\det \tilde{M}_z = 1, \qquad K_1(0) = K_4(0) = 1,$$

and, if the scheme is time-symmetric, then $K_1(z) = K_4(z)$. An essential role in the analysis is played by the stability polynomial, defined as

$$p(z) = \frac{1}{2}\mathrm{tr}\,\tilde{M}_z = \frac{1}{2}(K_1(z) + K_4(z)).$$

The eigenvalues of $\tilde{M}_z$ are the zeros of $\lambda^2 - 2p(z)\lambda + 1$ and they determine the stability of the given method: if z is such that $|p(z)| < 1$, then $\tilde{M}_z$ has complex conjugate eigenvalues of modulus 1 and the powers $\tilde{M}_z^n$, $n \geq 0$, remain bounded, whereas if $|p(z)| > 1$, then $\tilde{M}_z^n$ grows exponentially with n. Linear instability occurs when $+1$ or -1 is an eigenvalue with multiplicity 2 and $\tilde{M}_z$ is not diagonalizable.

The stability interval is defined as the longest interval $(0, z_*)$ such that $\tilde{M}_z^n$ is bounded for all the iterations n. In the particular case of the Störmer–Verlet method, one has $z_* = 2$. In fact, if we concatenate k steps of length h/k of the Störmer–Verlet method to (3.54), the resulting scheme is stable for $0 < z/k < 2$, or alternatively, its stability interval is $(0, 2k)$. This is, in fact, the longest stability interval one can achieve by considering any splitting method with $s = k$ stages [72]. In consequence, the Störmer–Verlet method may be applied with longer scaled time steps z/k than any other splitting method with k stages.

3.4 Splitting methods for special problems

Whereas the previous analysis is completely general, there are problems arising in applications leading to the equation (3.2), where the vector field possesses a special structure that allows us to simplify the treatment and design schemes without taking into account all the order conditions. This is usually the case when the system can be split into two parts of different natures or different algebraic structures. Some of them are reviewed in this section.

3.4.1 RKN splitting methods

Many differential equations of practical interest are of the form

$$\ddot{y} = g(y), \tag{3.57}$$

where $y \in \mathbb{R}^D$ and $g : \mathbb{R}^D \longrightarrow \mathbb{R}^D$. This is the case for Hamiltonian systems $H(q,p) = T(p) + V(q)$, where the kinetic energy $T(p)$ is quadratic in the momenta p, i.e., $T(p) = \frac{1}{2}p^T M^{-1} p$ for a constant invertible symmetric matrix M, and $V(q)$ is the potential. Then the equations of motion can be written in the form (3.57) with $y = q$, $g(y) = -M^{-1}\nabla V(y)$.

By transforming (3.57) into a first-order ODE system (of dimension $d = 2D$) in the new variables $x = (y, v)^T$, with $v = \dot{y}$, it is clear that the resulting equation

$$\dot{x} = \frac{d}{dt}\begin{pmatrix} y \\ v \end{pmatrix} = \begin{pmatrix} v \\ g(y) \end{pmatrix}$$

can be expressed as $\dot{x} = f_1(x) + f_2(x)$, with

$$f_1(y, v) = (v, 0)^T, \qquad f_2(y, v) = (0, g(y))^T, \tag{3.58}$$

and splitting methods of the form (3.40) can be applied, with the exact h-flows $\varphi_h^{[1]}$ and $\varphi_h^{[2]}$ given by

$$\varphi_h^{[1]} : \begin{pmatrix} y_0 \\ v_0 \end{pmatrix} \longmapsto \begin{pmatrix} y_0 + h\, v_0 \\ v_0 \end{pmatrix}, \qquad \varphi_h^{[2]} : \begin{pmatrix} y_0 \\ v_0 \end{pmatrix} \longmapsto \begin{pmatrix} y_0 \\ v_0 + h\, g(y_0) \end{pmatrix}. \tag{3.59}$$

As we have seen in subsection 2.4, the class of Runge–Kutta methods can be conveniently adapted to deal with equation (3.57) to get more efficient schemes, the so-called Runge–Kutta–Nyström methods. In fact, this is also the case for splitting methods: special schemes can be designed to improve the accuracy while reducing the computational cost with respect to the general composition (3.40). For analogy, they are sometimes called splitting methods of RKN-type. The key point here is that the differential operators F_1 and F_2 associated with (3.58),

$$F_1 = \sum_{i=1}^{d} v_i \frac{\partial}{\partial y_i}, \qquad F_2 = \sum_{i=1}^{d} g_i(y) \frac{\partial}{\partial v_i}, \tag{3.60}$$

verify (see Appendix A.1)

$$[F_1, F_2] = -\sum_{i=1}^{d} g_i(y) \frac{\partial}{\partial y_i} + \sum_{i,j=1}^{d} v_i \frac{\partial g_j}{\partial y_i} \frac{\partial}{\partial v_j},$$

and therefore

$$[F_2, [F_1, F_2]] = \sum_{j=1}^{d} \tilde{g}_j(y) \frac{\partial}{\partial v_j}, \tag{3.61}$$

with $\tilde{g}(y) = 2g'(y)g(y)$. Since $\tilde{g}(y)$ only depends on y, then $[F_2, [F_2, [F_1, F_2]] = 0$ identically. In other words, $F_{2212} = 0$ in (3.44), which introduces linear dependencies among higher-order terms in the expansion of $\log(\Psi(h))$ [243] and therefore contributes to a reduction in the number of order conditions when the order $r \geq 4$ (see [52] for a detailed analysis). The corresponding number d_r for $r \leq 11$ is given in Table 3.1. Thus, a consistent time-symmetric

RKN method of order 8 requires 16 order conditions, whereas a tenth-order scheme involves 41. Since the order conditions up to order three are identical as in the general case, the results for negative time steps still apply.

The class of problems for which this reduction in the order conditions is in fact more general than (3.58). It includes in particular the case where f_2 only depends on y and $f_1(y,v)$ is linear in v. For Hamiltonian systems, this generalization corresponds to $H(q,p) = T(q,p) + V(q)$, where $T(q,p)$ is quadratic in p. In that case, the flow associated with $T(q,p)$ should be easily computed for the RKN splitting methods to be advantageous.

Another situation where there is also a reduction in the number of order conditions corresponds to the time-dependent Schrödinger equation considered in Chapter 6, with $F_1 = \hat{T}$ and $F_2 = \hat{V}$, since the graded Lie algebra involved is isomorphic to the problem discussed above. In fact, it is conjectured in [240] (and checked up to order 20) that the case (3.58) (where $f_1(y,v) = v$) and the more general case (where $f_1(y,v)$ is linear in v but may depend on y) give rise to the same reduction in the order conditions.

3.4.2 Methods with commutators

Notice that the Lie derivative $[F_2,[F_1,F_2]]$ in (3.61) has the same form as F_2 in (3.60), but with a different function $\tilde{g}(y) = 2g'(y)g(y)$. In consequence, the corresponding h-flow reads

$$\varphi_h^{[212]}(y,v) = (y,\, v + h^3\tilde{g}(y))^T,$$

so that one might then incorporate this map into the splitting scheme to improve its efficiency. Of course, this makes sense if the gain in accuracy, stability, or any other favorable property compensates for the extra computational cost due to the presence of the additional flow. A popular fourth-order method belonging to this class is

$$\psi_h = \varphi_{b_1h}^{[2]} \circ \varphi_{a_1h}^{[1]} \circ \varphi_{b_2h}^{[2]} \circ \varphi_{d_2h}^{[212]} \circ \varphi_{b_2h}^{[2]} \circ \varphi_{a_1h}^{[1]} \circ \varphi_{b_1h}^{[2]}, \tag{3.62}$$

with

$$b_1 = \frac{1}{6}, \quad a_1 = \frac{1}{2}, \quad b_2 = \frac{1}{3}, \quad d_2 = \frac{1}{72}, \tag{3.63}$$

first proposed by [205] and [109]. Here and in the sequel we denote

$$\varphi_{d_jh}^{[212]}(y,v) = (y,\, v + d_jh^3\tilde{g}(y))^T,$$

for a given coefficient d_j. Moreover, in the RKN case $[F_2,[F_2,[F_1,F_2]]] \equiv 0$, which implies that the flows $\varphi_h^{[212]}$ and $\varphi_h^{[2]}$ commute. Hence, the three central terms in (3.62) can be merged into

$$\begin{aligned}\widetilde{\varphi}_{2b_2h,d_2h} &\equiv \varphi_{b_2h}^{[2]} \circ \varphi_{d_2h}^{[212]} \circ \varphi_{b_2h}^{[2]}(y,v) \\ &= (y,\, v + 2hb_2\, g(y) + h^3 d_2\, g'(y)g(y))^T.\end{aligned} \tag{3.64}$$

For Hamiltonian systems $H(p,q) = \frac{1}{2}p^T M^{-1} p + V(q)$, F_{212} is the operator associated to the Hamiltonian function $-(\nabla V)^T M^{-1} \nabla V$, depending only on q. Thus (3.64) is the h-flow of Hamiltonian

$$2b_2\, V(q) - d_2\, h^2 (\nabla V(q))^T M^{-1} \nabla V(q), \tag{3.65}$$

which reduces to the potential $V(q)$ when $b_2 = 1/2$ and $d_2 = 0$. This explains the term "splitting methods with modified potentials" frequently used in the literature [227, 284, 329]. The procedure can be generalized by considering "modified potentials" of even higher degree in h [49, 57]. Again, we can introduce additional free parameters into the scheme and construct more efficient integrators as long as the simultaneous evaluation of $g(y)$, $\frac{\partial g}{\partial y}(y)g(y)$, etc. is not substantially more expensive than the evaluation of $g(y)$ itself. This is the case, in particular, of the two-dimensional Kepler problem given by the Hamiltonian (1.57) with $\mu = 1$, i.e. $H = \frac{1}{2}(p_1^2 + p_2^2) - \frac{1}{r}$. Whereas one stage $\varphi_{b_i h}^{[2]} \circ \varphi_{a_i h}^{[1]}(q_k, p_k)$ can be carried out with one square root and 16 products/additions, the replacement of $\varphi_{b_i h}^{[2]}$ by $\widetilde{\varphi}_{b_j h, d_j h}$ only increases the cost by seven products/additions. Notice that for this problem

$$\nabla V(q)^T M^{-1} \nabla V(q) = \frac{1}{r^4}, \qquad \text{and} \qquad \nabla\left(\nabla V(q)^T M^{-1} \nabla V(q)\right) = -\frac{4}{r^6}(q_1, q_2)^T,$$

where the square root, $r = \sqrt{q_1^2 + q_2^2}$, has already been computed.

In addition to the reduction in the number of force evaluations, including flows associated with commutators has another advantage: since the coefficients a_i, b_i do not have to satisfy all the order conditions at order $r \geq 3$, the results for negative time steps do not apply here, and in fact methods of order greater than two do exist within this class. In addition to (3.62), other "forward" third and fourth-order methods (i.e., with all $a_j > 0$ and $b_j > 0$) involving second derivatives of the potential are known [285, 268, 269] and applied to systems where the presence of negative coefficients leads to severe stability problems [21]. Although achieving order six in general requires some negative coefficients [110, 19], it is indeed possible to construct a 6th-order processed method for cubic potentials with all a_j, b_j being positive [43].

3.4.3 Near-integrable systems

Very often in applications one has to deal with differential equations such as

$$\dot{x} = f_1(x) + \varepsilon f_2(x), \tag{3.66}$$

where $|\varepsilon| \ll 1$ and the h-flows $\varphi_h^{[1]}$, $\varphi_h^{[2]}$ corresponding to f_1 and εf_2, respectively, are readily available. In classical Hamiltonian mechanics, in particular, is rather common to have a Hamiltonian function H which is a small perturbation of an exactly integrable Hamiltonian H_1, that is, $H = H_1 + \varepsilon H_2$, with $0 < \varepsilon \ll 1$ [159, 271].

For this type of problem, splitting methods of the form (3.40),

$$\psi_h = \varphi^{[1]}_{a_{s+1}h} \circ \varphi^{[2]}_{b_s h} \circ \varphi^{[1]}_{a_s h} \circ \cdots \circ \varphi^{[1]}_{a_2 h} \circ \varphi^{[2]}_{b_1 h} \circ \varphi^{[1]}_{a_1 h},$$

are especially well adapted. On the one hand, the error is at most $\mathcal{O}(\varepsilon)$ and vanishes with ε, since in that case the scheme reproduces the exact solution. On the other hand, typically, $|\varepsilon| \ll h$ (or at least $|\varepsilon| \approx h$), so that one is mainly interested in eliminating error terms with small powers of ε, and its number grows as a polynomial in the order r, instead of exponentially. To see this point, it is illustrative to write the first terms of the operator obtained by applying the BCH formula to (3.43) (with F_2 replaced by εF_2):

$$\Psi(h) = \mathrm{e}^{h(F_1+\varepsilon F_2+E(h,\varepsilon))}, \tag{3.67}$$

with

$$\begin{aligned} E(h,\varepsilon) = {} & h\,\varepsilon\, v_{12}F_{12} + h^2\,\varepsilon\, v_{112}F_{112} + h^2\,\varepsilon^2\, v_{212}F_{212} + h^3\,\varepsilon\, v_{1112}F_{1112} \\ & + h^3\varepsilon^2 v_{1212}F_{1212} + h^3\varepsilon^3\, v_{2212}F_{2212} + \mathcal{O}(h^4) \end{aligned} \tag{3.68}$$

provided that the method is consistent. We see then that there is only one error term of order εh^k, namely the term $h^k \varepsilon v_{11\cdots 12}$ in (3.68). Analogously, it can be shown that there are $\lfloor \frac{1}{2}(k-1) \rfloor$ terms of order $\mathcal{O}(\varepsilon^2 h^k)$ and $\lfloor \frac{1}{6}(k-1)(k-2) \rfloor$ terms of order $\mathcal{O}(\varepsilon^3 h^k)$ [236].

In the treatment of splitting methods for near-integrable systems, it is convenient to introduce, following [236], the notion of *generalized order*. Thus, we say that ψ_h is of generalized order $(r_1, r_2, \ldots, r_m)$, with $r_1 \geq r_2 \geq \cdots \geq r_m$, if

$$\psi_h(x) - \varphi_h(x) = \mathcal{O}(\varepsilon h^{r_1+1} + \varepsilon^2 h^{r_2+1} + \cdots + \varepsilon^m h^{r_m+1}) \quad \text{as} \quad (h,\varepsilon) \longrightarrow (0,0).$$

Observe that r_1 is the order of consistency the method would have in the limit $\varepsilon \to 0$. With this notation, a method such that its local error is $\mathcal{O}(\varepsilon h^{2n+1} + \varepsilon^2 h^3)$ is said to be of generalized order $(2n, 2)$. In particular, the Strang scheme is of order $(2, 2)$.

As in the general case, besides applying the BCH formula, there are other strategies to get the generalized order conditions, i.e., the conditions that the coefficients a_i, b_i must satisfy for a splitting method to be of a prescribed (generalized) order $(r_1, r_2, \ldots, r_m)$. A general analysis in terms of a particular subset of multi-indices called Lyndon multi-indices is carried out in [41, 52].

3.5 Processing

The concept of *conjugacy* plays a fundamental role in the theory of dynamical systems [183]. Two flows φ_t and ϕ_t are (topologically) conjugate if there exists a homeomorphism[2] $\pi : \mathbb{R}^d \longrightarrow \mathbb{R}^d$ that satisfies

$$\phi_t(\pi(x)) = \pi(\varphi_t(x))$$

for all x. In that case, the two systems that give rise to the respective flows φ_t and ϕ_t are dynamically equivalent, in the sense that the map π takes the solution curves of one system to those of the other. In the context of numerical integration methods for systems of ODEs, where the h-flow φ_h of the original system is replaced by a map ψ_h depending on the small parameter h such that $\psi_h \approx \varphi_h$, conjugacy can be used to analyze the long-time behavior of the numerical error introduced by ψ_h [52]. The idea consists in analyzing how near ψ_h and φ_h are from being conjugate to each other and using that to estimate the long-term evolution of the errors.

The homeomorphism π can also be used to enhance the performance of a given splitting method ψ_h [284, 329, 238, 213] of order r, integrating instead the problem with the map

$$\hat{\psi}_h = \pi_h^{-1} \circ \psi_h \circ \pi_h, \tag{3.69}$$

where we have added a subscript to the conjugacy map to indicate its dependence on the step-size h. In this context it makes sense to choose $\pi_h(x) = x + \mathcal{O}(h^r)$. Application of n steps of the new integrator $\hat{\psi}_h$ leads to $\hat{\psi}_h^n = \pi_h^{-1} \circ \psi_h^n \circ \pi_h$. In words, the *pre-processor* π_h is applied only once, then the *kernel* ψ_h acts once per step, and, finally, the action of the *post-processor* or *corrector* π_h^{-1} is evaluated only when output is required. This technique then provides the (higher) accuracy of $\hat{\psi}_h$ at essentially the cost of the less accurate method ψ_h as long as the evaluation of π_h is computationally cheap compared to n evaluations of ψ_h.

One of the simplest examples of a processed integrator is provided by the Strang splitting. It can be written as

$$\begin{aligned} S_h^{[2]} &= \varphi_{h/2}^{[2]} \circ \varphi_h^{[1]} \circ \varphi_{h/2}^{[2]} = \varphi_{h/2}^{[2]} \circ \varphi_{-h}^{[2]} \circ \varphi_h^{[2]} \circ \varphi_h^{[1]} \circ \varphi_{h/2}^{[2]} \\ &= \varphi_{-h/2}^{[2]} \circ \chi_h \circ \varphi_{h/2}^{[2]} = \pi_h^{-1} \circ \chi_h \circ \pi_h, \end{aligned} \tag{3.70}$$

in terms of $\pi_h = \varphi_{h/2}^{[2]}$ and the Lie–Trotter method $\chi_h = \varphi_h^{[2]} \circ \varphi_h^{[1]}$. Hence, applying Lie–Trotter as the basic integrator with processing yields a second order of approximation. In terms of exponentials of operators, one has

$$\Psi(h) = \mathrm{e}^{\frac{h}{2}F_2}\, \mathrm{e}^{hF_1}\, \mathrm{e}^{\frac{h}{2}F_2} = \mathrm{e}^{\frac{h}{2}F_2} \left(\mathrm{e}^{hF_1}\mathrm{e}^{hF_2}\right) \mathrm{e}^{-\frac{h}{2}F_2}.$$

Instead of enhancing a previously existing integrator in this way, one may also design a processed method from scratch, that is, without requiring that ψ_h be an rth order approximation of φ_h. In the case of splitting methods, we

[2] That is, a one-to-one, onto, and continuous function with an inverse that is also continuous.

determine the h-parametric maps ψ_h and π_h as compositions of basic flows $\varphi^{[1]}_{a_j h}$ and $\varphi^{[2]}_{b_j h}$ with different sequences of coefficients a_j and b_j, such that the processed map $\hat{\psi}_h = \pi_h^{-1} \circ \psi_h \circ \pi_h$ is a good approximation of φ_h for a sufficiently small h. Typically, one requires that $\pi_h^{-1} \circ \psi_h \circ \pi_h(x) - \varphi_h(x) = \mathcal{O}(h^{r+1})$, so that the processed integrator is of order r [227, 56].

In any of the situations considered above (either analyzing the performance of a given splitting integrator with the help of a conjugacy map π_h, or enhancing an existing rth-order splitting integrator by processing or designing a splitting processing integrator from the very beginning), one needs to study the *effective order conditions* of ψ_h. These are the conditions on the parameters a_i, b_i that guarantee that there exists a processor map π_h such that $\pi_h^{-1} \circ \psi_h \circ \pi_h(x) - \varphi_h(x) = \mathcal{O}(h^{r+1})$, in which case the method ψ_h is said to be of *effective order* r [77]. A general treatment of the effective order conditions of several classes of numerical integrators, including splitting and composition methods, can be found in [45] and [46]. It is based on the series expansion (3.44) (for splitting) and (3.48) (for composition) of the formal logarithm of the Lie transformation $\Psi(h)$, and it is shown that the conditions for effective order r can be written in terms of the coefficients $v_{j_1 \cdots j_m}$ (splitting) and $w_{j_1 \cdots j_m}$ (composition).

Example 3.2. To illustrate this feature, we next obtain the effective order conditions of a processed method of order 4 whose kernel is the symmetric composition (3.47) of a basic first-order scheme χ_h and its adjoint. Let $\Psi(h) = \exp(K(h))$ and $\Pi(h) = \exp(P(h))$ be the Lie transformations associated to the kernel and the processor, respectively. Then, the operator associated with the processed method, $\hat{\Psi}(h) = \exp(\hat{K}(h))$, can be expressed as

$$\hat{\Psi}(h) = \Pi(h)\Psi(h)\Pi(h)^{-1} = \mathrm{e}^{P(h)}\mathrm{e}^{K(h)}\mathrm{e}^{-P(h)},$$

and thus, by applying equation (A.35), one has

$$\hat{K}(h) = \mathrm{e}^{\mathrm{ad}_{P(h)}} K(h) = K(h) + [P(h), K(h)] + \frac{1}{2}[P(h), [P(h), K(h)]] + \cdots. \tag{3.71}$$

If we assume that the operator $P(h)$ associated with the map π_h can be expanded in a similar way as the operator $K(h)$ associated with the kernel, we can write

$$\begin{aligned} K(h) &= h\, w_1 Y_1 + h^3(w_3 Y_3 + w_{12}[Y_1, Y_2]) + \mathcal{O}(h^5) \\ P(h) &= h\, p_1 Y_1 + h^2 p_2 Y_2 + h^3(p_3 Y_3 + p_{12}[Y_1, Y_2]) + \mathcal{O}(h^4). \end{aligned}$$

The coefficients of the kernel, α_j, have to be selected so that

$$w_1 = \sum_{i=1}^{2s} \alpha_i = 1, \qquad w_3 = \sum_{i=1}^{2s} \alpha_i^3 = 0, \tag{3.72}$$

whereas the processor has to satisfy

$$p_1 = 0, \qquad p_2 = w_{12}, \qquad p_3 = 0, \qquad p_{12} = 0.$$

In this way

$$K(h) = hY_1 + h^3 w_{12}[Y_1, Y_2] + \mathcal{O}(h^5), \qquad P(h) = h^2 w_{12} Y_2 + \mathcal{O}(h^4)$$

so that, upon substitution into (3.71), one has $\hat{K}(h) = hY_1 + \mathcal{O}(h^5) = hF + \mathcal{O}(h^5)$ and the processed integrator (3.69) is of conventional order 4. The expression of $w_{12}(\alpha)$ is given explicitly in (3.49). □

We thus see that a consistent ($w_1 = 1$) time-symmetric composition method of effective order 4 requires only one order condition for the kernel instead of two. The remaining conditions that make the whole scheme of order 4 can be fulfilled by the processor. One can proceed similarly for higher orders, both for composition and splitting methods. In fact, the analysis shows that many of the order conditions of the processed method $\hat{\psi}_h$ can be satisfied by π_h, so that ψ_h must fulfill a much reduced set of restrictions, also of lower complexity. Specifically, kernels (3.47) of effective order r only require $1 + c_r$ conditions instead of $c_1 + \cdots + c_r$, the total number of order conditions to achieve conventional order r. For a symmetric kernel, the number of effective order conditions to reach order $r = 2p$ reduces to $1 + (c_3 - c_2) + \cdots + (c_{2p-1} - c_{2p-2})$ instead of $1 + c_3 + \ldots + c_{2p-1}$ for a symmetric method of conventional order $r = 2p$.

When the kernel is time-symmetric, the processor leading to a minimum main order truncation error must be such that $\pi_{-h} = \pi_h$. With this requirement, the whole method is also time-symmetric, that is $\hat{\psi}_{-h}^{-1} = \hat{\psi}_h$ [45]. However, this "ideal" processor is in general very difficult to obtain and in practice will only be approximated. This is done again with a composition of the form

$$\widetilde{\pi}_h = \chi_{\gamma_{2m}h} \circ \chi^*_{\gamma_{2m-1}h} \circ \cdots \circ \chi_{\gamma_2 h} \circ \chi^*_{\gamma_1 h}, \tag{3.73}$$

with certain coefficients γ_j, $j = 1, \ldots, 2m$, chosen so that

$$\widetilde{\pi}_h = \pi_h + \mathcal{O}(h^q)$$

with $q \geq r$, the order of the method itself. In this way, the overall scheme

$$\widetilde{\pi}_h^{-1} \circ \psi_h \circ \widetilde{\pi}_h \tag{3.74}$$

has the same structure as the kernel, since

$$\widetilde{\pi}_h^{-1} = \chi_{-\gamma_1 h} \circ \chi^*_{-\gamma_2 h} \circ \cdots \circ \chi_{-\gamma_{2m-1}h} \circ \chi^*_{-\gamma_{2m}h}. \tag{3.75}$$

The resulting scheme (3.74), however, is no longer time-symmetric because the composition is not palindromic: $\widetilde{\pi}_{-h} \neq \widetilde{\pi}_h$, in contrast with $\pi_{-h} = \pi_h$ for the ideal processor. In addition, it involves negative coefficients, since

$$\gamma_1 + \cdots + \gamma_{2m} = 0$$

for consistency. This feature can be dealt with by the notion of *starter* [35], whereas the first drawback can be overcome by using in (3.74) the *adjoint* of $\widetilde{\pi}_h$ instead of its inverse,

$$\widetilde{\pi}_h^* = \chi_{\gamma_1 h} \circ \chi_{\gamma_2 h}^* \circ \cdots \circ \chi_{\gamma_{2m-1} h} \circ \chi_{\gamma_{2m} h}^*,$$

so that

$$\widetilde{\pi}_h^* = \widetilde{\pi}_{-h}^{-1} = \pi_{-h}^{-1} + \mathcal{O}(h^q) = \pi_h^{-1} + \mathcal{O}(h^q)$$

if the processor satisfies $\pi_{-h} = \pi_h$. In this way we propose to apply, for n steps, the scheme

$$\widetilde{\psi}_h^n \equiv \widetilde{\pi}_h^* \circ \psi_h^n \circ \widetilde{\pi}_h. \tag{3.76}$$

Since

$$(\widetilde{\psi}_h^n)^{-1} = \widetilde{\pi}_h^{-1} \circ (\psi_h^n)^{-1} \circ (\widetilde{\pi}_h^*)^{-1}$$

and

$$\widetilde{\psi}_{-h}^n = \widetilde{\pi}_{-h}^* \circ \psi_{-h}^n \circ \widetilde{\pi}_{-h} = \widetilde{\pi}_h^{-1} \circ \psi_{-h}^n \circ (\widetilde{\pi}_h^*)^{-1},$$

then the modified processed integrator (3.76) is also time-symmetric if ψ_h is time-symmetric. A similar procedure can be carried out for splitting methods [43].

The processing technique does indeed lead to efficient methods as long as frequent outputs are not required. Otherwise, the computational cost increases. To ameliorate this situation, in [45, 225] a technique has been developed for obtaining approximations to the post-processor at virtually no cost and without loss of accuracy. The key idea is to replace $\widetilde{\pi}_h^{-1}$ by a new map $\hat{\pi}_h^{-1} \simeq \widetilde{\pi}_h^{-1}$ obtained from the intermediate stages in the computation of $X_n = \psi_h(X_{n-1})$, with $X_0 = \widetilde{\pi}_h(x_0)$. More specifically, one replaces $x_n = \widetilde{\pi}_h^{-1}(X_n)$ by $\hat{x}_n = \sum_j \beta_j X_{n,j}$, where $X_{n,j}$ are the intermediate stages in the computation of the transformed variable, $\psi_h(X_{n-1})$, and β_j are coefficients that satisfy a system of linear equations. The error introduced by the approximation $\hat{\pi}_h^{-1}$ is of a purely local nature [45] and does not propagate along the evolution, contrary to the error committed by $\widetilde{\pi}_h$. The same procedure can be applied to the post-processor $\widetilde{\pi}_h^*$ instead.

3.6 Splitting methods for non-autonomous systems

So far we have focused on splitting methods for autonomous differential equations. The natural question is whether the same techniques can be applied when there is an explicit time dependence in the equation to integrate. The ideal situation would be that the methods designed for $\dot{x} = f(x)$ could also be used (maybe with only minor modifications) when we have $\dot{x} = f(t, x)$. In addition, one would like that the schemes previously considered before for special problems would still be valid in the non-autonomous case. All these issues are analyzed next.

Let us consider the general situation, corresponding to a system of the form

$$\dot{x} = f(t,x) = f_1(t,x) + \cdots + f_m(t,x), \qquad x(0) = x_0, \tag{3.77}$$

i.e., when the explicit time dependence is present in each part. Then we can take t as a new coordinate and transform (3.77) into an equivalent autonomous equation to which standard splitting algorithms can subsequently be applied. More specifically, equation (3.77) is equivalent to the enlarged system

$$\underbrace{\frac{d}{dt}\begin{pmatrix} x \\ x_{d+1} \end{pmatrix}}_{y} = \underbrace{\begin{pmatrix} 0 \\ 1 \end{pmatrix}}_{\hat{f}_0} + \underbrace{\begin{pmatrix} f_1(x_{d+1},x) \\ 0 \end{pmatrix}}_{\hat{f}_1(y)} + \cdots + \underbrace{\begin{pmatrix} f_m(x_{d+1},x) \\ 0 \end{pmatrix}}_{\hat{f}_m(y)}, \tag{3.78}$$

with $x_{d+1} \in \mathbb{R}$. If the resulting (autonomous) equations

$$\dot{y} = \hat{f}_i(y), \qquad i = 0, 1, \ldots, m$$

can be solved, then we may use any splitting method of the form (3.40), since x_{d+1} advances only with $\hat{f}_0$ and remains constant for the rest of the system.

We can do better for problems which are separable into just two parts,

$$\dot{x} = f(t,x) = f_1(t,x) + f_2(t,x), \qquad x(0) = x_0. \tag{3.79}$$

If t is taken as a new coordinate *twice*, and we write

$$\underbrace{\frac{d}{dt}\begin{pmatrix} x \\ x_{d+1} \\ x_{d+2} \end{pmatrix}}_{z} = \underbrace{\begin{pmatrix} f_1(x_{d+1},x) \\ 0 \\ 1 \end{pmatrix}}_{\hat{f}_1(z)} + \underbrace{\begin{pmatrix} f_2(x_{d+2},x) \\ 1 \\ 0 \end{pmatrix}}_{\hat{f}_2(z)}, \tag{3.80}$$

with $x_{d+1}, x_{d+2} \in \mathbb{R}$, then we can apply the same splitting schemes designed for autonomous systems separable into two pieces to

$$\dot{z} = \hat{f}_1(z), \qquad \dot{z} = \hat{f}_2(z).$$

This is so because x_{d+1} is constant when integrating the first equation, and x_{d+2} is constant when solving the second one. The procedure can be viewed as a generalization of the one proposed in [294] for Hamiltonian systems $H(t,q,p) = T(t,p) + V(t,q)$: by introducing two new coordinates q_{d+1}, q_{d+2} and their associated momenta, p_{d+1}, p_{d+2}, we instead deal with the formally autonomous Hamiltonian

$$\tilde{H}(q_{d+1}, q_{d+2}, q, p_{d+1}, p_{d+2}, p) = \big(T(p_{d+2}, p) + p_{d+1}\big) + \big(V(q_{d+1}, q) - q_{d+2}\big).$$

For the non-autonomous second order differential equation

$$\ddot{y} = g(t, y), \tag{3.81}$$

it is more convenient to split the system in the extended phase space as

$$\frac{d}{dt}\begin{pmatrix} y \\ v \\ y_{d+1} \end{pmatrix} = \begin{pmatrix} v \\ 0 \\ 1 \end{pmatrix} + \begin{pmatrix} 0 \\ g(y_{d+1}, y) \\ 0 \end{pmatrix},$$

since then we have an autonomous system with the same algebraic structure as those considered in subsection 3.4.1, so that RKN splitting methods (even including commutators) can also be used.

For Hamiltonian systems $H(t, q, p) = T(p) + V(t, q)$, this is equivalent to introducing a new coordinate $q_{d+1} = t$ and its associated momentum, $p_{d+1} = -H$, and consider the extended (autonomous) Hamiltonian function

$$\tilde{H}(q_{d+1}, q, p_{d+1}, p) = \big(T(p) + p_{d+1}\big) + V(q_{d+1}, q),$$

which is still quadratic in momenta, so that symplectic RKN methods can be used. Notice that $\tilde{H}$ is only linear in p_{d+1} and modified potentials only involve derivatives of the potential with respect to q, but not with respect to q_{d+1}, i.e., they do not require time derivatives. In this case the evolution for p_{d+1} is irrelevant and so there is no need to compute it.

Finally, if we take t as two new coordinates in the non-autonomous near-integrable system

$$\dot{x} = f(t, x) = f_1(t, x) + \varepsilon f_2(t, x), \qquad x(0) = x_0, \qquad |\varepsilon| \ll 1, \tag{3.82}$$

then the special structure of a near-integrable system is destroyed. A partial remedy consists in separating the system as

$$\frac{d}{dt}\begin{pmatrix} x \\ x_{d+1} \end{pmatrix} = \underbrace{\begin{pmatrix} f_1(x_{d+1}, x) \\ 1 \end{pmatrix}}_{\hat{f}_1} + \underbrace{\varepsilon\begin{pmatrix} f_2(x_{d+1}, x) \\ 0 \end{pmatrix}}_{\varepsilon \hat{f}_2}, \tag{3.83}$$

which requires us to numerically solve the non-autonomous unperturbed system.

It is important to remark that taking time as an additional coordinate is of interest only if the time dependence in f_i, $i = 1, \dots, m$ is cheap to compute. Otherwise the resulting algorithm may be computationally costly, since these functions have to be evaluated s times per time step. This drawback can be avoided by approximating the exact solution at each step by a composition of maps that somehow incorporate average values of the vector fields with different weights on the subinterval $[t_n, t_{n+1}]$ [37]. Specifically, the schemes read

$$\psi_h = \varphi_h^{[\hat{A}_{s+1}]} \circ \varphi_h^{[\hat{B}_s]} \circ \varphi_h^{[\hat{A}_s]} \circ \cdots \circ \varphi_h^{[\hat{A}_2]} \circ \varphi_h^{[\hat{B}_1]} \circ \varphi_h^{[\hat{A}_1]}, \tag{3.84}$$

where the maps $\varphi_h^{[\hat{A}_i]}$, $\varphi_h^{[\hat{B}_i]}$ are the exact 1-flows corresponding to the time-independent differential equations

$$\dot{x} = \hat{A}_i(x), \qquad \dot{x} = \hat{B}_i(x), \qquad i = 1, 2, \ldots \tag{3.85}$$

respectively, with

$$\hat{A}_i(x) \equiv h \sum_{j=1}^{k} \rho_{ij} f_1(\tau_j, x), \qquad \hat{B}_i(x) \equiv h \sum_{j=1}^{k} \sigma_{ij} f_2(\tau_j, x). \tag{3.86}$$

Here $\tau_j = t_n + c_j h$ and the (real) constants c_j, ρ_{ij}, σ_{ij} are chosen in such a way that ψ_h provides an approximation of order r. These methods have the additional advantage that, when applied to (3.79) with the time frozen, reproduce they the standard splitting (3.40), since $\sum_j \rho_{ij} = a_i$ and $\sum_j \sigma_{ij} = b_i$.

3.7 Some efficient splitting and composition methods

Constructing specific splitting and composition integrators requires numerically solving a system of polynomial equations to get the parameters of the methods. These equations may have: (i) no real solutions; (ii) a finite number of solutions or (iii) families of solutions depending on one or several parameters. The procedure may present severe technical difficulties: first, to find all the solutions (this can be a very difficult task for high order methods), and second, to choose the solution or solutions providing the "best" (in an appropriate sense) methods.

Given a set of integrators within the same class, we say that the most efficient method is the one that provides the desired accuracy with the lowest computational cost. Evidently, the best integration method will depend on the desired level of accuracy –since high-order integrators tend to be more efficient when high precision is required– as well as on the specific characteristics of the problem, such as near-integrability, the suitability for employing RKN-type methods, etc.

In this respect, it is quite common to compare the performance of different methods of the same order belonging to the same class of integrators when several solutions for the parameters are obtained. Sometimes it is even advantageous to consider more parameters than strictly necessary to solve the order conditions, although in this way the computational cost is increased.

We next present several relatively low-order methods that are used throughout the book, either in numerical examples or in the exercises. MATLAB codes for most of them are available on the accompanying website. For readers seeking more in-depth information, we recommend the recent review paper [52], which compiles references and coefficients for a wide range of splitting and composition schemes.

3.7.1 General (non-processed) methods

Symmetric composition of symmetric second-order methods: Orders 4 and 6. Starting from a basic symmetric second-order scheme $S_h^{[2]}$, we can achieve order r with the s-stage palindromic composition

$$\mathrm{SS}r_s: \quad \underbrace{S_{\alpha_1 h}^{[2]} \circ S_{\alpha_2 h}^{[2]} \circ \cdots \circ S_{\alpha_2 h}^{[2]} \circ S_{\alpha_1 h}^{[2]}}_{s \text{ stages}}. \tag{3.87}$$

The following schemes are among the most popular within this class. For clarity, we provide the sequence of coefficients for each method.

- $\mathrm{SS4}_3$: $(\alpha_1, \alpha_2, \alpha_1)$, with $\alpha_1 = \frac{1}{2-2^{1/3}}$, $\alpha_2 = 1 - 2\alpha_1$. This corresponds to the triple jump (3.31);
- $\mathrm{SS4}_5$: $(\alpha_1, \alpha_1, \alpha_2, \alpha_1, \alpha_1)$, with $\alpha_1 = \frac{1}{4-4^{1/3}}$, $\alpha_2 = 1 - 4\alpha_1$. It corresponds to the quintuple jump (3.33) with $k = 2$ [310];
- $\mathrm{SS6}_9$: $(\alpha_1, \alpha_2, \alpha_3, \alpha_4, \alpha_5, \alpha_4, \alpha_3, \alpha_2, \alpha_1)$, with coefficients [237]

$$\begin{aligned} \alpha_1 &= 0.1867 & \alpha_4 &= -0.8432656233877346 \\ \alpha_2 &= 0.5554970237124784 & \alpha_5 &= 1 - 2(\alpha_1 + \alpha_2 + \alpha_3 + \alpha_4) \\ \alpha_3 &= 0.1294669489134754. \end{aligned}$$

Splitting into two parts: Orders 2 and 4. The general form of an s-stage symmetric method of order r in this case is

$$\mathrm{S}r_s: \quad \underbrace{\varphi_{a_1 h}^{[1]} \circ \varphi_{b_1 h}^{[2]} \circ \varphi_{a_2 h}^{[1]} \circ \cdots \circ \varphi_{a_2 h}^{[1]} \circ \varphi_{b_1 h}^{[2]} \circ \varphi_{a_1 h}^{[1]}}_{s \text{ stages}}, \tag{3.88}$$

and two representative schemes correspond to the following sequences of coefficients:

- $\mathrm{S2}_2$: $(a_1, b_1, a_2, b_1, a_1)$, with $b_1 = \frac{1}{2}$, $a_2 = 1 - 2a_1$.

  The choice $a_1 = 0.19318332750378$ minimizes the leading error in energy and has positive coefficients [237, 268], while the choice $a_1 = (3 - \sqrt{3})/6 = 0.21132\ldots$ is more efficient for Hamiltonian Monte Carlo simulations [58].

- $\mathrm{S4}_6$: $(a_1, b_1, a_2, b_2, a_3, b_3, a_4, b_3, a_3, b_2, a_2, b_1, a_1)$ with [65]

$$\begin{aligned} a_1 &= 0.0792036964311956 & b_1 &= 0.2095151066133620 \\ a_2 &= 0.3531729060497740 & b_2 &= -0.1438517731798181 \\ a_3 &= -0.0420650803577195 & b_3 &= 1/2 - (b_1 + b_2) \\ a_4 &= 1 - 2(a_1 + a_2 + a_3). \end{aligned}$$

Composition method–adjoint: Order 4. Any scheme of the form (3.88) can be reformulated as a composition of a first-order method and its adjoint, according with the analysis of section 3.2.2. Thus, in particular, the previous method $S4_6$ reads

$$\chi_{\alpha_1 h} \circ \chi^*_{\alpha_2 h} \circ \chi_{\alpha_3 h} \circ \chi^*_{\alpha_4 h} \circ \chi_{\alpha_5 h} \circ \chi^*_{\alpha_6 h} \circ \chi_{\alpha_6 h} \circ \chi^*_{\alpha_5 h} \circ \chi_{\alpha_4 h} \circ \chi^*_{\alpha_3 h} \circ \chi_{\alpha_2 h} \circ \chi^*_{\alpha_1 h}$$

with coefficients

$$\begin{aligned} \alpha_1 &= 0.0792036964311956 & \alpha_4 &= -0.3667132690474261 \\ \alpha_2 &= 0.1303114101821661 & \alpha_5 &= 0.3246481886897060 \\ \alpha_3 &= 0.2228614958676080 & \alpha_6 &= 1/2 - (\alpha_1 + \cdots + \alpha_5). \end{aligned}$$

Runge–Kutta–Nyström methods: Order 4. The fourth-order scheme with $s = 6$ stages designed in [65] is particularly efficient for the problem (3.57) with the splitting (3.59). It corresponds to the composition

$$\text{NB4}_6: \quad \varphi^{[2]}_{b_1 h} \circ \varphi^{[1]}_{a_1 h} \circ \cdots \circ \varphi^{[1]}_{a_3 h} \circ \varphi^{[2]}_{b_4 h} \circ \varphi^{[1]}_{a_3 h} \circ \cdots \circ \varphi^{[1]}_{a_1 h} \circ \varphi^{[2]}_{b_1 h},$$

with coefficients

$$\begin{aligned} b_1 &= 0.0829844064174052 & a_1 &= 0.2452989571842710 \\ b_2 &= 0.3963098014983681 & a_2 &= 0.6048726657110800 \\ b_3 &= -0.0390563049223486 & a_3 &= 1/2 - (a_1 + a_2) \\ b_4 &= 1 - 2(b_1 + b_2 + b_3). \end{aligned}$$

We recall that, when used in Hamiltonian systems, $\varphi^{[2]}_h$ must be associated with the flow of the potential energy.

As representative of methods with modified potentials, we highlight the popular fourth-order scheme (3.62)

$$\text{MB4}_2: \quad \varphi^{[2]}_{b_1 h} \circ \varphi^{[1]}_{a_1 h} \circ \widetilde{\varphi}_{2b_2 h, d_2 h} \circ \varphi^{[1]}_{a_1 h} \circ \varphi^{[2]}_{b_1 h},$$

with $a_1 = \frac{1}{2}, b_1 = \frac{1}{6}, b_2 = \frac{1}{3}, d_2 = \frac{1}{72}$, and $\widetilde{\varphi}_{2b_2 h, d_2 h}$ given by (3.64).

Methods for near-integrable systems. For perturbed problems, we next collect a pair of schemes of generalized order $(s_1, s_2, \ldots)$ of the form

$$\text{NIA}(s_1, s_2, \ldots)_s: \quad \varphi^{[1]}_{a_1 h} \circ \varphi^{[2]}_{b_1 h} \circ \varphi^{[1]}_{a_2 h} \circ \cdots \circ \varphi^{[1]}_{a_2 h} \circ \varphi^{[2]}_{b_1 h} \circ \varphi^{[1]}_{a_1 h},$$

namely

- $\text{NIA}(4,2)_2 : (a_1, b_1, a_2, b_1, a_1)$, with $a_1 = \frac{3-\sqrt{3}}{6}$, $a_2 = 1 - 2a_1$, $b_1 = \frac{1}{2}$ [236].
- $\text{NIA}(10,6,4)_8 : (a_1, b_1, a_2, b_2, a_3, b_3, a_4, b_4, a_5, b_4, a_4, b_3, a_3, b_2, a_2, b_1, a_1)$ with [41]:

$$\begin{aligned} a_1 &= 0.0380944974224122 & b_1 &= 0.0958588808370752 \\ a_2 &= 0.1452987161169130 & b_2 &= 0.2044461531429988 \\ a_3 &= 0.2076276957255412 & b_3 &= 0.2170703479789911 \\ a_4 &= 0.4359097036515262 & b_4 &= 1/2 - (b_1 + b_2 + b_3) \\ a_5 &= 1 - 2(a_1 + a_2 + a_3 + a_4). \end{aligned}$$

We recall that $\varphi_h^{[2]}$ corresponds to the flow of the small perturbation part of the problem.

3.7.2 Processed methods

Symmetric composition of symmetric second-order methods: Order 6. We select the following processed method with a 13-stage symmetric kernel, ψ_h, and a 12-stage processor π_h [46]:

$$\text{PSS6}_{13}: \quad \underbrace{\left(S^{[2]}_{-\gamma_1 h} \circ \cdots \circ S^{[2]}_{-\gamma_{12} h}\right)}_{\pi_h^{-1}} \circ \underbrace{\left(S^{[2]}_{\alpha_{13} h} \circ \cdots \circ S^{[2]}_{\alpha_1 h}\right)}_{\psi_h} \circ \underbrace{\left(S^{[2]}_{\gamma_{12} h} \circ \cdots \circ S^{[2]}_{\gamma_1 h}\right)}_{\pi_h}$$

with coefficients

$$\begin{aligned}
&\alpha_1 = \alpha_2 = \alpha_3 = \alpha_4 && \gamma_1 = 1/10 \\
&\alpha_4 = 0.1256962887201060 && \gamma_2 = 0.2250802987611760 \\
&\alpha_5 = 0.1480706601149650 && \gamma_3 = 0.1912446945111610 \\
&\alpha_6 = -0.3508563708238280 && \gamma_4 = -0.2127637921948900 \\
&\alpha_7 = 1 - 2(\alpha_1 + \ldots + \alpha_6) && \gamma_5 = -0.0966015730658229 \\
&\alpha_8 = \alpha_6, \ldots, \alpha_{12} = \alpha_2, \alpha_{13} = \alpha_1 && \gamma_6 = -(\gamma_1 + \gamma_2 + \gamma_3 + \gamma_4 + \gamma_5) \\
& && \gamma_{6+i} = -\gamma_i, \; i = 1, \ldots, 6.
\end{aligned}$$

Composition method–adjoint: Order 4. With $s = 4$ and a symmetric kernel, we have the fourth-order scheme

$$\text{PS4}_4: \quad \underbrace{\left(\chi^*_{-\gamma_1 h} \circ \chi_{-\gamma_2 h} \circ \cdots \circ \chi^*_{-\gamma_7 h}\right)}_{\pi_h^{-1}}$$

$$\circ \underbrace{\left(\chi_{\alpha_8 h} \circ \chi^*_{\alpha_7 h} \circ \cdots \circ \chi_{\alpha_2 h} \circ \chi^*_{\alpha_1 h}\right)}_{\psi_h} \circ \underbrace{\left(\chi_{\gamma_7 h} \circ \cdots \circ \chi^*_{\gamma_2 h} \circ \chi_{\gamma_1 h}\right)}_{\pi_h}$$

with

$$\begin{aligned}
&\alpha_1 = \tfrac{6}{25} && \gamma_1 = 0.1171835753202670 \\
&\alpha_2 = \tfrac{-15+\sqrt{18069}}{300} && \gamma_2 = 0.4731269439352650 \\
&\alpha_3 = \tfrac{-15-\sqrt{18069}}{300} && \gamma_3 = -1.3516714399468860 \\
&\alpha_4 = \tfrac{1}{2} - (\alpha_1 + \alpha_2 + \alpha_3) = \tfrac{9}{25} && \gamma_4 = 1.3502981604903750 \\
&\alpha_{9-i} = \alpha_i, \; i = 1, 2, 3, 4 && \gamma_5 = -0.4530449481299280 \\
& && \gamma_6 = 0.0571927978097620 \\
& && \gamma_7 = -(\gamma_1 + \cdots + \gamma_6).
\end{aligned}$$

As already noted, this composition can also be used as a splitting method for problems which can be separated into two parts.

Runge–Kutta–Nyström methods: Order 4. An efficient fourth-order scheme of this class is formed by taking the Störmer–Verlet method with one modified potential as the kernel and an appropriate processor [56]. Specifically, the method reads

$$\text{PM4}_1: \quad \underbrace{\left(\varphi^{[2]}_{-y_1 h} \circ \varphi^{[1]}_{-z_1 h} \circ \varphi^{[2]}_{-y_2 h} \circ \varphi^{[1]}_{-z_2 h} \circ \varphi^{[2]}_{-y_3 h} \circ \varphi^{[1]}_{-z_3 h}\right)}_{\pi_h^{-1}}$$

$$\circ \underbrace{\left(\varphi^{[1]}_{h/2} \circ \widetilde{\varphi}_{h,h/24} \circ \varphi^{[1]}_{h/2}\right)}_{\psi_h} \circ \underbrace{\left(\varphi^{[1]}_{z_3 h} \circ \varphi^{[2]}_{y_3 h} \circ \varphi^{[1]}_{z_2 h} \circ \varphi^{[2]}_{y_2 h} \circ \varphi^{[1]}_{z_1 h} \circ \varphi^{[2]}_{y_1 h}\right)}_{\pi_h}$$

with coefficients

$$\begin{aligned} z_1 &= 0.8749306155955435 & y_1 &= 0.1859353996846055 \\ z_2 &= -0.2371066801510219 & y_2 &= 0.0731969797858114. \\ z_3 &= -0.5363539829039128 & y_3 &= -0.1576624269298081. \end{aligned}$$

The kernel of this method corresponds precisely to the scheme (1.62) illustrated in Chapter 1 on Hamiltonian systems. In this way, the second-order scheme (1.62) is conjugate to a fourth-order method, and this helps to explain its good behavior for the examples considered there.

Methods for near-integrable systems. The following composition provides a method of generalized order (7,6,4) which is particularly efficient in practice:

$$\text{PNI}(7,6,4)_3: \quad \underbrace{\left(\varphi^{[1]}_{-z_1 h} \circ \varphi^{[2]}_{-y_1 h} \circ \cdots \circ \varphi^{[1]}_{-z_6 h} \circ \varphi^{[2]}_{-y_6 h}\right)}_{\pi_h^{-1}}$$

$$\circ \underbrace{\left(\varphi^{[1]}_{a_1 h} \circ \varphi^{[2]}_{b_1 h} \circ \varphi^{[1]}_{a_2 h} \circ \varphi^{[2]}_{b_2 h} \circ \varphi^{[1]}_{a_2 h} \circ \varphi^{[2]}_{b_1 h} \circ \varphi^{[1]}_{a_1 h}\right)}_{\psi_h} \circ \underbrace{\left(\varphi^{[2]}_{y_6 h} \circ \cdots \circ \varphi^{[2]}_{y_1 h} \circ \varphi^{[1]}_{z_1 h}\right)}_{\pi_h}$$

with coefficients

$$\begin{aligned} a_1 &= 0.5600879810924619 & a_2 &= 1/2 - a_1 \\ b_1 &= 1.5171479707207228 & b_2 &= 1 - 2b_1 \\ z_1 &= 0.3346222298730800 & y_1 &= 1.6218101180868010 \\ z_2 &= 1.0975679907321640 & y_2 &= 0.0061709468110142 \\ z_3 &= 1.0380887460967830 & y_3 &= 0.8348493592472594 \\ z_4 &= 0.6234776317921379 & y_4 &= 0.0511253369989315 \\ z_5 &= 1.102753206303191 & y_5 &= 0.5633782670698199 \\ z_6 &= 0.0141183222088869 & y_6 &= 1/2. \end{aligned}$$

Again, the small part must be evolved with the $\varphi^{[2]}_h$ map.

3.8 Examples of application

We next illustrate how the splitting and composition methods collected above behave in practice on some simple examples.

Example 3.3. Our first illustration concerns the ABC flow system (2.31). The corresponding vector field can be split into three parts as

$$f = f^{[1]} + f^{[2]} + f^{[3]} = A \begin{pmatrix} 0 \\ \sin x \\ \cos x \end{pmatrix} + B \begin{pmatrix} \cos y \\ 0 \\ \sin y \end{pmatrix} + C \begin{pmatrix} \sin z \\ \cos z \\ 0 \end{pmatrix}. \quad (3.89)$$

As a basic symmetric second-order method, we take the composition

$$S_h^{[2]} = \varphi_{h/2}^{[1]} \circ \varphi_{h/2}^{[2]} \circ \varphi_h^{[3]} \circ \varphi_{h/2}^{[2]} \circ \varphi_{h/2}^{[1]},$$

but any other permutation can also be considered. Although composing N times $S_h^{[2]}$ requires in principle N evaluations of $\varphi_h^{[3]}$ and $2N$ evaluations of $\varphi_h^{[2]}$ and $\varphi_h^{[1]}$, an efficient implementation (by concatenating the last map with the first one at the following step) reduces the last number to $N+1$ evaluations of $\varphi_h^{[1]}$. Similar savings can be achieved in higher-order methods.

As a basic first-order method, we take $\chi_h = \varphi_h^{[1]} \circ \varphi_h^{[2]} \circ \varphi_h^{[3]}$, with adjoint $\chi_h^* = \varphi_h^{[3]} \circ \varphi_h^{[2]} \circ \varphi_h^{[1]}$. Notice that

$$\chi_{\alpha_{2j}h} \circ \chi^*_{\alpha_{2j-1}h} = \varphi^{[1]}_{\alpha_{2j}h} \circ \varphi^{[2]}_{\alpha_{2j}h} \circ \varphi^{[3]}_{(\alpha_{2j}+\alpha_{2j-1})h} \circ \varphi^{[2]}_{\alpha_{2j-1}h} \circ \varphi^{[1]}_{\alpha_{2j-1}h},$$

so that s-stage methods constructed either by composition of $S_h^{[2]}$ or $\chi_h \circ \chi_h^*$ require exactly the same number of evaluations of elementary flows $\varphi_h^{[i]}$. For this reason, the value of s will be used to measure the computational cost of both families of integrators.

We first take $A = 1$, $B = 2$, $C = 3$, initial conditions $(x_0, y_0, z_0) = (1, 2, 3)$ and integrate the system until the final time $t_{\mathrm{f}} = 20$ with time steps $h = t_{\mathrm{f}}/2^i$, $i = 2, 3, \ldots, 10$. Figure 3.1 shows, in a double logarithmic scale, the Euclidean norm of the error in the solution at t_{f} as a function of the number of stages s required (when using different values for the time step) for different symmetric-symmetric composition methods, namely the processed scheme $\mathrm{PSS6}_{13}$, and the non-processed $\mathrm{SS4}_5$ (squares) and $\mathrm{SS6}_9$ (diamonds). For comparison, we have also included the simple composition $\mathrm{SS4}_3$ and the fourth-order processed scheme $\mathrm{PS4}_3$. We observe that while higher-order methods require more computational effort per step, they ultimately prove more efficient when high accuracy is needed. In particular, the sixth-order processed method –despite using a relatively simple kernel – outperforms the non-processed $\mathrm{SS6}_9$ scheme in this regime.

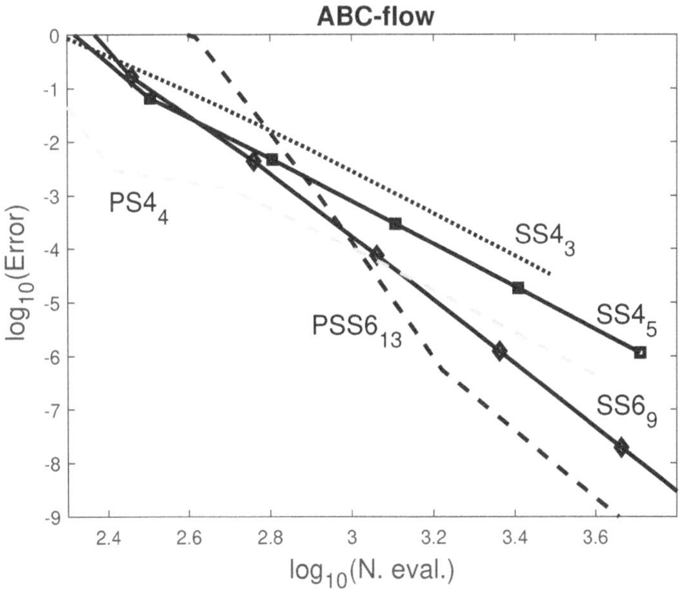

FIGURE 3.1: Euclidean norm of the error in the solution at $t_{\mathrm{f}} = 20$ vs. the number of stages for different processed and non-processed symmetric-symmetric composition methods applied to the ABC flow. Notice the improvement achieved by processed schemes.

To illustrate the possible advantages of using methods with more stages than the minimum necessary to satisfy the order conditions and also integrators based on composition of the basic first-order method and its adjoint, we repeat the experiment with the following fourth-order schemes: $SS4_3$, $SS4_5$, $S4_6$, and $PS4_4$. The corresponding efficiency diagram is depicted in Figure 3.2 (left). Then we take $A = 0$, so that system (3.89) can be seen as separable into just two parts and carry out again the integration. The corresponding results are shown in Figure 3.2 (right). We observe the good performance exhibited by the processed method and the six-stage method, both built for general separable problems. □

Example 3.4. For our second illustration, we take again the two-dimensional Kepler problem (1.57)–(1.58) with initial conditions given by (1.61) and eccentricity $e = 0.2$. We integrate until $t_{\mathrm{f}} = 200$ and measure the average error in energy and the average Euclidean norm of the error in positions and momenta for different values of the time step as a function of the number of force evaluations. For this problem it makes sense to use methods with modified potentials, since the extra cost due to its inclusion is marginal. If one map

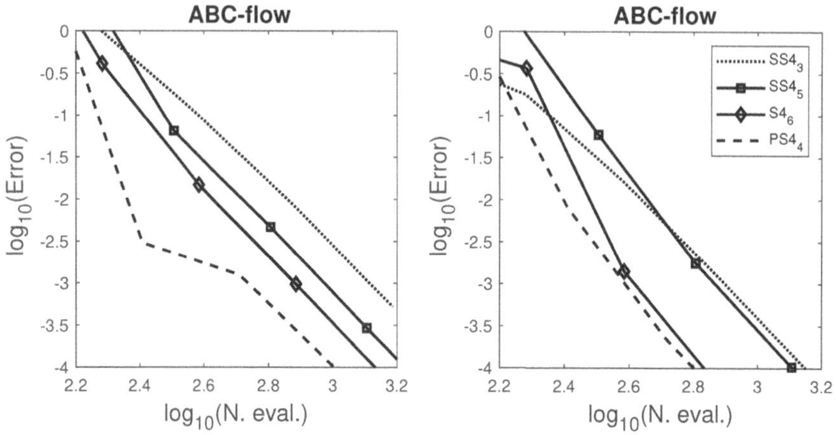

FIGURE 3.2: Error in the solution at the final time $t_f = 20$ vs. computational effort for the ABC-flow with different fourth-order schemes. Diagram on the right corresponds to the case $A = 0$.

containing the modified potential requires a cost K times the evaluation of the potential, then the corresponding curve should be displaced approximately $\log(2)(K-1) \simeq 0.3(K-1)$ units to the right (typically, $1 \le K \le 2$). The following methods are considered: $NB4_6$, $MB4_2$ and $PM4_1$. Figure 3.3 shows the results for the error in energy (left) and in phase space (right). □

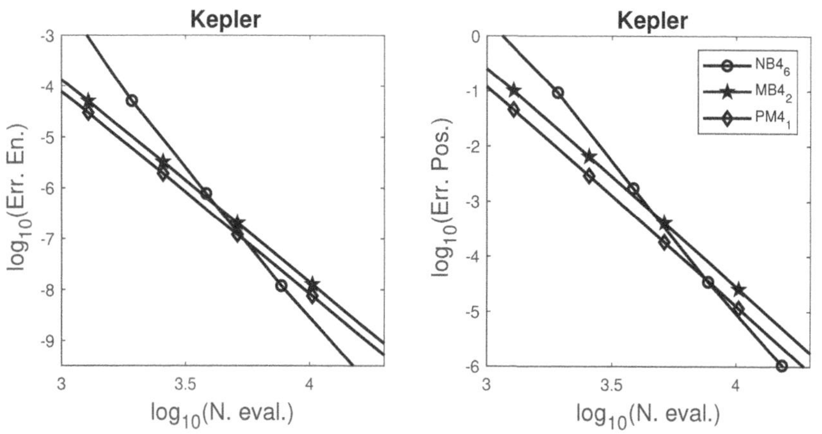

FIGURE 3.3: Average error in energy (left) and in the Euclidean norm of the error in positions and momenta (right) for different values of the time step vs. the number of force evaluations for the two-dimensional Kepler problem.

3.9 Exercises

1. Consider the two-stage one-parameter family of symmetric second-order splitting methods whose associated differential operator reads

$$\Psi_h = \mathrm{e}^{a_1 h F_1} \mathrm{e}^{h/2 F_2} \mathrm{e}^{a_2 h F_1} \mathrm{e}^{h/2 F_2} \mathrm{e}^{a_1 h F_1}$$

and $a_2 = 1 - 2a_1$. Show that

$$v_{112} = \frac{1}{24}(6a_1 - 1), \qquad v_{212} = -\frac{1}{12}(6a_1^2 - 6a_1 + 1)$$

in the corresponding expansion (3.44). Assuming that these two terms contribute equally to the error, find the value of a_1 which minimizes the quantity

$$\mathcal{E} = \sqrt{v_{112}^2 + v_{212}^2}.$$

Compare your result with the one provided for method $S2_2$, equation (3.88).

2. Suppose that for a given problem it holds that $\|F_1\| \ll \|F_2\|$ in the composition of Exercise 1. Find parameters such that the dominant term in the error vanishes, that is, $v_{212} = 0$. Repeat the exercise assuming now that $\|F_2\| \ll \|F_1\|$. The resulting methods are of generalized order (4,2) for near-integrable systems.

3. Given the two-stage one-parameter family of symmetric second-order composition methods

$$S2_2: \quad \chi_{\alpha_1 h} \circ \chi^*_{\alpha_2 h} \circ \chi_{\alpha_2 h} \circ \chi^*_{\alpha_1 h}, \tag{3.90}$$

with $\alpha_2 = 1/2 - \alpha_1$, suppose that $\|[Y_1, Y_2]\| \ll \|Y_3\|$ for a certain problem. Find the optimal value of α_1 that minimizes w_3 from the order conditions. Assuming now that $\|[Y_1, Y_2]\| \gg \|Y_3\|$, find the optimal value of α_1 that vanishes w_{12}.

4. The Lotka–Volterra model can be expressed as

$$\frac{d}{dt}\begin{pmatrix} u \\ v \end{pmatrix} = \begin{pmatrix} u(v-2) \\ 0 \end{pmatrix} + \begin{pmatrix} 0 \\ v(1-u) \end{pmatrix}, \tag{3.91}$$

and has $I(u,v) = \log u - u + 2\log v - v$ as a first integral (see e.g., [170]). Show that the change of variables $u = \mathrm{e}^q$, $v = \mathrm{e}^p$ transforms this problem into a separable Hamiltonian system with $H(q,p) = I(u(q), v(p)) = p - \mathrm{e}^p + q^2 - \mathrm{e}^q$. Solve the problem with initial conditions $(u(0), v(0)) = (3,3)$ until $t_{\mathrm{f}} = 100$ with the methods collected in section 3.7 adapted to this case (splitting into two parts and symmetric compositions of second-order symmetric schemes). Which is the most efficient method leading to an error smaller than $\mathcal{E} = 10^{-7}$ in the invariant $I(u, v)$ at the final time?

5. Repeat the numerical experiments of the previous exercise, but now with the splitting

$$\frac{d}{dt}\begin{pmatrix} u \\ v \end{pmatrix} = \begin{pmatrix} -2 & 0 \\ 0 & 1 \end{pmatrix}\begin{pmatrix} u \\ v \end{pmatrix} + \begin{pmatrix} uv \\ -uv \end{pmatrix}. \tag{3.92}$$

6. The equations defining the Lorenz system (2.30) can be split as

$$\frac{d}{dt}\begin{pmatrix} x \\ y \\ z \end{pmatrix} = \begin{pmatrix} -\sigma & \sigma & 0 \\ r & -1 & 0 \\ 0 & 0 & b \end{pmatrix}\begin{pmatrix} x \\ y \\ z \end{pmatrix} + \begin{pmatrix} 0 & 0 & 0 \\ 0 & 0 & -x \\ 0 & x & 0 \end{pmatrix}\begin{pmatrix} x \\ y \\ z \end{pmatrix}, \tag{3.93}$$

i.e., as linear plus a nonlinear part, and both are exactly solvable. Take $\sigma = 10$, $b = 8/3$, $r = 28$, initial conditions $(x_0, y_0, z_0) = (1, 2, 3)$ and integrate the system until $t_{\mathrm{f}} = 5$. Take as the exact solution at the final time the numerical result obtained with MATLAB with sufficiently high accuracy. Compare the performance of the fourth-order methods used in Figure 3.2 (right).

7. The Toda lattice Hamiltonian system

$$H = \frac{1}{2}\sum_{i=1}^{N} p_i^2 + \sum_{i=1}^{N-1} \left(e^{q_i - q_{i+1}} - 1\right) + \left(e^{q_N - q_1} - 1\right) \tag{3.94}$$

has several conserved quantities. One of them, $I = \sum_{i=1}^{N} p_i$, is exactly preserved by splitting methods. Take $N = 10$, initial conditions $q_1(0) = 0$, $p_1(0) = -1$, $q_i(0) = 0$, $p_i(0) = 1/9$, $i = 2, \ldots, 10$, and integrate until $t_{\mathrm{f}} = 100 \times 2\pi$. Compute the average error in energy for different time steps using the RKN splitting method NB4_6 and the scheme S4_6. Depict the efficiency diagram.

8. Consider again the Hamiltonian (3.94), but now with $N = 3$, i.e.,

$$H = \frac{1}{2}(p_1^2 + p_2^2 + p_3^2) + (e^{q_1 - q_2} - 1) + (e^{q_2 - q_3} - 1) + (e^{q_3 - q_1} - 1).$$

Compare the computational cost in terms of evaluations of exponentials and products/additions for one stage of a splitting method $\varphi_{a_i h}^{[1]} \circ \varphi_{b_i h}^{[2]}$ and obtain the extra cost introduced by replacing $\varphi_{b_i h}^{[2]}$ with the map $\widetilde{\varphi}_{2b_j h, d_j h}$.

9. The perturbed Kepler Hamiltonian

$$H = \frac{1}{2}(p_x^2 + p_y^2) - \frac{1}{r} - \frac{\varepsilon}{2r^3}\left(1 - \alpha\frac{3x^2}{r^2}\right), \tag{3.95}$$

with $r = \sqrt{x^2 + y^2}$, describes in first approximation the dynamics of a satellite moving into the gravitational field produced by a slightly oblate spheric planet. The motion takes place in a plane containing the symmetry axis of the planet when $\alpha = 1$, whereas $\alpha = 0$ corresponds to a plane perpendicular to that axis [246]. Take $\varepsilon = 0.001$, which approximately corresponds to a satellite moving under the influence of the Earth [202], $\alpha = 1$, and initial conditions $x = 1 - e$, $y = 0$, $p_x = 0$, $p_y = \sqrt{(1+e)/(1-e)}$, with $e = 0.2$. Determine numerically the trajectory up to the final time $t_f = 50 \cdot 2\pi$, and compute the mean error in energy with the Strang splitting with $H = T(p) + V(q)$ and $H = H_K(q,p) + \varepsilon V_I(q)$, where $H_K(q,p)$ denotes the unperturbed Kepler problem.

10. The non-linear damped harmonic oscillator with external force

$$m\ddot{x} = -k\,x + \epsilon\,x^2 - \alpha\dot{x} + f(t)$$

can be written as a first-order separable system as

$$\frac{d}{dt}\begin{pmatrix} x \\ v \end{pmatrix} = \begin{pmatrix} 0 & 1 \\ -\kappa & -\beta \end{pmatrix}\begin{pmatrix} x \\ v \end{pmatrix} + \begin{pmatrix} 0 \\ \varepsilon\,x^2 + \tilde{f}(t) \end{pmatrix},$$

where $\beta = \frac{\alpha}{m}$ and $\tilde{f}(t) = \frac{f(t)}{m}$. Show that it is possible to use Runge–Kutta–Nyström methods to solve this problem.

11. The motion of a charged particle in a constant magnetic field perturbed by k electrostatic plane waves is described by the time-dependent Hamiltonian [88]

$$H(q,p,t) = \frac{1}{2}p^2 + \frac{1}{2}q^2 + \varepsilon \sum_{i=1}^{k} \cos(q - \omega_i t). \tag{3.96}$$

Take $\varepsilon = 10^{-3}$, $k = 10$, $\omega_i = i/10$, $i = 1, \ldots, 10$, and integrate until $t_f = 20$. Compare the results obtained with the Strang splitting applied to the system written as $T(p) + V(t,q)$ and $H_0(p,q) + \varepsilon V_I(t,q)$. Take in both cases as step size $h = 1/5$.

12. Consider the van der Pol equation (2.34) split as follows ($y_1 = y, y_2 = \dot{y}$)

$$\frac{d}{dt}\begin{pmatrix} y_1 \\ y_2 \end{pmatrix} = \begin{pmatrix} 0 & 1 \\ -1 & 0 \end{pmatrix}\begin{pmatrix} y_1 \\ y_2 \end{pmatrix} + \begin{pmatrix} 0 \\ -a(1 - y_1^2)y_2 \end{pmatrix}, \tag{3.97}$$

both parts being exactly solvable. Take $y_1(0) = 1$, $y_2(0) = 0$, $a = \frac{1}{4}$ and compute numerically the exact solution at $t_f = 5$. Compare the performance of the second-order splitting method with the fourth-order methods used to solve Exercise 8. Repeat the numerical experiments for $a = \frac{1}{10}$ and $a = \frac{1}{100}$. What do you observe?

13. The motion of an electron in the Coulomb potential subjected to a constant magnetic field $b = (b_1, b_2, b_3)$ is described by [218]

$$\ddot{q} = -\frac{q}{r^3} + b \times \dot{q},$$

$q \in \mathbb{R}^3$, $r = \|q\|$. Denoting $p = \dot{q}$, we can write this equation either as

$$\frac{d}{dt}\begin{pmatrix} q \\ p \end{pmatrix} = \begin{pmatrix} 0 \\ -\frac{q}{r^3} \end{pmatrix} + \begin{pmatrix} p \\ \hat{b}\,p \end{pmatrix}, \tag{3.98}$$

or

$$\frac{d}{dt}\begin{pmatrix} q \\ p \end{pmatrix} = \begin{pmatrix} p \\ -\frac{q}{r^3} \end{pmatrix} + \begin{pmatrix} 0 \\ \hat{b}\,p \end{pmatrix}. \tag{3.99}$$

See Exercise 10 in Chapter 4 for a definition of the matrix $\hat{b}$. In the first case, both parts are exactly solvable (by means of the Rodrigues formula (4.72)), leading to Scovel's method (see [218, p. 95]), whereas the second can be considered as a perturbed Kepler problem. Take $q = (1, 1, 1)$, $p = (0, 0, 0)$ and compute the maximum error in energy, $H = \frac{1}{2}p^T p - \frac{1}{r}$, of the Strang splitting applied to both systems in the interval $t \in [0, 100]$. Repeat the experiment for $b = (0, 0, \varepsilon)$ with $\varepsilon = 1$ and $\varepsilon = \frac{1}{100}$. What do you observe?

Chapter 4

Other types of geometric numerical integrators

Although splitting and composition methods, together with some variants of Runge–Kutta schemes, constitute a significant class of geometric numerical integrators for ordinary differential equations, there are other types of methods known to possess very favorable qualitative properties when applied to certain systems appearing in applications. Here we will provide a brief survey of some of them.

Hamiltonian systems form one of the most common examples of ordinary differential equations to be analyzed from a "geometric integration" point of view, due to their very specific properties related to the symplecticity of the flow and the preservation of the energy and volume in phase space. As a matter of fact, the theory of Hamiltonian systems (fascinating in itself) offers new insight and opens new possibilities for constructing specifically oriented numerical schemes preserving their main features.

In this chapter we first continue with our (necessarily concise) treatment of Hamiltonian dynamical systems, this time reviewing the fundamental role played by generating functions of canonical transformations and the Hamilton–Jacobi equation. This allows us to introduce in a natural way a new class of symplectic integrators obtained by appropriate truncations of cleverly chosen generating functions (section 4.1). Next, the connection with the Lagrangian formulation is established through Hamilton's variational principle and Legendre transforms. In this way, yet another possibility of designing geometric integrators can be envisaged: the so-called variational integrators, which verify by construction a discrete analogue of Hamilton's principle and thus inherit at the discrete level several properties the continuous system has, especially when symmetries are present (section 4.2).

The flow of a Hamiltonian system preserves, in particular, volume in phase space. There are other systems that, not being Hamiltonian, also possess this property: in fact, any divergence-free vector field defines a volume-preserving flow. It makes sense, then, to analyze whether conventional integrators share this property at the discrete level and eventually to construct specific methods adapted to this situation. This will be the subject of section 4.3, whereas in section 4.4 we review some particular examples of Lie group integrators, both in the linear and non-linear cases. These include schemes based on the Magnus

DOI: 10.1201/9781003527626-4

expansion, Crouch-Grossmann methods, and the class of positivity-preserving integrators.

Extrapolation methods are also widely used to integrate initial value problems when high accuracy is desired, especially when a time-symmetric second-order integrator is used as a basic scheme. If we take the Strang splitting as the basic scheme and carry out polynomial extrapolation, we end up with a particular linear combination of splitting methods, and thus, by following the same logic as in Chapter 2, it is relevant to analyze how geometric properties are then preserved. This issue is addressed in section 4.5.

4.1 Symplectic methods based on generating functions

4.1.1 Crash course on Hamiltonian dynamics II: Canonical transformations and generating functions

As we saw in Chapter 1, the flow originated by Hamilton's equations of motion defines a symplectic transformation, and this property puts very specific restrictions on the solutions of such systems. Symplectic (or canonical) transformations play indeed a central role in Hamiltonian theory [159]. In particular, they allow one to find coordinates where the equations of motion take a particularly simple form.

Given a Hamiltonian function $H(q,p,t)$, with equations of motion

$$\dot{q}_i = \frac{\partial H}{\partial p_i}, \qquad \dot{p}_i = -\frac{\partial H}{\partial q_i}, \qquad i = 1, \ldots, d,$$

the (invertible) mapping from the coordinates and momenta (q_i, p_i) to the new set (Q_i, P_i) defined by

$$Q_i = Q_i(q,p,t), \qquad P_i = P_i(q,p,t), \qquad i = 1, \ldots, d \tag{4.1}$$

is called *symplectic* if its Jacobian matrix M is a symplectic matrix for all values of (q,p,t), i.e.,

$$M^T J M = J \qquad \text{or} \qquad M J M^T = J, \tag{4.2}$$

where J is the canonical matrix (1.34). Denoting, as in Chapter 1, $x = (q,p)^T$ and $X = (Q,P)^T$, the entries of M are given by $M_{ij} = \partial X_i / \partial x_j$. The set of all $2d \times 2d$ symplectic matrices forms the Lie group $\mathrm{Sp}(2d)$.

To get some additional insight into symplectic mappings, consider the Poisson bracket of the different X_i with each other. Using equation (1.35), we get

$$\begin{aligned} \{X_i, X_j\} &= \sum_{m,n} \frac{\partial X_i}{\partial x_m} J_{mn} \frac{\partial X_j}{\partial x_n} = \sum_{m,n} M_{im} J_{mn} M_{jn} \\ &= \sum_{m,n} M_{im} J_{mn} (M^T)_{nj} = (M J M^T)_{ij}, \end{aligned}$$

so that applying the symplectic condition (4.2), we arrive at

$$\{X_i, X_j\} = (MJM^T)_{ij} = J_{ij} = \{x_i, x_j\}. \tag{4.3}$$

In consequence, a necessary and sufficient condition for the map (4.1) to be symplectic is that it preserves the fundamental Poisson brackets (1.36). This in turn is equivalent to the condition that the map must preserve the Poisson bracket Lie algebra of all dynamical variables [16]. Thus, in particular, in the new variables (Q_i, P_i), there must exist a function $K(Q, P, t)$ such that the resulting equations of motion are also in Hamiltonian form,

$$\dot{Q}_i = \frac{\partial K}{\partial P_i}, \qquad \dot{P}_i = -\frac{\partial K}{\partial Q_i}, \qquad i = 1, \ldots, d.$$

In other words, the function K plays the role of the Hamiltonian in the new coordinate set, independently of the particular H considered [159]. Most notably, H, K and the two coordinate systems are connected by a relation of the form

$$\sum_{i=1}^{d} p_i dq_i - H dt = \sum_{i=1}^{d} P_i dQ_i - K dt + dF, \tag{4.4}$$

where the left- and right-hand sides are differential forms and $F(q, p, t)$ is any function of the phase space variables with continuous second derivatives [16, 159]. Moreover, the new and old Hamiltonian functions are related through [274]

$$K(Q, P, t) = H(q, p, t) + \frac{\partial F}{\partial t}, \tag{4.5}$$

whereas if time does not enter explicitly into the transformation, one has the same Hamiltonian function,

$$K(Q, P) = H(q(Q, P), p(Q, P)).$$

In that case, the 1-form $p_i dq_i - P_i dQ_i$ is clearly an exact differential:

$$p_i dq_i - P_i dQ_i = dF(q, p). \tag{4.6}$$

The converse is also true [16, 170]. Here and in the sequel, we adopt the sum over repeated indices notation so that $\sum_{i=1}^{d} P_i dQ_i$ is denoted simply by $P_i dQ_i$, and so on.

The function F acts then as a connection point between both sets of canonical variables and is called appropriately the *generating function* of the canonical transformation. It indeed specifies the equations of the map, in the sense that the transformation can be entirely reconstructed from F, as we next show.

Suppose that, in a neighborhood of some point (q_0, p_0) in phase space, one can take the set (q, Q) as independent coordinates. Equivalently, assume that

$$\det \frac{\partial(q, Q)}{\partial(q, p)} = \det \frac{\partial Q}{\partial p} \neq 0.$$

The function F in (4.6) can be expressed locally in these coordinates, $F(q,p) \equiv F_1(q,Q)$, so that

$$p_i dq_i - P_i dQ_i = \frac{\partial F_1}{\partial q_i} dq_i + \frac{\partial F_1}{\partial Q_i} dQ_i.$$

Since q_i and Q_i are independent, one arrives at

$$p_i = \frac{\partial F_1}{\partial q_i}, \qquad P_i = -\frac{\partial F_1}{\partial Q_i}, \qquad i = 1, \ldots, d.$$

These are relations defining p_i as functions of q_j, Q_j. Inverting them, we can express Q_j in terms of q_i, p_i. Next, we replace these expressions for Q_j into the second set of equations to give P_j as functions of q_i, p_i and thus the transformation is completed. Finally, we invert these relations to express (q,p) in terms of (Q,P) and substitute in $H(q,p)$ to get K, that is,

$$K(Q,P) = H(q(Q,P), p(Q,P)).$$

We see then that the canonical transformation (4.1) in $\mathbb{R}^{2d}$ is given entirely by *one* function of $2d$ variables.

There are, however, some canonical transformations that cannot be generated by a function of the type $F_1(q,Q)$ just analyzed. This is the case, in particular, of the identity transformation, since q_i and $Q_i = q_i$ are certainly dependent. One may then consider other types of generating functions. Suppose in particular that the coordinates q and the new momenta P constitute a set of independent local coordinates in $\mathbb{R}^{2d}$. In other words, $\det(\partial(q,P)/\partial(q,p)) = \det(\partial P/\partial p) \neq 0$. Then (4.6) can be written as

$$p_i dq_i + Q_i dP_i = d(Q_i P_i + F),$$

so that $Q_i P_i + F$, expressed in terms of (q,P), is also a generating function, which we call F_2:

$$F_2(q,P) = Q_i P_i + F(q,p).$$

Proceeding as before, we find

$$p_i = \frac{\partial F_2}{\partial q_i}, \qquad Q_i = \frac{\partial F_2}{\partial P_i}, \qquad i = 1, \ldots, d. \tag{4.7}$$

The first relations are to be solved for P_i in terms of q_j, p_j, and these have to be inserted into the second ones to get Q_i.

If we take instead p_i, Q_i as $2d$ independent coordinates, we obtain a third kind of generating function $F_3(p,Q) = q_i p_i - F(q,p)$ and the transformation can be constructed in the same way from the relations

$$q_i = -\frac{\partial F_3}{\partial p_i}, \qquad P_i = -\frac{\partial F_3}{\partial Q_i}. \tag{4.8}$$

There are, of course, other possibilities of combining variables q, p, Q, P to get some set of $2d$ independent coordinates. The particular choice

$$\frac{Q_i + q_i}{2}, \qquad \frac{P_i + p_i}{2}, \qquad i = 1, \ldots, d$$

is useful to construct numerical schemes. In this case, it is a simple exercise to show that (4.6) can be written as

$$-(P-p)_i\, d(Q+q)_i + (Q-q)_i\, d(P+p)_i = 2\, dF_4$$

for some function $F_4\big((Q+q)/2, (P+p)/2\big)$ [170]. The transformation can be obtained from this fourth type of generating function F_4 by the relations

$$\begin{aligned} Q_i &= q_i + (\partial_2 F_4((Q+q)/2, (P+p)/2))_i \\ P_i &= p_i - (\partial_1 F_4((Q+q)/2, (P+p)/2))_i \end{aligned}$$

where ∂_1 (respectively, ∂_2) denotes the derivative with respect to the first argument (resp., the second). In a more compact way, we can write

$$X = x + J\, \nabla F_4((X+x)/2), \tag{4.9}$$

where $x = (q,p)^T$, $X = (Q,P)^T$ and J is the canonical matrix (1.34).

Example 4.1. Consider the generating function of the second type $F_2(q,P) = q_i P_i + \tau H(q,P)$ for a certain parameter τ. Then (4.7) results in

$$p_i = P_i + \tau \frac{\partial}{\partial q_i} H(q,P), \qquad Q_i = q_i + \tau \frac{\partial}{\partial P_i} H(q,P).$$

By setting $\tau = h$ and identifying (q,p) with (q_n, p_n) and (Q,P) with (q_{n+1}, p_{n+1}), respectively, we get the symplectic Euler method (1.41). Notice that a simple characterization of the symplecticity of scheme (1.41) is obtained in this way.

More generally, consider an s-stage Runge–Kutta method (2.10) whose coefficients satisfy the relations $b_i\, a_{ij} + b_j a_{ji} = b_i\, b_j$ for all i, j. As we have seen in section 2.2, such a method defines a symplectic transformation $(q,p) \mapsto (Q,P)$ when applied to a Hamiltonian system, namely

$$\begin{aligned} Q &= q + h \textstyle\sum_{i=1}^{s} b_i \nabla_p H(\tilde{Q}_i, \tilde{P}_i), & \tilde{Q}_i &= q + h \textstyle\sum_{j=1}^{s} a_{ij} \nabla_p H(\tilde{Q}_j, \tilde{P}_j), \\ P &= p - h \textstyle\sum_{i=1}^{s} b_i \nabla_q H(\tilde{Q}_i, \tilde{P}_i), & \tilde{P}_i &= p - h \textstyle\sum_{j=1}^{s} a_{ij} \nabla_q H(\tilde{Q}_j, \tilde{P}_j). \end{aligned} \tag{4.10}$$

It was established by Lasagni in 1988 [211] (see also [170]) that

$$F_2(q,P) = h \sum_{i=1}^{s} b_i H(\tilde{Q}_i, \tilde{P}_i) - h^2 \sum_{i,j=1}^{s} b_i a_{ij} \nabla_q H(\tilde{Q}_i, \tilde{P}_i)^T \nabla_p H(\tilde{Q}_j, \tilde{P}_j) \tag{4.11}$$

constitutes a generating function for this transformation. □

Since the exact flow of a Hamiltonian system is symplectic, one may try to construct a generating function F for the canonical transformation from the coordinates and momenta $(q(t), p(t))$ at time t to the "new" coordinate set defined by the $2d$ initial values $(q_0, p_0) \equiv (Q, P)$. Once the canonical transformation is constructed in the usual way, we have formally obtained the solution of the problem in the form $q = q(q_0, p_0, t)$, $p = p(q_0, p_0, t)$. We can ensure that the new variables are constant in time simply by requiring that the transformed Hamiltonian K be identically zero. Therefore, in accordance with the existing relationship between the Hamiltonians and the generating function (4.5), F must satisfy the equation

$$H(q,p) + \frac{\partial F}{\partial t} = 0. \tag{4.12}$$

This basic relation adopts different forms depending on the partial set of independent variables chosen. Thus, if one takes F as a function of the "old" coordinates q and the "new" constant momenta P, $F_2(q, P, t)$, then (4.12) becomes

$$H(q, \nabla_q F_2) + \frac{\partial F_2}{\partial t} = 0. \tag{4.13}$$

This nonlinear first-order partial differential equation is called the *Hamilton–Jacobi equation*, and the solution, usually denoted by S, allows one to integrate Hamilton's equations of motion [16, 69, 159].

Example 4.2. In the particular case of the fourth kind of generating function $F_4((Q+q)/2, (P+p)/2)$, by introducing the variables $w = (x+X)/2$, $W = J^{-1}(X - x)$ in terms of $x = (q,p)^T$ and $X = (Q,P)^T$, we can write the canonical transformation (4.9) as $W = \nabla_w F_4(w,t)$ and

$$x = w - \frac{1}{2} JW = w - \frac{1}{2} J\nabla_w F_4(w,t).$$

Inserting this expression into (4.12) results in the following formula for the Hamilton–Jacobi equation:

$$H(w - \frac{1}{2} J\nabla_w F_4(w,t)) + \frac{\partial F_4}{\partial t}(w,t) = 0. \tag{4.14}$$

□

4.1.2 Symplectic integrators based on generating functions

The previous formalism of generating functions allows one in principle to construct symplectic methods by solving approximately the Hamilton–Jacobi equation near the identity transformation and use the corresponding approximation S to determine a canonical transformation for the step $t_n \mapsto t_{n+1} = t_n + h$,

$$(q = q_n, p = p_n) \longmapsto (q_{n+1}, p_{n+1}) = (Q, P). \tag{4.15}$$

This, in fact, is the starting point taken by Feng Kang and his collaborators [141, 143, 145, 330] (see also [142] for a detailed treatment), as well as Channell and Scovel [103], to construct symplectic integrators. In this procedure, one first expands the generating function as

$$S(q, P, h) = G_0(q, P) + hG_1(q, P) + h^2 G_2(q, P) + h^3 G_3(q, P) + \cdots, \tag{4.16}$$

where $G_0(q, P) = q_i P_i$ generates the identity transformation; then insert this expression into the Hamilton–Jacobi equation (4.13) and finally compare equal powers of h. If this procedure is carried out up to order h^r, a set of formulas is obtained that allows one to construct (implicitly) a symplectic method of order r. It is worth noticing that for $r \geq 2$ the method requires, in addition, computing higher derivatives of $H(q, p)$ and that the resulting methods are implicit, even for Hamiltonians of the form $H(q, p) = T(p) + V(q)$ [103, 142].

The requirement of computing higher derivatives of the Hamiltonian can be relaxed, however, by considering special constructions for the generating functions. Thus, in particular, in [250], the following generating function (of the fourth type considered before) is proposed

$$S(w, h) = h \sum_{i=1}^{s} b_i H\big(w + hc_i J \nabla_w H(w)\big)$$

and the coefficients b_i, c_i are determined so that it agrees with the solution of the Hamilton–Jacobi equation (4.14) up to a certain order. Methods of order 4 and 5 have been constructed by following this approach. On the other hand, in [230] the following generating function is proposed

$$S(w, h) = h \sum_{i=1}^{s} b_i H(Y_i) + h^2 \sum_{i,j=1}^{s} \beta_{ij} \nabla H(Y_i)^T J H(Y_j) \tag{4.17}$$

depending on the real numbers b_i, α_{ij} and $\beta_{ij}(= -\beta_{ji})$, with

$$Y_i = w + hJ \sum_{j=1}^{s} \alpha_{ij} \nabla H(Y_j).$$

Taking $s = 4$, it is then possible to construct a time-symmetric sixth-order scheme $x_{n+1} = \psi_h(x_n)$ of the form

$$x_{n+1} = x_n + J \nabla S\big((x_n + x_{n+1})/2, h\big)$$

with $x_n = (q_n, p_n)^T$, which is implicit in x_{n+1}, although the function S can be computed explicitly. Notice that the generating function (4.17) constitutes a generalization of (4.11), corresponding to a symplectic Runge–Kutta method.

4.2 Variational integrators

4.2.1 Crash course on Hamiltonian dynamics III: Hamilton's principle and Lagrangian formulation

Much emphasis is placed on Hamiltonian systems in this book. As a matter of fact, according to [170], "Hamiltonian systems form the most important class of ordinary differential equations in the context of Geometric Numerical Integration," mainly due to their wide range of applications. In the study of Hamiltonian dynamics, there are several domains closely interconnected: the Hamiltonian function and its associated equations of motion, together with the symplecticity of the flow; the first-order partial differential (Hamilton–Jacobi) equation, whose solutions allow one to construct generating functions and symplectic transformations (section 4.1.1); and the variational principles and the Lagrangian formulation, which turns out to be more convenient in covariant relativistic theories [231]. In this section we will briefly summarize the main features of this Lagrangian formulation, and we will see how it leads in a natural way to the construction of yet another family of symplectic methods.

Although in modern treatments Hamiltonian systems are introduced first and then the Lagrangian function and the corresponding Euler–Lagrange equations are derived later [2, 234, 274], the classical approach proceeds just in the reverse way: Lagrange's equations are obtained from a principle of critical action (Hamilton's principle), and then the Legendre transform leads to Hamilton's equations of motion [16, 159, 271].

In Lagrangian mechanics, one has a differentiable manifold (the configuration space) and a function on its tangent bundle (the Lagrangian) [16]. Let $\mathbf{Q}$ be the configuration space, with generalized coordinates $q = (q_1, q_2, \ldots, q_d)^T$ and generalized velocities $\dot{q} = (\dot{q}_1, \ldots, \dot{q}_d)^T$. Let us assume for simplicity a Lagrangian function of the form

$$L(q, \dot{q}) = \frac{1}{2}\dot{q}^T M \dot{q} - V(q), \tag{4.18}$$

where M is a symmetric positive-definite mass matrix and V is a potential function. We form the action function by integrating L along a curve $q(t)$ and then compute variations of the action while holding the endpoints of the curve $q(t)$ fixed so that

$$\delta \int_0^T L(q(t), \dot{q}(t))\, dt = 0. \tag{4.19}$$

This is, roughly speaking, the idea of *Hamilton's principle of critical action* [159]. Computing the variation in (4.19), we get

$$\int_0^T \left(\frac{\partial L}{\partial q_i} \delta q_i + \frac{\partial L}{\partial \dot{q}_i} \delta \dot{q}_i \right) dt = \int_0^T \left(\frac{\partial L}{\partial q_i} - \frac{d}{dt} \left(\frac{\partial L}{\partial \dot{q}_i} \right) \right) \delta q_i \, dt = 0,$$

where we have used integration by parts for the second term and the boundary conditions at the end points of the time interval, $\delta q_i(T) = \delta q_i(0) = 0$. Requiring that this hold for all such variations, then

$$\frac{\partial L}{\partial q_i} - \frac{d}{dt}\frac{\partial L}{\partial \dot{q}_i} = 0, \qquad i = 1, \ldots, d, \tag{4.20}$$

i.e., we get the well-known *Euler–Lagrange equations* [159, 271]. For the particular Lagrangian (4.18), equations (4.20) reduce to

$$M\ddot{q} = -\nabla V(q),$$

that is, Newton's second law. The Lagrangian formulation can be extended to more general systems, including problems with constraints and non-autonomous systems [16]. Of course, the flow is symplectic, and symmetries of the Lagrangian function automatically lead to conservation laws (of momentum, angular momentum, etc.) by virtue of Noether's theorem [16, 159]. In addition, Hamilton's equations are shown to be equivalent to Euler–Lagrange equations.

To see this feature in more detail, let us introduce

$$p_i = \frac{\partial L}{\partial \dot{q}_i} \tag{4.21}$$

and assume that the transformation $(q_i, \dot{q}_j) \longmapsto (q_i, p_j)$ is invertible. By definition, the *Legendre transform* of $L(q, \dot{q})$ with respect to $\dot{q}$ is the function [16]

$$H(q, p) = p_i \dot{q}_i - L(q, \dot{q}), \tag{4.22}$$

where $\dot{q}$ is expressed in terms of p by means of (4.21), which depends on the parameters q. The function (4.22) is then the Hamiltonian of the system.

By applying the chain rule to (4.22) and taking into account (4.21), we get

$$\frac{\partial H}{\partial p_i} = \dot{q}_i + p_j \frac{\partial \dot{q}_j}{\partial p_i} - \frac{\partial L}{\partial \dot{q}_j}\frac{\partial \dot{q}_j}{\partial p_i} = \dot{q}_i.$$

Likewise,

$$\frac{\partial H}{\partial q_i} = p_j \frac{\partial \dot{q}_j}{\partial q_i} - \frac{\partial L}{\partial q_i} - \frac{\partial L}{\partial \dot{q}_j}\frac{\partial \dot{q}_j}{\partial q_i} = -\frac{\partial L}{\partial q_i} = -\dot{p}_i,$$

where the last equality follows directly from (4.20). In consequence, the system of Euler–Lagrange equations (4.20) is equivalent to the system of $2d$ first-order equations

$$\dot{q}_i = \frac{\partial H}{\partial p_i}, \qquad \dot{p}_i = -\frac{\partial H}{\partial q_i},$$

where $H(q,p) = p_i \dot{q}_i - L(q,\dot{q})$ is the Legendre transform of the Lagrangian function viewed as a function of $\dot{q}$ [16, 159].

Example 4.3. For the Lagrangian function (4.18), one has

$$p_i = \partial L / \partial \dot{q}_i = (M\dot{q})_i,$$

so that $\dot{q}_i = (M^{-1}p)_i$ and the Hamiltonian is given by

$$H(q,p) = p^T M^{-1} p - L(q, M^{-1}p) = p^T M^{-1} p - \frac{1}{2} p^T M^{-1} p + V(q) = T(p) + V(q),$$

the total energy of the mechanical system. □

4.2.2 An introduction to variational integrators

The basic idea behind variational integrators is to discretize the previous variational formulation of a given problem. In the case of conservative systems, it is Hamilton's principle of critical action (4.19) that is discretized, and an analogue of the continuous procedure is applied for obtaining a map which approximates the exact trajectory corresponding to the solution of the (continuous) Euler–Lagrange equations (4.20) [219, 235].

The variational method for deriving integrators means that the resulting algorithms automatically inherit distinctive features of the exact flow. In particular, they are symplectic by construction, the momenta associated with symmetries of the system are exactly preserved, and they present a good behavior with respect to the propagation of the error in energy with time.

The starting point is to consider an approximation of the action integral, thus giving the *discrete Lagrangian*

$$L_h(q_n, q_{n+1}, h) \approx \int_{t_n}^{t_{n+1}} L(q(t), \dot{q}(t))\, dt,$$

where $q(t)$ is the exact solution of the Euler–Lagrange equations (4.20) joining q_n and q_{n+1} in the given time interval. Notice that the discrete Lagrangian is a function of two positions, q_n and q_{n+1}, and the time step $h = t_{n+1} - t_n$, instead of the position q and velocity $\dot{q}$.

Next, one considers a discrete curve of points $\{q_n\}_{n=0}^N$ and evaluates the discrete action along this sequence by adding up the discrete Lagrangian on each adjacent pair, i.e.,

$$S_h\left(\{q_n\}_{n=0}^N\right) = \sum_{n=0}^{N-1} L_h(q_n, q_{n+1}, h)$$

and finally computes, as in the continuous case, variations of this action sum with the boundary points q_0 and q_N held fixed. Then

$$\begin{aligned}\delta S_h(\{q_n\}) &= \sum_{n=0}^{N-1} (D_1 L_h(q_n, q_{n+1}, h)\delta q_n + D_2 L_h(q_n, q_{n+1}, h)\delta q_{n+1}) \\ &= \sum_{n=1}^{N-1} \Big(D_2 L_h(q_{n-1}, q_n, h) + D_1 L_h(q_n, q_{n+1}, h)\Big)\delta q_n \\ &\quad + D_1 L_h(q_0, q_1, h)\delta q_0 + D_2 L_h(q_{N-1}, q_N, h)\delta q_N,\end{aligned}$$

where a discrete integration by parts and a rearrangement of the summation have been applied [222]. Here and in the sequel, $D_i L_h$ stands for the derivative of L_h with respect to the ith argument. Requiring now that the variations of the action be zero for any choice of δq_n such that $\delta q_0 = \delta q_N = 0$, we get the *discrete Euler–Lagrange equations*

$$D_2 L_h(q_{n-1}, q_n, h) + D_1 L_h(q_n, q_{n+1}, h) = 0, \tag{4.23}$$

which must hold for any n. Taking initial conditions (q_0, q_1), equations (4.23) define a recursive procedure for computing the sequence $\{q_n\}_{n=0}^{N}$. In this way, the map $(q_n, q_{n+1}) \longmapsto (q_{n+1}, q_{n+2})$ gives us an integrator for the flow defined by the continuous Euler–Lagrange equations.

Example 4.4. Suppose we approximate the action integral $\int_{t_n}^{t_{n+1}} L(q(t), \dot{q}(t))dt$ by the simple rectangle rule, where the velocity is replaced by $(q_{n+1} - q_n)/h$. Then the discrete Lagrangian corresponding to (4.18) reads

$$L_h(q_n, q_{n+1}, h) = h\frac{1}{2}\left(\frac{q_{n+1} - q_n}{h}\right)^T M \left(\frac{q_{n+1} - q_n}{h}\right) - hV(q_n),$$

so that

$$\begin{aligned}D_2 L_h(q_{n-1}, q_n, h) &= M\left(\frac{q_n - q_{n-1}}{h}\right) \\ D_1 L_h(q_n, q_{n+1}, h) &= -M\left(\frac{q_{n+1} - q_n}{h}\right) - h\nabla V(q_n),\end{aligned}$$

and finally, the discrete Euler–Lagrange equations are

$$M\left(\frac{q_{n+1} - 2q_n + q_{n-1}}{h^2}\right) = -\nabla V(q_n),$$

i.e., we recover the Störmer–Verlet scheme in the form (1.50) for Newton's equation $M\ddot{q} = -\nabla V(q)$. This provides yet another proof of the symplectic character of the Störmer–Verlet method. □

In general, any integrator that is a solution of the discrete Euler–Lagrange equations for some discrete Lagrangian is called a *variational integrator*. It is indeed possible to rewrite a variational integrator in a position-momentum form rather than involving two positions. To do that, one first defines the momentum at the nth step as

$$p_n = D_2 L_h(q_{n-1}, q_n, h) = -D_1 L_h(q_n, q_{n+1}, h), \tag{4.24}$$

where the last equality follows from the discrete Euler–Lagrange equations (4.23). With this definition, we can write the integrator as

$$p_n = -D_1 L_h(q_n, q_{n+1}, h), \qquad p_{n+1} = D_2 L_h(q_n, q_{n+1}, h). \tag{4.25}$$

Given the initial condition (q_0, p_0), one solves implicitly the first equation of (4.25) to find q_1 and then determines p_1 from the second equation. Repeating this procedure, we get the sequence of pairs (q_i, p_i), $i = 1, 2, \ldots$, which, by construction, preserves the canonical symplectic form. Therefore, the numerical flow is symplectic [235, 325]. Notice that the sequence $\{q_n\}_{n=0}^N$ obtained in this way satisfies the discrete Euler–Lagrange equations (4.23) for all n, due to the definition of the momentum (4.24).

The order of accuracy of the variational integrator depends, of course, on the order of approximation of the discrete Lagrangian to the continuous action integral. The discrete Lagrangian is of order r if

$$L_h(q_n, q_{n+1}, h) = \int_{t_n}^{t_{n+1}} L(q(t), \dot{q}(t))dt + \mathcal{O}(h^{r+1}),$$

where $q(t)$ is the unique solution of the continuous Euler–Lagrange equations with $q(t_n) = q_n$ and $q(t_{n+1}) = q_{n+1}$. It can be shown that if L_h is of order r, then the corresponding variational integrator is also of order r, i.e., $q_n = q(nh) + \mathcal{O}(h^{r+1})$ [325]. Therefore, high-order variational integrators require constructing discrete Lagrangians by using more accurate approximation rules of the corresponding action integral. In addition, if the discrete Lagrangian is such that $L_h(q_n, q_{n+1}, h) = -L_h(q_{n+1}, q_n, -h)$, for all n, then the resulting variational integrator is automatically time-symmetric and therefore of even order [325].

Example 4.5. Consider the following discrete Lagrangian

$$L_h(q_n, q_{n+1}, h) = \frac{h}{2}\left(\frac{q_{n+1} - q_n}{h}\right)^T M \left(\frac{q_{n+1} - q_n}{h}\right) - hV\big((1-\alpha)q_n + \alpha q_{n+1}\big),$$

depending on a parameter $\alpha \in [0, 1]$. The corresponding discrete Euler–Lagrange equations (4.23) are in this case

$$\begin{aligned} M\left(\frac{q_{n+1} - 2q_n + q_{n-1}}{h^2}\right) &= -(1-\alpha)\nabla V\big((1-\alpha)q_n + \alpha q_{n+1}\big) \\ &\quad - \alpha \nabla V\big((1-\alpha)q_{n-1} + \alpha q_n\big) \end{aligned}$$

and the position-momentum form (4.25) is given by

$$\begin{aligned} p_n &= M\left(\frac{q_{n+1} - q_n}{h}\right) + (1-\alpha)h\nabla V\big((1-\alpha)q_n + \alpha q_{n+1}\big) \\ p_{n+1} &= M\left(\frac{q_{n+1} - q_n}{h}\right) - \alpha h \nabla V\big((1-\alpha)q_n + \alpha q_{n+1}\big). \end{aligned}$$

Notice that the resulting method is implicit and of order one. When $\alpha = 1/2$, we recover the well-known implicit midpoint rule (1.54): the method is symmetric and of second order. □

The class of symplectic partitioned Runge–Kutta methods also constitutes a particular subset of variational integrators. In the context of Lagrangian mechanics a partitioned Runge–Kutta scheme is defined by the map $(q_n, p_n) \longmapsto (q_{n+1}, p_{n+1})$ obtained through

$$\begin{aligned}
q_{n+1} &= q_n + h\sum_{j=1}^{s} b_j \dot{Q}_j & p_{n+1} &= p_n + h\sum_{j=1}^{s} \tilde{b}_j \dot{P}_j \\
Q_i &= q_n + h\sum_{j=1}^{s} a_{ij} \dot{Q}_j & P_i &= p_n + h\sum_{j=1}^{s} \tilde{a}_{ij} \dot{P}_j \qquad i = 1, \ldots, s \\
P_i &= \frac{\partial L}{\partial \dot{q}}(Q_i, \dot{Q}_i) & \dot{P}_i &= \frac{\partial L}{\partial q}(Q_i, \dot{Q}_i) \qquad i = 1, \ldots, s,
\end{aligned} \tag{4.26}$$

the internal stages being the points (Q_i, P_i). If the coefficients satisfy

$$\begin{aligned}
b_i \tilde{a}_{ij} + \tilde{b}_j a_{ji} &= b_i \tilde{b}_j \qquad i, j = 1, \ldots, s \\
b_i - \tilde{b}_i &= 0 \qquad i = 1, \ldots, s
\end{aligned} \tag{4.27}$$

then it is possible to construct a discrete Lagrangian generating the method as follows [235]. Given points (q_n, q_{n+1}), equations (4.26) implicitly define p_n, p_{n+1}, Q_i, P_i, $\dot{Q}_i$, $\dot{P}_i$ for $i = 1, \ldots, s$ as functions of (q_n, q_{n+1}). It turns out, then, that

$$L_h(q_n, q_{n+1}, h) = h\sum_{i=1}^{s} b_i L(Q_i, \dot{Q}_i) \tag{4.28}$$

leads to a position-momentum map $(q_n, p_n) \longmapsto (q_{n+1}, p_{n+1})$, which is precisely the partitioned Runge–Kutta method (4.26) [308]. In this way the method, with the requirements (4.27), is symplectic. Moreover, since the discrete Lagrangian (4.28) is linear, the integrator inherits linear symmetries of the continuous Lagrangian L, and in particular quadratic momentum maps.

Variational integrators are particularly useful for dealing with systems possessing constraints. Suppose for simplicity that these constraints are represented by a function g, which is zero for all configurations. We can work in the space of full configurations by using Lagrange multipliers to enforce that $g(q) = 0$. In that case the discretization proceeds by taking variations of the action, including these Lagrange multipliers λ_k,

$$\delta \sum_{n=0}^{N-1} \Big(L_h(q_n, q_{n+1}, h) + \lambda_{n+1} \cdot g(q_{n+1}) \Big) = 0$$

resulting in the constrained discrete Euler–Lagrange equations

$$\begin{aligned}
&D_2 L_h(q_{n-1}, q_n, h) + D_1 L_h(q_n, q_{n+1}, h) = -\lambda_n \cdot \nabla g(q_n) \\
&g(q_{n+1}) = 0
\end{aligned}$$

which are subsequently solved for λ_n and q_{n+1} [325].

As in the continuous case, the discrete version of Noether's theorem guarantees that symmetries of the system lead naturally to conservation laws for the numerical flow defined by the variational integrator [235].

In addition to the previous Lagrangian setting, there are other approaches to generate variational integrators, for instance, those based on the Galerkin approximation (which involves the replacement of the infinite-dimensional function space where the Lagrangian is defined by a suitable finite-dimensional space), the characterization of the discrete Lagrangian in terms of solutions of the Hamilton–Jacobi equation, etc. [220]. On the other hand, Hamiltonian variational integrators can also be directly formulated from a discretization of the Hamilton variational principle in phase space, which allows one to extend the framework of variational integrators to systems with degenerate Hamiltonians [221].

4.3 Volume-preserving methods

As we have pointed out several times, the flow φ_t of a Hamiltonian system preserves volume in phase space, $\mathrm{vol}(\varphi_t(D)) = \mathrm{vol}(D)$, for any bounded open set D and for any t such that $\varphi_t(x)$ exists, with $x \in D$. This property (known as Liouville's theorem) can be derived from the transformation formula for multiple integrals [149], since $\mathrm{vol}(D) = \int_D dx$ and $|\det(\varphi_t'(x))| = 1$ for symplectic transformations. By the same token, every symplectic integration method applied to a Hamiltonian system automatically inherits this volume-preserving property.

Preservation of volume is of paramount importance in many dynamical systems arising in physical applications, such as particle tracking in incompressible fluid flows, in perturbations of Hamiltonian systems, and in the discretization of wave equations [142, 241]. Volume-preserving integrators also play a fundamental role in the Hamiltonian (or hybrid) Monte Carlo method in molecular simulations, data science, and statistics [72, 265]. We will return to this issue in Chapter 7.

Any differential equation $\dot{x} = f(x)$ whose vector field f is divergence-free, i.e., such that

$$\mathrm{div} f(x) = \sum_{i=1}^{d} \frac{\partial f_i(x)}{\partial x_i} = 0, \tag{4.29}$$

produces a volume-preserving flow. This can be easily shown from the differential equation satisfied by the derivative of the flow along the trajectory, $X(t) \equiv \frac{\partial \varphi_t(x)}{\partial x}$,

$$\dot{X} = A(t)X, \qquad X(0) = I, \tag{4.30}$$

known as the variational equation. Here $A(t) = f'(x(t))$ represents the Jacobian matrix of f evaluated at $x(t) = \varphi_t(x_0)$. A well-known result from the theory of linear differential equations establishes that [114]

$$\frac{d}{dt} \det X = \mathrm{tr} A(t) \det X.$$

In our case $\mathrm{tr} A(t) = \mathrm{div} f(x(t))$. In consequence, $\det X(t) = 1$ if and only if f is divergence-free. We note in passing that for a Hamiltonian system, one has $f = J\nabla H$ and therefore $\mathrm{div} f(x) = 0$.

Since volume preservation constitutes the characteristic geometric feature of divergence-free dynamical systems, it is worth analyzing how standard integrators behave with respect to this property and eventually consider specific methods for this class of systems.

Example 4.6. Consider first, for simplicity, the implicit midpoint rule $x_{n+1} = x_n + hf\left((x_{n+1} + x_n)/2\right)$ applied to a linear system $\dot{X} = A\,X$, $X(0) = I$, where the $d \times d$ matrix A is such that $\mathrm{tr} A = 0$. As pointed out before, $\det X(t) = 1$ and the system is volume-preserving. The numerical approximation reads

$$X_{n+1} = \left(I - \frac{h}{2} A\right)^{-1} \left(I + \frac{h}{2} A\right) X_n,$$

and so the numerical flow is also volume-preserving if

$$\det \left(\frac{\partial X_{n+1}}{\partial X_n} \right) = 1 = \frac{\det(I + hA/2)}{\det(I - hA/2)}.$$

Denoting by $P(\lambda)$ the characteristic polynomial of the matrix A, $P(\lambda) = \det(A - \lambda I)$, the above condition can be written as

$$1 = \frac{P(-2/h)}{(-1)^d P(2/h)}$$

and so the implicit midpoint rule is volume-preserving if $P(\lambda) = (-1)^d P(-\lambda)$. A straightforward calculation shows that this condition is automatically satisfied if $d = 2$, whereas for $d = 3$ it is required in addition that $\det A = 0$, and more conditions have to be imposed when $d > 3$ [142].

The same argument can be applied in the general case, just replacing A by the Jacobian matrix of the vector field [142]. In consequence, the implicit midpoint rule is always volume-preserving for divergence-free systems of dimension 2, but *not* in general (except if the system is Hamiltonian, in which case it is symplectic and hence volume-preserving). □

As a matter of fact, this result can be generalized to any symplectic method: when the dimension $d = 2$, area-preserving maps are automatically symplectic, so any symplectic method is volume-preserving, even if the system is not Hamiltonian. When $d \geq 3$ the situation is different: all standard methods are generally not volume-preserving, even for linear systems. This is a particular consequence of the following result [142, 144]:

Theorem 3 *Let $R(z)$ be a differentiable function defined in a neighborhood of $z = 0$ in $\mathbb{C}$ and satisfying that $R(0) = 1$ and $R'(0) = 1$. Then, for $d \geq 3$, $\det R(A) = 1$ for all $d \times d$ matrices with $\mathrm{tr} A = 0$ if and only if $R(z) = \exp(z)$.*

In other words, there are no consistent analytic approximations to the exponential function mapping the set of traceless matrices to matrices with unit determinant other than the exponential itself. Since explicit (respectively, implicit) Runge–Kutta methods provide polynomial (respectively, rational) approximations to the exponential, we cannot expect that they provide volume-preserving numerical flows. In consequence, new ways need to be explored to construct methods preserving the phase-space volume for generic divergence-free vector fields.

One such approach, due to Feng and Shang [144], constitutes in fact a particular application of splitting methods. Following the treatment in [241, 244], we start with a divergence-free system of ordinary differential equations, i.e., equations

$$\dot{x}_i = f_i(x), \qquad i = 1, \ldots, d \tag{4.31}$$

verifying (4.29). Given an arbitrary point $\tilde{x}$, we can insert

$$f_d = f_d(\tilde{x}) + \int_{\tilde{x}}^{x_d} \frac{\partial f_d(x)}{\partial x_d} dx_d = f_d(\tilde{x}) - \int_{\tilde{x}}^{x_d} \left(\sum_{i=1}^{d-1} \frac{\partial f_i(x)}{\partial x_i} \right) dx_d$$

into (4.31) to get

$$\begin{aligned} \dot{x}_1 &= f_1(x) \\ &\vdots \\ \dot{x}_{d-1} &= f_{d-1}(x) \\ \dot{x}_d &= f_d(\tilde{x}) - \sum_{i=1}^{d-1} \int_{\tilde{x}}^{x_d} \frac{\partial f_i(x)}{\partial x_i} dx_d. \end{aligned} \tag{4.32}$$

The point $\tilde{x}$ has to be conveniently chosen (for instance, such that $f(\tilde{x}) = 0$). Now system (4.32) is split as the sum of $d-1$ vector fields,

$$\begin{aligned} \dot{x}_i &= 0 \qquad i \neq j, d \\ \dot{x}_j &= f_j(x) \\ \dot{x}_d &= f_d(\tilde{x})\delta_{j,d-1} - \int_{\tilde{x}}^{x_d} \frac{\partial f_j(x)}{\partial x_j} dx_d \end{aligned} \tag{4.33}$$

for $j = 1, \ldots, d-1$ (the symbol $\delta_{i,j}$ corresponds to the Kronecker delta). It is worth remarking that every subsystem (4.33) is divergence-free and corresponds indeed to a two-dimensional Hamiltonian system

$$\dot{x}_j = \frac{\partial H_j}{\partial x_d}, \qquad \dot{x}_d = -\frac{\partial H_j}{\partial x_j},$$

with $H_j = -f_d(\tilde{x})\delta_{j,d-1}x_j + \int_{\tilde{x}}^{x_d} f_j(x)dx_d$. Here the x_i, $i \neq j, d$, have to be treated as fixed parameters. The idea is then to solve all these two-dimensional problems either exactly or approximately with a symplectic integrator $\psi_h^{[j]}$ (which in any case is volume-preserving in $\mathbb{R}^d$) and construct an integrator for f by composition: $\psi_h = \psi_h^{[1]} \circ \psi_h^{[2]} \circ \cdots \circ \psi_h^{[d-1]}$ [241].

Example 4.7. The following 3D Volterra system

$$\dot{x} = x(y-z), \qquad \dot{y} = y(z-x), \qquad \dot{z} = z(x-y)$$

is an example of a volume-preserving system. Split the system into a set of two-dimensional Hamiltonian systems.
Solution. We first obtain the Hamiltonian functions from $f_1 = x(y-z)$ and $f_2 = y(z-x)$ as

$$H_1 = \int_0^z x(y-s)ds = xz(y - \frac{1}{2}z), \qquad H_2 = \int_0^z y(s-x)ds = yz(\frac{1}{2}z - x),$$

respectively. Then the system can be split as the sum of the vector fields associated to each Hamiltonian as follows:

$$\frac{d}{dt}\begin{pmatrix} x \\ y \\ z \end{pmatrix} = \underbrace{\begin{pmatrix} x(y-z) \\ 0 \\ -z\left(y - \frac{1}{2}z\right) \end{pmatrix}}_{f^{[1]}} + \underbrace{\begin{pmatrix} 0 \\ y(z-x) \\ -z\left(\frac{1}{2}z - x\right) \end{pmatrix}}_{f^{[2]}}. \tag{4.34}$$

In this case, the solution of both systems can be written in closed form, and any splitting method can be used as a method. Alternatively, one can solve each system using, for example, implicit symplectic methods to build high-order methods using appropriate compositions. □

Other techniques exist to generate volume-preserving integrators. Among them, we can mention in particular the approach in [277] and the volume-preserving schemes obtained from generating functions (analogously to symplectic integrators) [142]. Moreover, the conditions to be satisfied for the existence of volume-preserving Runge–Kutta schemes have also been analyzed [142, 170, 23].

4.4 Lie group methods

Lie groups play a fundamental role in many branches of science, since they constitute the natural framework to express symmetries originating in the very formulation of the models. Thus, in Hamiltonian mechanics, the set of symplectic maps defined on the phase space forms the symplectic group, a particular instance of a Lie group. Very often, the solution of the initial value problem $\dot{x} = f(t, x)$, $x(0) = x_0$, evolves in a differentiable manifold acted upon transitively by a Lie group, i.e., in a *homogeneous space* [267]. Examples of homogeneous spaces include spheres, tori, isospectral orbits, and also Lie groups themselves. Under such circumstances, it is certainly advantageous to design integration methods such that the corresponding numerical flow respects this structure, i.e., is defined in the same Lie group as the continuous flow [121]. These are called *Lie group methods*, and an exhaustive survey is available in the literature collecting the main aspects of the subject [194]. Rather than considering the problem in its greatest generality, in this section we will restrict ourselves to the simpler (but still very important in applications) situation of time-dependent differential equations defined on matrix Lie groups.

4.4.1 Linear matrix equations: Magnus expansion

Linear systems of differential equations with varying coefficients of the form

$$\frac{dX}{dt} = A(t)X, \qquad X(0) = I, \tag{4.35}$$

where $X(t)$ and $A(t)$ are (sufficiently smooth real or complex) $d \times d$ matrices, appear time and again in many branches of science and technology, ranging from atomic and molecular physics to geometric control in mechanical systems. The variational equation (4.30) is just an example of its ubiquity.

From the general theory of differential equations, (4.35) is closely related to the linear homogeneous system

$$\frac{dx}{dt} = A(t)x, \qquad x(0) = x_0 \tag{4.36}$$

with $x \in \mathbb{C}^d$, in the sense that $x(t) = X(t)x_0$ is a solution of (4.36) if $X(t)$ verifies (4.35). Equivalently, (4.35) is the evolution equation for the fundamental matrix of system (4.36) [114].

In spite of its apparent simplicity, obtaining closed-form solutions for (4.35) is only possible in very particular situations. If $d = 1$, i.e., in the scalar case, the problem is trivial: the solution reduces to a quadrature and an ordinary exponential evaluation:

$$X(t) = \exp\left(\int_0^t A(s)ds\right). \tag{4.37}$$

For $d > 1$, (4.37) is still the solution if for any pair of values of t, t_1 and t_2, one has $A(t_1)A(t_2) = A(t_2)A(t_1)$, which is true for a constant A. More generally, this is also the case when the matrices $A(t)$ and $\int_0^t A(s)ds$ commute [114].

In the general non-commutative case, the approach followed by Magnus in [229] was to express the solution $X(t)$ of (4.35) as the exponential of a certain matrix $\Omega(t)$,

$$X(t) = \exp \Omega(t). \tag{4.38}$$

By substituting (4.38) into (4.35), one can derive the differential equation satisfied by the exponent Ω [194]:

$$\dot{\Omega} = d\exp_{\Omega}^{-1}(A(t)), \qquad \Omega(0) = 0, \tag{4.39}$$

where

$$d\exp_{\Omega}^{-1}(A) \equiv \frac{\mathrm{ad}_{\Omega}}{\mathrm{e}^{\mathrm{ad}_{\Omega}} - I} A = \sum_{k=0}^{\infty} \frac{B_k}{k!} \mathrm{ad}_{\Omega}^{k}(A). \tag{4.40}$$

Here B_k are the Bernoulli numbers [3], ad^k is a shorthand for an iterated commutator,

$$\mathrm{ad}_{\Omega}^{0} A = A, \qquad \mathrm{ad}_{\Omega}^{k+1} A = [\Omega, \mathrm{ad}_{\Omega}^{k} A],$$

and $[\Omega, A] = \Omega A - A\Omega$. Integration of (4.39) by iteration gives

$$\Omega(t) = \int_0^t A(t_1)dt_1 - \frac{1}{2}\int_0^t \left[\int_0^{t_1} A(t_2)dt_2, A(t_1)\right] dt_1 + \cdots$$

and in general an infinite series for Ω,

$$\Omega(t) = \sum_{k=1}^{\infty} \Omega_k(t), \qquad \text{with} \qquad \Omega_k(0) = 0, \tag{4.41}$$

whose terms are increasingly complex expressions involving time-ordered integrals of iterated commutators of A evaluated at different times. Although several procedures have been proposed over the years to obtain explicit expressions of Ω_k for any k (recurrence relations [203], techniques based on binary trees [195], combinatorial techniques applied to iterated integrals [249, 321, 4], etc.), the resulting expressions are not unique (due to the Jacobi identity and other identities appearing at higher orders), and very often not all the terms are independent. This problem is addressed in [13], where an explicit formula for Ω_k is obtained as an iterated integral of a linear combination of $(k-1)!$ right-nested independent commutators of A. Specifically, one has

$$\begin{aligned}
\Omega_1(t) &= \int_0^t A(t_1)dt_1, \\
\Omega_k(t) &= \frac{1}{k} \sum_{\sigma \in \mathfrak{S}_{k-1}} (-1)^{d_b} \frac{1}{\binom{k-1}{d_b}} \int_0^t dt_1 \int_0^{t_1} dt_2 \cdots \int_0^{t_{k-1}} dt_k \\
&\quad [A(t_{\sigma(1)}), [A(t_{\sigma(2)}) \cdots [A(t_{\sigma(k-1)}), A(t_k)] \cdots]], \qquad k \geq 2,
\end{aligned} \tag{4.42}$$

where the sum is extended over the $(k-1)!$ permutations $\sigma \in \mathfrak{S}_{k-1}$ of $\{1, 2, \ldots, k-1\}$, and d_b is the number of descents of the permutation σ. We recall that σ has a descent in i if $\sigma(i) > \sigma(i+1)$, $i = 1, \ldots, n-1$.

Equations (4.38) and (4.41) constitute the so-called *Magnus expansion* for the solution of (4.35), whereas the infinite series (4.41) with (4.42) is known as the *Magnus series* [54].

Constructing the solution of (4.35) as $\exp(\Omega(t))$ presents some advantages when $A(t)$ belongs to some matrix Lie (sub)algebra $\mathfrak{g}$ for all t, so that $X(t)$ evolves in the matrix Lie group $\mathcal{G}$ having $\mathfrak{g}$ as its corresponding Lie algebra. See Appendix A.3 for a review of Lie groups and Lie algebras. As all terms in the Magnus series are constructed as sums of multiple integrals of nested commutators, it is clear that Ω and indeed any approximation to it obtained by truncation will also be in $\mathfrak{g}$. Finally, its exponential will be in $\mathcal{G}$.

For instance, suppose that (4.35) is obtained after discretizing in space the time-dependent Schrödinger equation. Then $\mathfrak{g} = \mathfrak{u}(n)$ is the set of all skew-Hermitian matrices, $\mathcal{G}$ is the matrix Lie group of all complex unitary matrices, and the approximate solutions obtained with the Magnus expansion are unitary matrices by construction.

Another important property of the Magnus expansion is that *it actually converges* [255, 256]. More precisely, if equation (4.35) is defined in a Hilbert space and $A(t)$ is a linear operator, then the Magnus series (4.41) with Ω_k given by (4.42) converges in the interval $t \in [0, T)$ such that $\int_0^T \|A(s)\|_2 ds < \pi$ and the sum $\Omega(t)$ satisfies $\exp \Omega(t) = X(t)$ [91]. Here $\|A(t)\|_2$ denotes the 2-norm of the operator $A(t)$.

4.4.2 Numerical schemes based on the Magnus expansion

The Magnus expansion has found extensive use in mathematical physics, quantum chemistry, control theory, quantum computing, etc., as a tool to construct explicitly analytic approximations to the exact solution (see [53, 54] and references therein). These approximate solutions are fairly accurate inside the convergence domain, especially when high-order terms in the Magnus series can be computed.

One should notice, however, the increasing complexity of the terms Ω_k (4.42) in the Magnus series. Although in some cases these expressions can be computed explicitly beyond $k = 2$ (for instance, when the elements of A and its commutators are polynomial or trigonometric functions), in general one has to turn to appropriate numerical techniques. Although several attempts were made before, it was in the pioneering work [195] where Iserles and Nørsett carried out the first systematic study of the Magnus expansion with the aim of constructing numerical integration algorithms for linear problems.

The process of rendering the Magnus expansion a practical numerical integration algorithm involves several stages: dividing the time interval of interest into subintervals of length h, truncating appropriately the Magnus series (4.41), and computing (or at least conveniently approximating) the multidimensional integrals appearing in the expression of Ω_k. The most striking fact in this respect is that just by evaluating $A(t)$ at the nodes of a univariate quadrature, it is possible to approximate all the multivariate integrals [195].

Depending on the particular quadrature rule used, one ends up with a different numerical scheme. The reader is referred to [54, 194] for a detailed analysis of the construction process. Here we only collect some of the most effective methods of orders 2 and 4 within this class.

Order 2. The simplest method is obtained by approximating the integral appearing in Ω_1 either by the midpoint rule

$$X_{n+1} = \mathrm{e}^{hA(t_n+h/2)} X_n, \tag{4.43}$$

or by the trapezoidal rule

$$X_{n+1} = \exp\left(\frac{h}{2}(A(t_n) + A(t_{n+1})\right) X_n.$$

Order 4. Choosing the Gauss–Legendre quadrature rule,

$$A_1 = A\left(t_n + \left(\frac{1}{2} - \frac{\sqrt{3}}{6}\right) h\right), \qquad A_2 = A\left(t_n + \left(\frac{1}{2} + \frac{\sqrt{3}}{6}\right) h\right) \tag{4.44}$$

we get the fourth-order scheme

$$\begin{aligned} \Omega^{[4]}(h) &= \frac{h}{2}(A_1 + A_2) - h^2\frac{\sqrt{3}}{12}[A_1, A_2] \\ X_{n+1} &= \exp(\Omega^{[4]}(h))X_n. \end{aligned} \tag{4.45}$$

Alternatively, approximating $\int_{t_n}^{t_n+h} A(s)ds$ with the Simpson rule, we have instead

$$\Omega^{[4]}(h) = \frac{h}{6}(A_1 + 4A_2 + A_3) - \frac{h^2}{72}[A_1 + 4A_2 + A_3, A_3 - A_1], \tag{4.46}$$

where

$$A_1 = A(t_n), \qquad A_2 = A(t_n + h/2), \qquad A_3 = A(t_n + h). \tag{4.47}$$

Although apparently more A evaluations are necessary in (4.46), this is not the case, since A_3 can be reused at the next integration step.

Although originally designed for a bounded operator $A(t)$, it has been shown that Magnus integrators still retain their full order of convergence when applied to the Schrödinger equation with $A(t) = -i(\Delta + \hat{V}(t))$ and a smooth time-dependent potential $\hat{V}(t)$ [184] for sufficiently regular solutions, and uniformly with respect to the space discretization.

There are, however, several issues related to the presence of commutators in the integrator. Thus, computing the action of iterated commutators on vectors can be very costly due to the number of matrix-vector products required. This is particularly relevant when considering the Schrödinger equation in two- and three-space dimensions [22].

There is, however, an alternative that circumvents these difficulties while still retaining the favorable properties of Magnus integrators: the so-called commutator-free quasi-Magnus (CFQM) methods. Here the basic idea is to replace the single exponential characteristic of Magnus integrators by a composition of several exponentials involving linear combinations of the values of the operator A at certain nodes, c_k, of a quadrature rule. The general form reads

$$X_{n+1} = \mathrm{e}^{h\,B_J} \cdots \mathrm{e}^{h\,B_1}\, X_n$$

$$c_k \in [0,1]\,, \qquad A_k = A(t_n + c_k h)\,, \qquad k \in \{1,\ldots,K\},$$

$$B_j = a_{j1}\,A_1 + \cdots + a_{jK}\,A_K\,, \qquad j \in \{1,\ldots,J\}.$$

Particular examples of CFQM methods are the fourth-order schemes [64]

$$X_{n+1} = \exp\left(\frac{h}{2}(\beta A_1 + \alpha A_2)\right) \exp\left(\frac{h}{2}(\alpha A_1 + \beta A_2)\right) X_n, \tag{4.48}$$

with $\alpha = \frac{1}{2} + \frac{\sqrt{3}}{3}$, $\beta = \frac{1}{2} - \frac{\sqrt{3}}{3}$, A_1, A_2 given in (4.44), and

$$X_{n+1} = \exp\left(\frac{h}{2}\left(-\frac{1}{6}A_1 + \frac{2}{3}A_2 + \frac{1}{2}A_3\right)\right) \exp\left(\frac{h}{2}\left(\frac{1}{2}A_1 + \frac{2}{3}A_2 - \frac{1}{6}A_3\right)\right) X_n, \tag{4.49}$$

where now A_1, A_2, A_3 are given in (4.47).

A detailed treatment and specific schemes up to order six can be found in [60, 61] and up to order eight in [9], whereas it is proved in [42] that CFQM methods applied to the Schrödinger equation with a time-dependent potential $\hat{V}(t)$ are unconditionally stable in the underlying Hilbert space and retain full order of convergence under low regularity requirements on the initial state. On the other hand, it was established in [61] that CFQM integrators employed for the numerical solution of parabolic-type equations are well-defined and stable only if their coefficients satisfy the positivity condition $\sum_{k=1}^{K} a_{jk} > 0$, $j = 1,\ldots,J$. Furthermore, it is shown in [186] that this condition cannot be fulfilled by methods of order greater than four.

Example 4.8. Consider the Mathieu equation

$$\ddot{x} + (\omega + \epsilon\cos(t))x = 0. \tag{4.50}$$

Write the equation as a system of first-order equations, taking as initial conditions $x(0) = 1$, $\dot{x}(0) = 0$, with $\epsilon = 1$. Solve numerically the system on the time interval $t \in [0, 20\pi]$ when $\omega = 1 + k\Delta\omega$, $k = 0, 1, 2, \ldots, 240$ with $\Delta\omega = 0.1$, i.e., for values of ω in the interval $\omega \in [1, 25]$. Use the standard 4th-order RK method (2.7) and the fourth-order Magnus integrators (4.45) and (4.48) with time step $h = \pi/20$ (so that all methods require the same

number of time-dependent function evaluations). Compare the numerical results with the exact solution at the final time (obtained also numerically to high accuracy) for each value of ω.
Solution. Figure 4.1 shows the results obtained. The superiority of the Magnus integrators manifests clearly as ω grows, i.e., when the problem becomes highly oscillatory. □

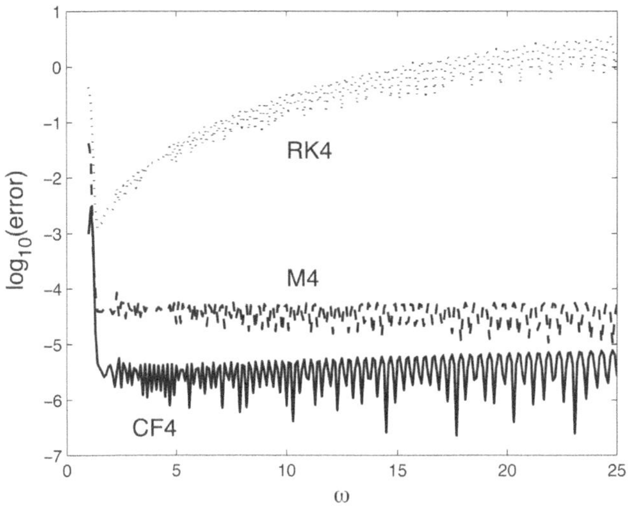

FIGURE 4.1: Error at the final time for the numerical integration of the Mathieu equation (4.50) for different values of ω using the standard fourth-order RK method (RK4), the fourth-order Magnus integrator using the Gaussian quadrature rule (4.45) (M4), and the CFQM scheme (4.48) (denoted as CF4).

4.4.3 Nonlinear matrix equations in Lie groups

The nonlinear time-dependent differential equation

$$\dot{X} = A(t, X)X, \qquad X(0) = X_0 \in \mathcal{G}, \tag{4.51}$$

defined in a matrix Lie group $\mathcal{G}$ appears in relevant physical fields such as rigid body mechanics, in the calculation of Lyapunov exponents ($\mathcal{G} \equiv \mathrm{SO}(n)$), and other problems arising in Hamiltonian dynamics ($\mathcal{G} \equiv \mathrm{Sp}(n)$). In fact, it can be shown that every differential equation evolving on a matrix Lie group $\mathcal{G}$ can be written in the form (4.51) [194].

As in the linear case, the solution of (4.51) can be represented by

$$X(t) = \exp(\Omega(t))X_0, \tag{4.52}$$

where $\Omega \in \mathfrak{g}$ now satisfies the differential equation

$$\dot{\Omega} = d\exp_{\Omega}^{-1}\left(A(t, \mathrm{e}^{\Omega}X_0)\right), \qquad \Omega(0) = 0, \tag{4.53}$$

and the $d\exp_{\Omega}^{-1}$ operator has been defined in (4.40). This is, in fact, the starting point to design integration methods that provide numerical approximations lying in $\mathcal{G}$ for all t. In doing so, at least two different strategies are possible: (i) to apply a Runge–Kutta method to (4.53) rather than to the original equation (4.51). This results in the so-called Runge–Kutta–Munthe-Kaas class of methods [257, 258]. (ii) To solve (4.53) by iteration, as in the linear case, thus giving

$$\begin{aligned} \Omega^{[m]}(t) &= \int_0^t d\exp_{\Omega^{[m-1]}(s)}^{-1} A(s, \mathrm{e}^{\Omega^{[m-1]}(s)} X_0) ds \qquad (4.54) \\ &= \int_0^t \sum_{k=0}^{\infty} \frac{B_k}{k!} \mathrm{ad}_{\Omega^{[m]}(s)}^k A(s, \mathrm{e}^{\Omega^{[m-1]}(s)} X_0) ds, \qquad m \geq 1 \end{aligned}$$

and construct explicit approximations by truncating appropriately the $d\exp^{-1}$ operator and analyzing the time dependence of $\Omega^{[m]}(t)$. In this way one has a nonlinear version of the Magnus expansion [93].

In the Runge–Kutta–Munthe-Kaas (RKMK) class of methods , the solution of (4.53) is approximated at each time step with a classical (implicit or explicit) Runge–Kutta method. The series defining the $d\exp^{-1}$ operator,

$$d\exp_u^{-1}(v) = \sum_{k=0}^{\infty} \frac{B_k}{k!} \mathrm{ad}_u^k v = v - \frac{1}{2}[u, v] + \frac{1}{12}[u, [u, v]] + \cdots,$$

can be truncated because it is used only in cases where the argument $u = \mathcal{O}(h)$. In this way an approximation $v \approx \Omega(t_n + h)$ is constructed, and the numerical solution is defined by $X_{n+1} = \exp(v) X_n$. More specifically, with an rth-order Runge-Kutta method with s stages, one can write the corresponding RKMK scheme in the following form, where h is the step size and a_{ij}, b_i are the Runge-Kutta parameters [258, 259]:

$$\begin{aligned} u_i &= \sum_{j=1}^{s} a_{ij} \, \mathtt{dexpinv}(u_j, k_j, r-1) \\ k_i &= h A(t_n + c_i h, \mathrm{e}^{u_i} X_n), \qquad i = 1, \ldots, s \\ v &= \sum_{i}^{s} b_i \, \mathtt{dexpinv}(u_i, k_i, r) \\ X_{n+1} &= \exp(v) X_n. \end{aligned}$$

Here $\mathtt{dexpinv}(u, v, r)$ denotes an rth-order approximation to $\mathrm{dexp}_u^{-1}(v)$, i.e., $d\exp_{tu}^{-1}(v) - \mathtt{dexpinv}(tu, v, r) = \mathcal{O}(t^{r+1})$ for all $u, v \in \mathfrak{g}$ at $t = 0$.

Particular examples of methods within this class are the following Lie group versions of familiar low-order integrators:

- Explicit Euler method (order 1)

$$X_{n+1} = \exp(hA(t_n, X_n)) X_n.$$

- Implicit midpoint rule (order 2)

$$v = hA(t_n + \frac{h}{2}, \mathrm{e}^v X_n), \qquad X_{n+1} = \exp(v) X_n.$$

- Trapezoidal rule (order 2, implicit)

$$\begin{aligned} u_1 &= hA(t_n, X_n) \\ u_2 &= hA(t_n + h, \mathrm{e}^{\frac{1}{2}(u_1+u_2)} X_n) \\ v &= \frac{1}{2}(u_1 + u_2) \\ X_{n+1} &= \exp(v) X_n. \end{aligned} \tag{4.55}$$

Higher-order methods can be designed along the same lines by introducing appropriate transformations and strategies to reduce the number of commutators required by the algorithm [260, 95].

The starting point for designing numerical methods based on the Magnus expansion for the nonlinear case (4.51) is equation (4.54). It is clear that

$$\Omega^{[1]}(t) = \int_0^t A(s, X_0) ds = \Omega(t, X_0) + \mathcal{O}(t^2), \tag{4.56}$$

whereas the truncation

$$\Omega^{[m]}(t) = \sum_{k=0}^{m-2} \frac{B_k}{k!} \int_0^t \mathrm{ad}^k_{\Omega^{[m-1]}(s)} A(s, \mathrm{e}^{\Omega^{[m-1]}(s)} X_0)) ds, \qquad m \geq 2 \tag{4.57}$$

once inserted in (4.52), provides an explicit approximation $X^{[m]}(t)$ for the solution of (4.51) that is correct up to terms $\mathcal{O}(t^{m+1})$ [93]. In addition, $\Omega^{[m]}(t)$ reproduces exactly the sum of the first m terms in the Ω series of the usual Magnus expansion for the linear equation $\dot{X} = A(t)X$. It makes sense, then, to regard $\exp(\Omega^{[m]}(t, X_0))X_0$ as an explicit Magnus expansion for the nonlinear equation (4.51).

Now, by replacing the integrals appearing in (4.56)-(4.57) by appropriate quadratures, we get another family of numerical schemes that, by construction, are Lie group methods. In particular, if we approximate $\Omega^{[1]}(t_n + h)$ in (4.56) by $\Omega^{[1]}(t_n + h) = hA(t_n, X_n) + \mathcal{O}(h^2)$, this results in the explicit second-order scheme

$$\begin{aligned} v_1 &\equiv \frac{h}{2} \left(A(t_n, X_n) + A(t_n + h, \mathrm{e}^{hA(t_n, X_n)} X_n) \right) = \Omega^{[2]}(t_n + h) + \mathcal{O}(h^3) \\ X_{n+1} &= \mathrm{e}^{v_1} X_n, \end{aligned} \tag{4.58}$$

which is precisely the two-stage RKMK method with Butcher tableau

$$\begin{array}{c|cc} 0 & & \\ 1 & 1 & \\ \hline & \frac{1}{2} & \frac{1}{2} \end{array}$$

The same procedure can be carried out at higher orders, consistently discretizing the integrals appearing in $\Omega^{[m]}(t_n + h)$ for $m > 2$ [93]. In general, and due to its iterative character, in a Magnus method of order r, unlike a conventional RKMK scheme, the internal stages provide approximations to the exact solution up to order $r - 1$. For this reason the new methods require, in general, more computational effort per time step, but on the other hand, variable step size and other techniques can be incorporated in a natural way into the algorithm, thus improving its overall efficiency [175].

4.4.4 Crouch–Grossman methods

Another class of Lie group integrators was proposed in [121] (see also [170, sect. IV.8] for more details). The idea is to appropriately modify Runge–Kutta schemes to render approximations lying in $\mathcal{G}$ by construction. Given the autonomous differential equation

$$\dot{X} = A(X)X, \qquad X(0) = X_0 \in \mathcal{G}, \tag{4.59}$$

and $b_i, a_{i,j}$, $(i, j = 1, \ldots, s)$ real numbers, an explicit s-stage Crouch–Grossman method is defined by

$$\begin{array}{rcllr} X^{(1)} & = & X_n, & K_1 = A(X^{(1)}), & \\ X^{(j)} & = & e^{ha_{j,j-1}K_{j-1}} \cdots e^{ha_{j,1}K_1} X_n, & K_j = A(X^{(j)}), & 2 \le j \le s \\ X_{n+1} & = & e^{hb_s K_s} \cdots e^{hb_1 K_1} X_n. & & \end{array} \tag{4.60}$$

As usual, the order conditions to be satisfied by the coefficients $b_i, a_{i,j}$, $(i, j = 1, \ldots, s)$ can be found by comparing the Taylor series expansions of the exact and the numerical solutions. It has been shown that the order conditions for classical Runge–Kutta methods form a subset of those for the Grouch–Grossman methods [270].

In particular, from the coefficients of Runge's method ($s = 2, a_{2,1} = 1, b_1 = b_2 = 1/2$) the following second-order method is obtained:

$$\begin{array}{rcll} X^{(1)} & = & X_n, & K_1 = A(X^{(1)}) \\ X^{(2)} & = & e^{\frac{h}{2}K_1} X_n, & K_2 = A(X^{(2)}), \\ X_{n+1} & = & e^{\frac{h}{2}K_2} e^{\frac{h}{2}K_1} X_n. & \end{array} \tag{4.61}$$

Several 3-stage methods of order 3 are designed in [121], whereas in [270] a fourth-order scheme with $s = 5$ stages is presented.

Example 4.9. Let us consider the motion of a free rigid body described by the Euler equations and written in matrix form as

$$\frac{d}{dt}\begin{pmatrix} x_1 \\ x_2 \\ x_3 \end{pmatrix} = \begin{pmatrix} 0 & x_3/I_3 & -x_2/I_2 \\ -x_3/I_3 & 0 & x_1/I_1 \\ x_2/I_2 & -x_1/I_1 & 0 \end{pmatrix}\begin{pmatrix} x_1 \\ x_2 \\ x_3 \end{pmatrix}.$$

Here the vector $x = (x_1, x_2, x_3)^T$ represents the angular momentum in the body frame, and I_1, I_2, I_3 are the principal moments of inertia. The norm and the kinetic energy of the system are, respectively,

$$M = x_1^2 + x_2^2 + x_3^2, \qquad H = \frac{1}{2}\left(\frac{x_1^2}{I_1} + \frac{x_2^2}{I_2} + \frac{x_3^2}{I_3}\right)$$

and are preserved along the evolution. In our numerical test we take $(I_1, I_2, I_3) = (5, 1, 1/5)$, initial condition $x_0 = (\cos(1.1), 0, \sin(1.1))^T$, step size $h = 1/10$ and integrate for $t \in [0, 10]$ with the second-order Heun method (1.19) (H2) and second-order Crouch–Grossman (CG2) scheme (4.61).

The resulting trajectory is depicted in the left panel of Figure 4.2. Specifically, the solid line stands for the exact solution (computed numerically to high accuracy). It evolves in the intersection of the sphere $M = const$ with the ellipsoid $H = const$. Circles correspond to the numerical trajectory obtained by CG2: it lies in the sphere but slightly moves from the ellipsoid. Finally, the numerical approximation rendered by the Heun method (stars) clearly separates from the sphere, and in fact the error grows exponentially for times $t > 10$.

Figure 4.2 (right) illustrates the time evolution of the error in the kinetic energy H for both schemes. While the error initially grows also linearly for the Heun method, it becomes exponential in the interval $t \in [10, 15]$. In terms of the norm M, Heun exhibits similar behavior, whereas CG2 maintains accuracy up to round-off error. □

4.4.5 Positivity-preserving methods

Differential equations $\dot{x} = f(x)$ modeling population dynamics must satisfy positivity and very often also mass preservation, that is,

$$\begin{array}{lll} \text{(i)} & x_i(0) \geq 0 \quad \Rightarrow \quad x_i(t) \geq 0 \ \forall t, \ i = 1, \ldots, d & \textit{(positivity preservation)} \\ \text{(ii)} & \mathbf{1}^T x(t) = \mathbf{1}^T x(0), \quad \text{with} \quad \mathbf{1} = (1, \ldots, 1)^T & \textit{(mass conservation)} \end{array}$$

Both of them are automatically fulfilled if the differential equation can be written as

$$\dot{x} = A(x)x,$$

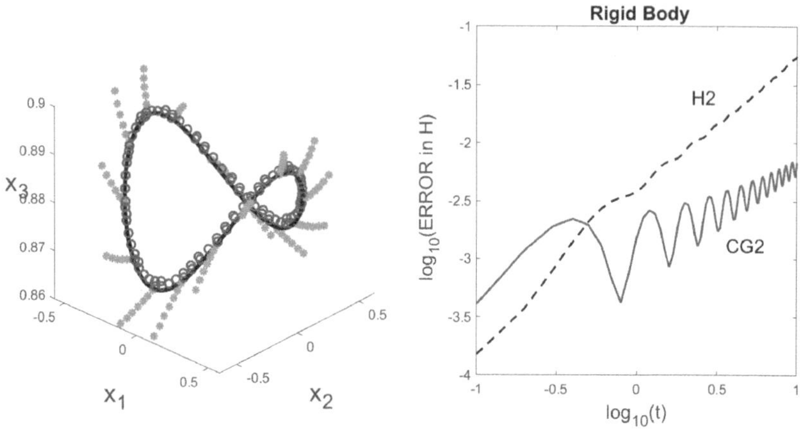

FIGURE 4.2: Left: trajectory of the rigid body from Example 4.9 (solid line) for $t \in [0, 10]$ and the numerical results obtained with a time step $h = 1/10$ with the Heun method (stars, H2) and the second-order Crouch–Grossman method (4.61) (circles, CG2). Right: Error in the kinetic energy of both methods along the numerical integration.

where A is a (graph Laplacian) matrix such that

- if $x_i \geq 0$, $i = 1, \ldots, d$, then $A(x)_{k,\ell} \geq 0$ when $k \neq \ell$, and $A_{k,k} \leq 0$, for $k, \ell = 1, \ldots, d$ (to guarantee positivity);
- $\sum_{k=1}^{d} A_{k,\ell} = 0$ for $\ell = 1, \ldots, d$ (to ensure mass conservation).

In such situations, it is expedient to employ a numerical integrator that preserves these properties at the discrete level during simulations. However, a serious limitation arises with two widely used classes of methods, namely Runge–Kutta and multistep schemes: if they unconditionally preserve positivity, then they cannot be better than first order [67].

It turns out, however, that splitting methods as well as the previous Lie group methods are able to overcome this order barrier. To see this point, consider again the problem in the extended phase space

$$\begin{aligned} \dot{y} &= A(z)y, \qquad y(0) = y_0 = x_0, \\ \dot{z} &= A(y)z, \qquad z(0) = z_0 = x_0, \end{aligned} \tag{4.62}$$

where $x(t) = y(t) = z(t)$. Now, we apply the Strang splitting to the system (4.62), that is

$$\begin{aligned} y_{1/2} &= \mathrm{e}^{\frac{1}{2}hA(z_0)} y_0, \\ z_1 &= \mathrm{e}^{hA(y_{1/2})} z_0, \\ y_1 &= \mathrm{e}^{\frac{1}{2}hA(z_1)} y_{1/2}, \\ x_1 &= \frac{1}{2}(y_1 + z_1). \end{aligned} \tag{4.63}$$

Both y_1 and z_1 are second-order approximations to the exact solution $x(h)$ preserving mass and positivity. This can be easily checked by realizing that $\mathbf{1}^T A = 0$ and so $\mathbf{1}^T \mathrm{e}^A x = \mathbf{1}^T x$. On the other hand, for $h > 0$, if $(z_0)_i = (x_0)_i \geq 0$, $i = 1, \ldots, d$, then $(\mathrm{e}^{hA(z_0)})_{jk} \geq 0$ for all j, k and thus $y_{1/2} \geq 0$, $i = 1, \ldots, d$. A similar reasoning applies to z_1 and y_1, so that x_1 preserves mass and positivity.

Moreover, it is known that the graph Laplacian has a single eigenvalue that is zero, with all other eigenvalues located in the left half of the complex plane. As a result, the mass- and positivity-preserving method (4.63) converges to a steady-state solution, which coincides with the exact steady state of the system.

It is worth noticing that the second-order Lie group methods (4.55), (4.58) and (4.61) are also mass- and positivity-preserving. However, if $h < 0$, positivity is no longer preserved with this approach, and in fact there appears to be an order barrier at second order when using Lie group integrators (see [63] for a more detailed treatment).

Example 4.10. The Robertson chemical reaction system is a popular model to check numerical schemes for stiff equations [174]. It consists of three reactions: $A \longrightarrow B$ (slow), $B + B \longrightarrow C + B$ (very fast) and $B + C \longrightarrow A + C$ (fast), which can be modeled by the differential system

$$\frac{d}{dt}\begin{pmatrix} x_1 \\ x_2 \\ x_3 \end{pmatrix} = \begin{pmatrix} -0.04 & 10^4\, x_3 & 0 \\ 0.04 & -3\cdot 10^7\, x_2 - 10^4\, x_3 & 0 \\ 0 & 3\cdot 10^7\, x_2 & 0 \end{pmatrix}\begin{pmatrix} x_1 \\ x_2 \\ x_3 \end{pmatrix}, \qquad (4.64)$$

formulated in terms of the concentrations $x = (x_1, x_2, x_3)$ of the components A, B, C, respectively. Since the matrix $A(x)$ is a graph Laplacian, then there is mass preservation, and if the initial conditions are positive, the solution remains positive for all times. This is also true for the second-order Lie group method (4.63), even when this problem is very stiff.

Notice, however, that this is no longer the case if the system is written in the equivalent form

$$\frac{d}{dt}\begin{pmatrix} x_1 \\ x_2 \\ x_3 \end{pmatrix} = \begin{pmatrix} -0.04 & 0 & 10^4\, x_2 \\ 0.04 & -3\cdot 10^7\, x_2 & -10^4\, x_2 \\ 0 & 3\cdot 10^7\, x_2 & 0 \end{pmatrix}\begin{pmatrix} x_1 \\ x_2 \\ x_3 \end{pmatrix}, \qquad (4.65)$$

since now the corresponding matrix $A(x)$ has negative off-diagonal elements, although the scheme is still mass-preserving, positivity is not guaranteed.

To illustrate the mass- and positivity-preserving scheme (4.63), we take the system (4.64) with initial condition $x_0 = [1, 0, 0]^T$ and step size $h = \frac{0.3}{N}$ for a given N on the time interval $t \in [0, 0.3]$ and also compare with the classical 4th-order RK method. Figure 4.3 (left) shows the x_2-component of the exact solution (solid line), together with the numerical approximations. $\mathrm{RK4}_{180}$ corresponds to the Runge–Kutta method with $N = 180$ (grey line with stars), whereas $\mathrm{PP2}_{100}$ and $\mathrm{PP2}_{10}$ stand for scheme (4.63) with $N = 100$ and $N = 10$, respectively.

From the graph, it is clear that $\mathrm{RK4}_{180}$ not only provides low accuracy results, but these are also qualitatively wrong, since the approximation takes negative values, whereas $\mathrm{PP2}_{100}$ provides the correct solution after a short transition. We have included the numerical result obtained with a significantly larger time step to illustrate the fact that the method not only preserves mass and positivity but also converges to the steady-state solution.

Finally, Figure 4.3 (right) shows the relative error in the 2-norm of the solution at the final time $t_{\mathrm{f}} = 0.3$ as a function of the time step used. The reference solution for measuring the error is numerically computed with high accuracy and is given by

$$(x_1, x_2, x_3) = (0.9886739393,\ 0.0000344771,\ 0.0112915834).$$

□

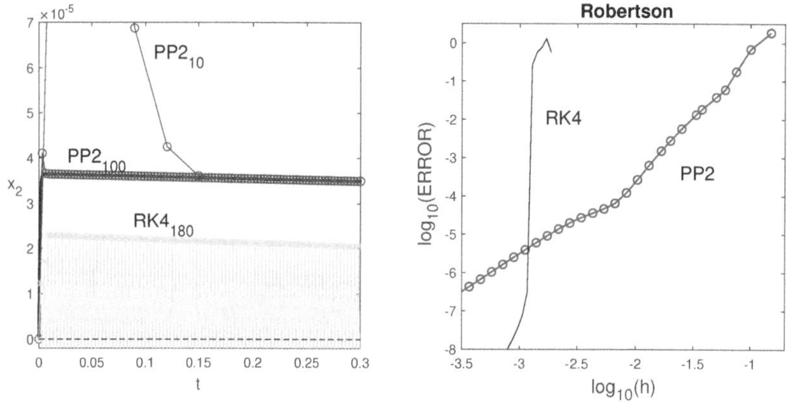

FIGURE 4.3: Left: solution for the component x_2 for the Robertson problem (4.64) (solid line) and the numerical solutions obtained with time steps $h = \frac{0.3}{N}$ by the classical Runge-Kutta method with $N = 180$ (grey line with stars, $\mathrm{RK4}_{180}$), the second-order positivity preserving method (4.63) with $N = 100$ ($\mathrm{PP2}_{100}$), and with $N = 10$ ($\mathrm{PP2}_{10}$). Right: relative error in the 2-norm of the solution x at $t_{\mathrm{f}} = 0.3$ versus the time step used with RK4 and PP2.

4.5 Linear combinations of splitting methods

In addition to composing a low-order basic method, it is also possible to construct a high-order approximation by considering linear combinations. Thus, it is well known that polynomial extrapolation gives rise to highly efficient schemes for solving initial value problems when the local error of the basic discretization method has an asymptotic expansion containing only even powers of h. This is the case for the midpoint rule and the Störmer–Verlet method.

We recall that in polynomial extrapolation, one starts with a basic low-order integration method, which is applied with different time steps h (e.g., $h/2, h/3, \ldots$). Then, by an appropriate combination of the results, one obtains a new scheme that approximates the exact solution to a higher order with essentially no additional cost [304]. More specifically, suppose $S_h^{[2]}$ denotes a time-symmetric second-order scheme, $\{k_1, k_2, \ldots, k_m\}$ is a given set of m integer numbers, and we form the linear combination

$$\phi_h \equiv \sum_{i=1}^{m} c_i \left(S_{h/k_i}^{[2]}\right)^{k_i}, \tag{4.66}$$

where $\left(S_{h/2}^{[2]}\right)^2 \equiv S_{h/2}^{[2]} \circ S_{h/2}^{[2]}$, etc. Then, ϕ_h furnishes an approximation of order $2m$ if the coefficients satisfy the linear equations

$$G_0 = \sum_{i=1}^{m} c_i = 1, \qquad G_{2+2j} = \sum_{i=1}^{m} \frac{c_i}{k_i^{2+2j}} = 0, \qquad j = 0, 1, \ldots, m-2,$$

with solutions

$$c_i = \prod_{j=1(\neq i)}^{m} \frac{k_i^2}{k_i^2 - k_j^2}.$$

In particular, if one takes the harmonic sequence $k_i = i$, the linear combination ϕ_h produces an extrapolation method of order $2m$ with only $m(m+1)/2$ evaluations of the basic map $S_h^{[2]}$ [171, 113]. For orders 4, 6 and 8, one has, respectively

$$\begin{aligned}
\phi_h^{[4]} &= -\frac{1}{3} S_h^{[2]} + \frac{4}{3}\left(S_{h/2}^{[2]}\right)^2, \\
\phi_h^{[6]} &= \frac{1}{24} S_h^{[2]} - \frac{16}{15}\left(S_{h/2}^{[2]}\right)^2 + \frac{81}{40}\left(S_{h/3}^{[2]}\right)^3, \\
\phi_h^{[8]} &= -\frac{1}{360} S_h^{[2]} + \frac{16}{45}\left(S_{h/2}^{[2]}\right)^2 - \frac{729}{280}\left(S_{h/3}^{[2]}\right)^3 + \frac{1024}{315}\left(S_{h/4}^{[2]}\right)^4.
\end{aligned} \tag{4.67}$$

Schemes of the form (4.66) are also referred to as multi-product expansions (MPE) [113] and have been shown to be competitive with other integration schemes on a number of problems, especially when high accuracy is required [111, 113], even when a fixed step size is used. On the other hand, they do not belong to the class of structure-preserving methods: the exact preservation of geometric properties of the solution is lost in general when one considers linear combinations of approximations obtained with different time steps. Generally speaking, linear combinations provide elements that do not belong to the Lie group where the differential equation is defined. Nevertheless, it can be shown that the numerical solution furnished by an MPE of order $2m$ still preserves geometric properties up to order $2m+1$. In particular, if it is applied to a Hamiltonian system, then the method preserves symplecticity up to order $2m+1$ [59].

Example 4.11. To illustrate this feature, we apply the Störmer–Verlet scheme to the simple harmonic oscillator with Hamiltonian $H = \frac{1}{2}(p^2+q^2)$:

$$S_h^{[2]} = \mathrm{e}^{\frac{h}{2}A}\mathrm{e}^{hB}\mathrm{e}^{\frac{h}{2}A} = \begin{pmatrix} 1 & h/2 \\ 0 & 1 \end{pmatrix}\begin{pmatrix} 1 & 0 \\ -h & 1 \end{pmatrix}\begin{pmatrix} 1 & h/2 \\ 0 & 1 \end{pmatrix},$$

and from the previous MPE $\phi_h^{[4]} = -\frac{1}{3}S_h^{[2]} + \frac{4}{3}\left(S_{h/2}^{[2]}\right)^2$. Let $M_h = \mathrm{e}^{h(A+B)}$ be the exact solution (3.12). Then, it is straightforward to verify that

$$\phi_h^{[4]} - M_h = \begin{pmatrix} 0 & -\frac{1}{480} \\ -\frac{1}{120} & 0 \end{pmatrix} h^5 + \mathcal{O}(h^6)$$

and

$$\det\left(\phi_h^{[4]}\right) = 1 - \frac{h^6}{288}.$$

In other words, the extrapolation scheme $\phi_h^{[4]}$ is of order 4, but it preserves the symplectic character of the system up to order 5. □

In general, if one starts with a basic time-symmetric geometric method $S_h^{[2r]}$ of order $2r$, the corresponding linear combination (4.66) provides a scheme of order $2r+2\ell$, $\ell = 1,\dots,r$, which, quite remarkably, preserves the geometric properties of the system (e.g., symplecticity for Hamiltonian problems) up to order $4r+1$ or higher [55, 102]. This can be shown by appropriately rewriting the differential operator associated with (4.66) as a product of exponentials of operators.

Multi-product expansions can be generalized by considering the linear combination

$$\phi_h^{[m,k]} = \sum_{i=1}^{m} c_i \prod_{j=1}^{k} S_{a_{ij}h}^{[2]}, \tag{4.68}$$

where, as before, $S_h^{[2]}$ is a second-order time-symmetric integrator and the condition $\sum_{i=1}^m c_i = \sum_{j=1}^k a_{ij} = 1$ is taken for consistency. The presence of additional coefficients may be used for various purposes. In particular, one could increase the order of preservation of geometric properties of the continuous system or reduce the most significant contributions to the truncation error. In contrast, MPEs use all available coefficients to increase the order of the method as much as possible. Notice that multi-product expansions of the form (4.68) are well suited for parallelization: each composition can be evaluated sequentially by individual processors, after which the resulting outputs are combined.

Methods of this form have been analyzed in [33, 59], where new fourth- and sixth-order schemes are presented that show a better efficiency than MPEs for a number of problems.

4.6 Exercises

1. Find the canonical transformation generated by $F_1(q,Q) = \alpha q^2 \cot Q$, with α constant. Obtain the Hamiltonian in the new coordinate system (Q,P) corresponding to the one-dimensional harmonic oscillator

$$H(q,p) = \frac{1}{2m}p^2 + \frac{1}{2}kq^2.$$

   Choose α to make K independent of Q and hence get the motion of the system in each representation.

2. Consider the Hamiltonian $H(q,p)$ and a symplectic change of coordinates given by

$$Q = Q(q,p), \qquad P = P(q,p).$$

   Show that the system in the new coordinates is also Hamiltonian with Hamiltonian $\hat{H}(Q,P) = H(q(Q,P), p(Q,P))$, i.e.,

$$\dot{Q} = \frac{\partial \hat{H}}{\partial P}, \qquad \dot{P} = -\frac{\partial \hat{H}}{\partial Q}.$$

3. Verify explicitly that the function (4.11) is indeed a generating function of the symplectic transformation (4.10).

4. Show explicitly that the midpoint rule applied to the linear system $\dot{X} = AX$, where the 3×3 matrix A is such that $\mathrm{tr}A = 0$, is volume-preserving if and only if $\det A = 0$.

5. Write explicitly the terms Ω_2 and Ω_3 of the Magnus expansion (4.41) from the general expression (4.42).

6. Repeat *Example 4.7* for the nonlinear 3D Volterra system

$$\dot{x} = x(By^2 - Cz^2), \qquad \dot{y} = y(Cz^2 - Ax^2), \qquad \dot{z} = z(Ax^2 - By^2)$$

   with A, B, C constants.

7. Split the ABC system (2.31) into a set of two-dimensional Hamiltonian systems in such a way that splitting methods render volume-preserving approximations.

8. The Pauli matrices are defined by

$$\sigma_1 = \begin{pmatrix} 0 & 1 \\ 1 & 0 \end{pmatrix}, \quad \sigma_2 = \begin{pmatrix} 0 & -i \\ i & 0 \end{pmatrix}, \quad \sigma_3 = \begin{pmatrix} 1 & 0 \\ 0 & -1 \end{pmatrix}. \tag{4.69}$$

Given $\boldsymbol{a} = (a_1, a_2, a_3) \in \mathbb{R}^3$, $\boldsymbol{\sigma} = (\sigma_1, \sigma_2, \sigma_3)$ and $\boldsymbol{a} \cdot \boldsymbol{\sigma} = a_1\sigma_1 + a_2\sigma_2 + a_3\sigma_3$, show that

$$(\boldsymbol{a}\cdot\boldsymbol{\sigma})(\boldsymbol{b}\cdot\boldsymbol{\sigma}) = \boldsymbol{a}\cdot\boldsymbol{b}\, I + i(\boldsymbol{a}\times\boldsymbol{b})\cdot\boldsymbol{\sigma}, \qquad [\boldsymbol{a}\cdot\boldsymbol{\sigma}, \boldsymbol{b}\cdot\boldsymbol{\sigma}] = 2i(\boldsymbol{a}\times\boldsymbol{b})\cdot\boldsymbol{\sigma} \tag{4.70}$$

and

$$\exp(i\boldsymbol{a}\cdot\boldsymbol{\sigma}) = \cos(a)\, I + i\frac{\sin(a)}{a}\boldsymbol{a}\cdot\boldsymbol{\sigma}, \tag{4.71}$$

where $a = \|\boldsymbol{a}\| = \sqrt{a_1^2 + a_2^2 + a_3^2}$. Show that (4.71) is a unitary transformation.

9. Consider equation (4.35) with $A(t) = -i\boldsymbol{a}(t)\cdot\boldsymbol{\sigma}$ and build explicitly the fourth-order Magnus integrators (4.45) for $\boldsymbol{a}(t) = (\mathrm{e}^{-t}, \cos(t), \sin(2t))$, but taking

$$\Omega^{[4]}(h) = \Omega_1(t) + \Omega_2(t),$$

with the expressions of $\Omega_1(h), \Omega_2(h)$ obtained in Exercise 6 and the results from Exercise 8.

10. (Rodrigues formula). Given $b = (b_1, b_2, b_3) \in \mathbb{R}^3$, show that

$$\mathrm{e}^{t\hat{b}} = I + \frac{\sin(t\beta)}{\beta}\hat{b} + \frac{1}{2}\left(\frac{\sin(t\beta/2)}{\beta/2}\right)^2 \hat{b}^2, \tag{4.72}$$

with

$$\hat{b} = \begin{pmatrix} 0 & -b_3 & b_2 \\ b_3 & 0 & -b_1 \\ -b_2 & b_1 & 0 \end{pmatrix}$$

and $\beta = \sqrt{b_1^2 + b_2^2 + b_3^2}$.

Hint: Write $\hat{b}^3$ in terms of $I, \hat{b}$ and $\hat{b}^2$. Denote by $X(t), Y(t)$ the left- and right-hand sides of the equation. Then, compute $\dot{X}(t), \dot{Y}(t)$ and show that they satisfy the same differential equations with the same initial conditions.

11. Given the Lagrangian function $L(q, \dot{q}) = \frac{1}{2}\dot{q}^T M\dot{q} - V(q)$, take the interpolating polynomial to connect q_n with q_{n+1} as

$$\tilde{q}(t) = q_n + t\frac{q_{n+1} - q_n}{h}, \qquad \text{so that} \quad \dot{\tilde{q}}(t) = \frac{q_{n+1} - q_n}{h}.$$

Substitute into the Lagrangian and use the midpoint rule to approximate the integral of the new Lagrangian in order to obtain the following second-order discrete Lagrangian

$$L_h(q_n, q_{n+1}, h) = h\frac{1}{2}\left(\frac{q_{n+1} - q_n}{h}\right)^T M \left(\frac{q_{n+1} - q_n}{h}\right) - hV(\frac{q_{n+1} + q_n}{2}).$$

Apply the discrete Euler–Lagrange equations (4.23) to obtain the associated symplectic integrator. Compare the results with the implicit midpoint rule RK method (2.16).

12. The *Susceptible–Infected–Recovered (SIR) model* describes the temporal epidemic evolution in terms of three variables for the population: $S(t)$ (Susceptible), $I(t)$ (Infected) and $R(t)$ (Recovered). It is usually assumed that the total population does not change during the infection period, and S, I and R denote the fractions with respect to the total population, so that $S(t) + I(t) + R(t) \equiv 1$. This model was proposed in [200] and can be written as

$$\frac{d}{dt}\begin{pmatrix} S \\ I \\ R \end{pmatrix} = \begin{pmatrix} -R_0 I & 0 & 0 \\ R_0 I & -1 & 0 \\ 0 & 1 & 0 \end{pmatrix}\begin{pmatrix} S \\ I \\ R \end{pmatrix}, \tag{4.73}$$

where $R_0 > 0$ is the basic reproduction number. This problem has mass- and positivity-preservation (the columns of the matrix sum to zero and the non-diagonal elements are positive). Take $R_0 = 1$ and integrate for the time interval $t \in [0, 100]$ with initial conditions $x_0 = [S, I, R]_0^T = [1-10^{-2}, 10^{-2}, 0]^T$. Integrate the system with method (4.63) with time steps $h = \frac{100}{2^{10}} 2^i$, $i = 1, 2, \ldots, 10$, and check that the method unconditionally preserves mass and positivity. Repeat the numerical experiment using the fourth-order standard Runge–Kutta method and find the smallest value of i for which any component turns negative along the numerical integration. Repeat the computations with $R_0 = 3$.

13. Given the rigid body problem studied in *Example 4.9*, decompose it as

$$\dot{x} = (R_1 + R_2 + R_3)x,$$

where $x = (x_1, x_2, x_3)^T$ and R_j are the following rotation matrices:

$$R_1 = \begin{pmatrix} 0 & 0 & 0 \\ 0 & 0 & x_1/I_1 \\ 0 & -x_1/I_1 & 0 \end{pmatrix}, \qquad R_2 = \begin{pmatrix} 0 & 0 & -x_2/I_2 \\ 0 & 0 & 0 \\ x_2/I_2 & 0 & 0 \end{pmatrix},$$

$$R_3 = \begin{pmatrix} 0 & x_3/I_3 & 0 \\ -x_3/I_3 & 0 & 0 \\ 0 & 0 & 0 \end{pmatrix}.$$

Apply the symmetric second-order method

$$S_h^{[2]} = \varphi_{h/2}^{[R_1]} \circ \varphi_{h/2}^{[R_2]} \circ \varphi_h^{[R_3]} \circ \varphi_{h/2}^{[R_2]} \circ \varphi_{h/2}^{[R_1]},$$

with the same time interval and time step as in *Example 4.9*, and compute the error in H. Compare with the result obtained in the right panel of Figure 4.2. Here $\varphi_{a_j h}^{[R_j]} = \exp(a_j h R_j)$.

14. Repeat Exercise 13, but now with the splitting

$$\frac{d}{dt}\begin{pmatrix} x_1 \\ x_2 \\ x_3 \end{pmatrix} = \begin{pmatrix} J_1 x_2 x_3 \\ 0 \\ 0 \end{pmatrix} + \begin{pmatrix} 0 \\ J_2 x_1 x_3 \\ 0 \end{pmatrix} + \begin{pmatrix} 0 \\ 0 \\ J_3 x_1 x_2 \end{pmatrix},$$

where

$$J_1 = \frac{1}{I_3} - \frac{1}{I_2}, \qquad J_2 = \frac{1}{I_1} - \frac{1}{I_3}, \qquad J_3 = \frac{1}{I_2} - \frac{1}{I_1}.$$

Notice that the flow corresponding to each part can also be computed explicitly. Compare the results obtained.

15. Decompose the vector field of the Robertson chemical reaction system examined in *Example 4.10* into three parts as

$$\frac{d}{dt}\begin{pmatrix} x_1 \\ x_2 \\ x_3 \end{pmatrix} = \begin{pmatrix} -0.04 x_1 \\ 0.04 x_1 \\ 0 \end{pmatrix} + \begin{pmatrix} 0 \\ -3 \cdot 10^7\, x_2^2 \\ 3 \cdot 10^7\, x_2^2 \end{pmatrix} + \begin{pmatrix} 10^4\, x_2 x_3 \\ -10^4\, x_2 x_3 \\ 0 \end{pmatrix}$$

and apply the symmetric second-order integrator used in Exercise 13 in the time interval $t \in [0, 0.3]$ with time steps $h = \frac{0.3}{10^k}$, $k = 1, 2, 3, 4$. Compare with the results shown in Figure 4.3. Observe that, because each individual reaction is integrated exactly, mass is conserved to machine precision. Moreover, positivity is unconditionally preserved, provided that all reactions are advanced forward in the time direction, i.e., when employing splitting methods with positive coefficients.

16. Consider the matrix differential equation $\dot{x} = A(x)x$, where $A(x)$ is a skew-symmetric matrix. Then the exact flow is given by an orthogonal transformation. Suppose that $S_h^{[2n]}$ is a (matrix) map preserving orthogonality, and take the linear combination

$$\phi_h = c_1 S_h^{[2n]} + c_2 \big(S_{h/2}^{[2n]}\big)^2$$

with $c_1 = -1/(2^{2n} - 1)$, $c_2 = 1 - c_1$. Show that ϕ_h is of order $2n + 2$, requires three evaluations of the basic method per step, and verifies $\phi_h^T \phi_h = I + \mathcal{O}(h^{4n+2})$. Particularize to the case $n = 3$.

Chapter 5

Long-time behavior of geometric integrators

5.1 Introduction. Examples

Much insight into the long-time behavior of splitting methods (including preservation or nearly preservation of invariants of the continuous system and structures in the phase space) can be gained by applying *backward error analysis*.

This concept arises in several branches of numerical analysis. Generally speaking, given a problem $\mathcal{P}$ with true solution $\mathcal{S}$, when a suitable numerical solver is applied, one ends up with an approximate solution $\tilde{\mathcal{S}}$. *Forward error analysis* consists, then, in estimating an appropriate distance between $\tilde{\mathcal{S}}$ and $\mathcal{S}$. *Backward error analysis*, on the other hand, consists in showing explicitly that $\tilde{\mathcal{S}}$ is indeed the *exact* solution of a problem $\tilde{\mathcal{P}}$ which is somehow close to $\mathcal{P}$ [83, 289]. Whereas the importance of backward error analysis has been recognized long ago in areas like numerical linear algebra [318], it is fair to say that only recently has it acquired a similar status in the study of long-time numeral integration of differential equations, thanks to the fundamental contributions of, among others, Sanz-Serna [288], Murua [261], Hairer [167] and Reich [280].

The reason is easy to grasp. In long-time integrations, the conclusion provided by any forward error analysis is that errors are very large, regardless of the numerical method being used. This is particularly true for systems possessing a chaotic regime, where exact solutions diverge from each other due to the existence of a positive Lyapunov exponent [327]. Therefore, any numerical method produces an output $\tilde{\mathcal{S}}$ far away from the exact solution $\mathcal{S}$. In this sense, any integrator is unsatisfactory, a conclusion that is not supported in practice, since numerical simulations have proved to be very helpful when analyzing these systems [324]. By the same token, classical error bounds are pointless in long-time integrations aimed to ascertain qualitative properties of the systems involved. Backward error analysis, in contrast, can show that a numerical simulation of a given problem provides the exact solution of a slightly perturbed system, and thus, by analyzing this perturbed system, extract valid conclusions about qualitative features of the original problem.

DOI: 10.1201/9781003527626-5

Before entering into a general formulation of the theory and its main results, it is worthwhile to illustrate some of the techniques involved by considering again the simple harmonic oscillator (1.2) ($k = m = 1$) and the numerical solution provided by the symplectic Euler-VT scheme (1.10) for the first step, namely

$$x_1 = M(h)x_0 = \begin{pmatrix} q_1 \\ p_1 \end{pmatrix} = \begin{pmatrix} 1-h^2 & h \\ -h & 1 \end{pmatrix} \begin{pmatrix} q_0 \\ p_0 \end{pmatrix}. \tag{5.1}$$

The exact solution (3.12), in accordance with the general treatment of Appendix A.1, can be expressed as $x(h) = \exp(hL_{X_H})[\mathrm{Id}](x_0)$, where X_H is the Hamiltonian vector field associated with the Hamiltonian $H(q,p) = \frac{1}{2}(p^2+q^2)$ (c.f. Appendix A.3.2).

We can in principle express the approximation (5.1) as $x_1 = \mathrm{e}^{hY(h)}x_0$, i.e., as the exponential of a certain matrix $Y(h)$ depending on the step size h. Specifically

$$Y(h) = \frac{1}{h}\log M(h) = \frac{1}{h}\sum_{m\geq 1}\frac{(-1)^{m-1}}{m}(M(h)-I)^m = \begin{pmatrix} -C(h) & D(h) \\ -D(h) & C(h) \end{pmatrix},$$

where

$$\begin{aligned} C(h) &= \frac{h}{2} + \frac{h^3}{12} + \frac{h^5}{60} + \frac{h^7}{280} + \cdots = \frac{2}{\sqrt{4-h^2}}\arcsin(h/2) \\ D(h) &= 1 + \frac{h^2}{6} + \frac{h^4}{30} + \frac{h^6}{140} + \cdots = \frac{4}{h\sqrt{4-h^2}}\arcsin(h/2) \end{aligned}$$

for $|h| < 2$. More compactly,

$$\begin{aligned} Y(h) &= \frac{2}{h\sqrt{4-h^2}}\arcsin(h/2)\begin{pmatrix} -h & 2 \\ -2 & h \end{pmatrix} \\ &= \frac{2}{h\sqrt{4-h^2}}\arcsin(h/2)\Big(2A + 2B + hC\Big) = A + B + \mathcal{O}(h), \end{aligned}$$

where matrices A, B and C are given in (A.33). Taking into account the existing isomorphism between the Lie algebra spanned by A, B, C and the Lie algebra spanned by the Hamiltonian vector fields X_U, X_V, X_W of (A.31), it is clear that the numerical solution (5.1) can be expressed as

$$x_1 = \exp\Big(g(h)(hL_{X_U} + hL_{X_V} + h^2 L_{X_W})\Big)[\mathrm{Id}](x_0) = \exp\Big(g(h)hL_{X_{\tilde{H}}}\Big)[\mathrm{Id}](x_0),$$

where

$$g(h) = \frac{2}{h\sqrt{4-h^2}}\arcsin(h/2), \qquad U = \frac{1}{2}p^2, \qquad V = \frac{1}{2}q^2, \qquad W = -q\,p.$$

In consequence, (5.1) *is* the exact solution at $t = h$ of the *perturbed* Hamiltonian system

$$\begin{aligned}\tilde{H}(q,p,h) &= g(h)(2U + 2V + hW) = \frac{2\arcsin(h/2)}{h\sqrt{4-h^2}}(p^2 + q^2 - h\,p\,q) \\ &= \frac{1}{2}(p^2+q^2) - \frac{h}{2}q\,p + h^2\frac{1}{12}(p^2+q^2) + \cdots \qquad (5.2)\end{aligned}$$

for $|h| < 2$. In other words, the numerical approximation (5.1), which is only of first order for the exact trajectories of the Hamiltonian $H(q,p) = \frac{1}{2}(p^2+q^2)$, *is in fact the exact solution at* $t = h, 2h, \ldots$ *of the perturbed Hamiltonian* (5.2).

If one takes instead the numerical solution (1.11) provided by the symplectic Euler-TV method and applies the same procedure, one arrives at the same conclusion, this time with the perturbed Hamiltonian $\tilde{H}(q,p,h) = g(h)(p^2 + q^2 + h\,p\,q)$.

How does this property manifest in practice? Neglecting the factor $g(h)$ (which is just a number for a fixed step size h), it is easy to verify that $\tilde{H}$ is invariant under the symplectic transformation (5.1). In other words, the successive iterated points obtained by the Euler-VT scheme all lie on the ellipse $p^2 - h\,p\,q + q^2 = \text{const}$, which differs from the original invariant of the continuous system, $p^2 + q^2 = \text{const}$, by $\mathcal{O}(h)$. In consequence, the error in the determination of the energy is bounded along the evolution. The existence of this backward error interpretation has therefore direct implications for the qualitative behavior of the numerical solution, as well as for its global error.

We note in passing that the differential equation *(ii)* in *Example 1.2* is obtained by considering $\tilde{H}(q,p,h)$ in (5.2) up to $\mathcal{O}(h)$.

The previous treatment can be generalized to other linear problems that, once formulated in appropriate coordinates, are expressed as a system of uncoupled harmonic oscillators, as we have seen in subsection 3.3.4. If the splitting method

$$x_{n+1} = e^{ha_{s+1}A}\, e^{hb_sB} \cdots e^{ha_2A}\, e^{hb_1B}\, e^{ha_1A} x_n$$

is applied to the harmonic oscillator $\dot{q} = p$, $\dot{p} = -\omega^2 q$ with step size $h < 2/\omega$, then the resulting map $\tilde{M}_z$ given by (3.56) can be expressed as

$$\tilde{M}_z = \begin{pmatrix} \cos\theta_z & \gamma_z \sin\theta_z \\ -\gamma_z^{-1}\sin\theta_z & \cos\theta_z \end{pmatrix},$$

where θ_z and γ_z are real functions such that $p(z) = \cos\theta_z$ and $K_2(z) = -\gamma_z^2 K_3(z)$ and $\theta_{-z} = -\theta_z$, $\gamma_{-z} = -\gamma_z$ if the splitting method is time-symmetric. It is then straightforward to verify that the map $\tilde{M}_z$ is precisely the z-flow of the modified Hamiltonian [58, 72]

$$\tilde{H}(q,p) = \frac{\theta_z}{2z}\left(\gamma_z p^2 + \frac{1}{\gamma_z} q^2\right).$$

Two relevant conclusions can thus be extracted from this analysis: (i) the solution furnished by the numerical scheme verifies a modified differential equation of the original one; (ii) in the case of symplectic integrators, this modified equation can be derived from a Hamiltonian which constitutes a perturbation of the original system.

5.2 Modified equations

The preceding discussion can be extended to arbitrary nonlinear ordinary differential equations and any consistent integrator, although the infinite series defining the modified equation (or the modified Hamiltonian $\tilde{H}$) associated with the numerical solution does not converge in general. Consider the initial value problem

$$\dot{x} = f(x), \qquad x(0) = x_0 \tag{5.3}$$

and a numerical integrator ψ_h producing the sequence of approximations $x_0, x_1, x_2, \ldots, x_n, \ldots$ at $t = 0, h, 2h, \ldots, nh, \ldots$. In backward error analysis, one looks for another differential equation

$$\dot{\tilde{x}} = f_h(\tilde{x}) \tag{5.4}$$

whose vector field is defined as a formal series in powers of h,

$$f_h(\tilde{x}) \equiv f(\tilde{x}) + h f^{[2]}(\tilde{x}) + h^2 f^{[3]}(\tilde{x}) + \cdots \tag{5.5}$$

and such that $x_n = \tilde{x}(t_n)$, with $t_n = nh$ [165, 170]. In this way, by analyzing the difference of the vector fields $f(x)$ and $f_h(x)$, it is possible to extract useful information about the qualitative behavior of the numerical solution and the global error $e_n = x_n - x(t_n) = \tilde{x}(nh) - x(t_n)$. Since the series is only asymptotic, to make the formalism rigorous, we have to give bounds on the coefficient functions $f^{[j]}(x)$ of the modified equation so as to determine an optimal truncation index, and finally, we must estimate the difference between the numerical solution x_n and the exact solution $\tilde{x}(t_n)$ of the modified equation.

The modified equation associated with a particular one-step method ψ_h can be constructed as follows. First, we assume that $\psi_h(x)$ is a consistent method that can be expanded as

$$\psi_h(x) = x + hf(x) + h^2 d_2(x) + h^3 d_3(x) + \cdots, \tag{5.6}$$

where $d_j(x)$ are known functions involving f and its derivatives. Second, expanding the solution of the modified equation (5.4) into a Taylor series and denoting for simplicity $x \equiv \tilde{x}(t)$ for a fixed t, we get

$$\tilde{x}(t+h) = x + hf_h(x) + \frac{h^2}{2} f_h' f_h(x) + \frac{h^3}{3!}(f_h''(f_h, f_h)(x) + f_h' f_h' f_h(x)) + \cdots, \tag{5.7}$$

where $f'(x)$ denotes the Jacobian matrix of f and $f''(f,f)(x)$ is a shorthand notation for the vector whose components are $f(Hf_i)f$, Hf_i being the Hessian of the i-component of the vector f. Inserting the series (5.5) into (5.7) and collecting terms in powers of h, one arrives at

$$\begin{aligned}\tilde{x}(t+h) &= x + hf(x) + h^2\left(f^{[2]}(x) + \frac{1}{2}f'f(x)\right)\\ &+ h^3\left(f^{[3]}(x) + \frac{1}{2}((f^{[2]})'f(x) + f'f^{[2]}(x)) + \frac{1}{3!}(f''(f,f)(x) + f'f'f(x))\right)\\ &+ \mathcal{O}(h^4),\end{aligned} \tag{5.8}$$

so that a comparison with (5.6) allows one to get the recursive relations determining $f^{[j]}(x)$ as

$$\begin{aligned} f^{[2]}(x) &= d_2(x) - \frac{1}{2}f'f(x),\\ f^{[3]}(x) &= d_3(x) - \frac{1}{2}((f^{[2]})'f(x) + f'f^{[2]}(x)) - \frac{1}{3!}(f''(f,f)(x) + f'f'f(x)),\end{aligned} \tag{5.9}$$

and so on. A systematic procedure to construct modified equations up to any order for one-step methods that can be formally expanded into B-series has been devised in [167, 261].

Example 5.1. As an illustration, let us consider the explicit Euler method $x_{n+1} = x_n + hf(x_n)$. It is then clear that $d_1 = d_2 = \cdots = 0$ in (5.6), and so f_h in the corresponding modified equation (5.4) reads

$$f_h(x) = f(x) + hf^{[2]}(x) + \mathcal{O}(h^2) = f(x) - \frac{h}{2}f'f(x) + \mathcal{O}(h^2). \tag{5.10}$$

In the particular case of the simple harmonic oscillator ($k = m = 1$), one has

$$f = \begin{pmatrix} p \\ -q \end{pmatrix}, \qquad f^{[2]} = -\frac{1}{2}\begin{pmatrix} 0 & 1 \\ -1 & 0 \end{pmatrix}\begin{pmatrix} p \\ -q \end{pmatrix} = \frac{1}{2}\begin{pmatrix} q \\ p \end{pmatrix}$$

and so it results in the equation (i) of *Example 1.2* in Chapter 1. If, on the other hand, the mathematical pendulum (1.12) is considered ($k = 1$), then $f = (p, -\sin q)^T$ and $f^{[2]} = (1/2)(\sin q, p\cos q)^T$, so that one ends up with the equation (i) in *Example 1.4*. □

Example 5.2. If we take instead the implicit midpoint rule $x_{n+1} = x_n + hf((x_n + x_{n+1})/2)$, a straightforward calculation shows that the first terms in the corresponding map (5.6) are given by

$$d_2(x) = \frac{1}{2}f'f(x), \qquad d_3(x) = \frac{1}{8}f''(f,f)(x) + \frac{1}{4}f'f'f(x).$$

Therefore, in the modified equation we have

$$f^{[2]}(x) = 0, \qquad f^{[3]}(x) = -\frac{1}{24} f''(f,f)(x) + \frac{1}{12} f'f'f(x). \tag{5.11}$$

Applying the same procedure to the simple pendulum, $f = (p, -\sin q)^T$, then

$$f'f'f = \begin{pmatrix} -p\cos q \\ \cos q \, \sin q \end{pmatrix}, \qquad f''(f,f) = \begin{pmatrix} 0 \\ p^2 \, \sin q \end{pmatrix}$$

and finally, the modified equation, up to terms of order $\mathcal{O}(h^3)$, reads

$$\dot{q} = p - \frac{h^2}{12} p \, \cos q, \qquad \dot{p} = -\sin q + \frac{h^2}{24}(\sin 2q - p^2 \, \sin q). \tag{5.12}$$

It is worth noticing that:

(a) the perturbation terms in the modified equation are of size $\mathcal{O}(h^2)$ in this case, since $f^{[2]} = 0$. Moreover, it can be shown that there are no $\mathcal{O}(h^3)$ terms in the expansion;

(b) the modified equation (5.12) is a Hamiltonian system, with Hamiltonian function

$$\tilde{H}(q,p,h) = \frac{1}{2}p^2 + (1 - \cos q) + \frac{h^2}{48}(-2p^2 \cos q + \cos 2q - 1), \tag{5.13}$$

since the midpoint rule is a symplectic method. Nevertheless, and in contrast with the original system $H(q,p) = \frac{1}{2}p^2 + (1 - \cos q) = T(p) + V(q)$, the modified Hamiltonian (5.13) is no longer separable into coordinates and momenta. □

All these features constitute, indeed, general properties of the modified equation, as we next state. The corresponding proofs can be found in, e.g., [170].

- **Adjoint methods**. Recall that the adjoint of a numerical method ψ_h, denoted by ψ_h^*, is defined through $\psi_h^* = \psi_{-h}^{-1}$. If $f_j(h)$ are the functions appearing in the series (5.5) of the modified equation corresponding to the integrator ψ_h, then the functions $(f^{[j]})^*(h)$ of the modified equation associated to ψ_h^*,

$$f_h^*(\tilde{x}) \equiv f(\tilde{x}) + h(f^{[2]})^*(\tilde{x}) + h^2(f^{[3]})^*(\tilde{x}) + \cdots,$$

verify that $(f^{[j]})^*(\tilde{x}) = (-1)^{j+1} f^{[j]}(\tilde{x})$. Therefore

$$f_h^*(\tilde{x}) = f(\tilde{x}) - h f^{[2]}(\tilde{x}) + h^2 f^{[3]}(\tilde{x}) + \cdots.$$

- **Time-symmetric methods**. For a time-symmetric method, one has $\psi_h^* = \psi_h$, so that $(f^{[j]})^*(x) = f^{[j]}(x)$. In consequence, $f^{[2j]}(x) = 0$ in the corresponding modified equation. In other words, the modified differential equation of a time-symmetric method only contains even powers of h.

- **Volume-preserving methods**. For volume-preserving methods applied to a divergence-free dynamical system, the modified equation is also divergence-free.

- **Symplectic methods**. For symplectic methods applied to a Hamiltonian system with a smooth Hamiltonian function H, the modified differential equation (5.4)–(5.5) is (locally) Hamiltonian: there are smooth functions $H_j : \mathbb{R}^{2d} \longrightarrow \mathbb{R}$ for $j = 2, 3, \ldots$, such that $f^{[j]}(x) = J\nabla H_j(x)$, where J is the canonical symplectic matrix (1.34). In consequence, there exists a modified Hamiltonian of the form

$$\tilde{H}(q,p) = H(q,p) + hH_2(q,p) + h^2H_3(q,p) + h^3H_4(q,p) + \cdots \quad (5.14)$$

such that the modified differential equation is given by

$$\dot{q} = \nabla_p \tilde{H}(q,p), \qquad \dot{p} = -\nabla_q \tilde{H}(q,p).$$

Of course, if the method is of order r, say, then $H_i = 0$ for $i \leq r$ in (5.14). In other words, the modified Hamiltonian has the form $\tilde{H} = H + h^r H_{r+1} + \cdots$. It can be shown [170, p. 345] that if the symplectic method ψ_h has a generating function

$$F(q,P,h) = hF_1(q,P) + h^2F_2(q,P) + h^3F_3(q,P) + \cdots$$

such that the functions $F_j(q,P)$ are defined on an open set D, then the modified differential equation is Hamiltonian with a Hamiltonian of the form (5.14), and the functions $H_j(q,p)$ are defined and smooth on D. In particular, for the Störmer–Verlet method (3.27) applied to the Hamiltonian $H(q,p) = T(p) + V(q)$, one has

$$\tilde{H} = T + V - h^2\left(\frac{1}{24}V_{qq}(T_p,T_p) + \frac{1}{12}T_{pp}(V_q,V_q)\right) + \mathcal{O}(h^4). \quad (5.15)$$

Example 5.3. The Lie formalism is also very useful to analyze numerical integrators using backward error analysis. Consider, for instance, the first step of the explicit Euler method, $x_1 = x_0 + hf(x_0)$. It can also be formally written as

$$x_1 = (\mathrm{Id} + hL_f\mathrm{Id})(x_0) = \exp(hL_{\hat{f}})[\mathrm{Id}](x_0) = \hat{\varphi}_h(x_0),$$

where

$$L_{\hat{f}} = \log(\mathrm{Id} + hL_f)) = L_f - \frac{1}{2}L_f^2 + \frac{1}{3}L_f^3 + \cdots$$

and the qualitative properties of the method can be studied through the associated vector field, $\hat{f}$. Clearly, $L_{\hat{f}}^2$ does not belong to the Lie algebra of vector fields, and therefore, the explicit Euler method does not preserve the underlying geometric structure associated with the original system. □

5.3 Modified equations of splitting methods

The example above shows the type of insight the Lie formalism developed in Appendices A.1–A.2 and applied in section 3.3 to obtain the order conditions of splitting and composition methods can provide in this setting. In the following, we particularize the treatment to splitting methods and show how the Baker–Campbell–Hausdorff (BCH) formula allows us to construct formally the modified vector field associated with the numerical integrator and, from here, to derive the corresponding modified differential equation if necessary.

To illustrate the general procedure, let us consider first the application to equation (5.3) when $f(x) = f_1(x) + f_2(x)$ of the Lie–Trotter splitting method $\chi_h = \varphi_h^{[2]} \circ \varphi_h^{[1]}$. As we stated in section 3.3, the differential operator associated with the exact h-flow can formally be expressed as

$$\Phi^h = \exp(hL_f) \equiv \exp(hF) = \exp(h(L_{f_1} + L_{f_2})) \equiv \exp(h(F_1 + F_2)),$$

whereas the corresponding differential operator associated to the numerical flow of χ_h is

$$\Psi(h) = \exp(hF_1)\, \exp(hF_2). \tag{5.16}$$

Applying now the BCH formula to (5.16), we can formally write $\Psi(h) = \exp(Z(h))$, where the first terms of $Z(h)$ can be obtained from the general expression (A.38) by replacing X and Y by the operators F_1 and F_2, respectively:

$$Z(h) = h(F_1 + F_2) + \frac{h^2}{2} F_{12} + \frac{h^3}{12}(F_{112} - F_{212}) - \frac{h^4}{24} F_{1212} + \mathcal{O}(h^5), \tag{5.17}$$

where the expression of $F_{j_1 \cdots j_m}$ is given in (3.45). We notice at once the following: (i) the operator $h^{-1}Z(h)$ thus obtained can be considered as the formal Lie derivative associated with the numerical flow of the integrator χ_h; (ii) this operator is an infinite series involving iterated Lie brackets of F_1 and F_2, and thus lies in the same Lie algebra as F_1 and F_2; in consequence, the numerical solution inherits the properties of the exact flow associated with this feature (e.g., Hamiltonian, volume-preserving, etc.); (iii) since χ_h is only first order, $Z(h) - h(F_1 + F_2) = \mathcal{O}(h^2)$ and (iv) it is indeed possible from the expression of $Z(h)$ to construct explicitly the modified equation associated with χ_h term by term.

Let us analyze in more detail this last point. From the expression of the Lie derivatives F_1 and F_2, equation (A.28), we get

$$F_{12} = \sum_j (F_{12})_j \frac{\partial}{\partial x_j}, \quad F_{112} = \sum_j (F_{112})_j \frac{\partial}{\partial x_j}, \quad F_{212} = \sum_j (F_{212})_j \frac{\partial}{\partial x_j},$$

and so on, with (see eq. (A.7))

$$\begin{aligned}
(F_{12})_j &= \sum_{i=1}^d \left((f_1)_i \frac{\partial (f_2)_j}{\partial x_i} - (f_2)_i \frac{\partial (f_1)_j}{\partial x_i} \right) = (f_1, f_2)_j \\
(F_{112})_j &= \sum_{i=1}^d \left((f_1)_i \frac{\partial (F_{12})_j}{\partial x_i} - (F_{12})_i \frac{\partial (f_1)_j}{\partial x_i} \right) = (f_1, (f_1, f_2))_j \\
(F_{212})_j &= \sum_{i=1}^d \left((f_2)_i \frac{\partial (F_{12})_j}{\partial x_i} - (F_{12})_i \frac{\partial (f_2)_j}{\partial x_i} \right) = (f_2, (f_1, f_2))_j.
\end{aligned}$$

In consequence, one has for f_h in the corresponding modified differential equation (5.4) associated to the integrator $\chi_h = \varphi_h^{[2]} \circ \varphi_h^{[1]}$ the expression

$$f_h(x) = f_1(x) + f_2(x) + \frac{h}{2}(f_1, f_2)(x) + \frac{h^2}{12}\Big((f_1, (f_1, f_2))(x) - (f_2, (f_1, f_2))(x) \Big) + \mathcal{O}(h^3).$$

If we work instead with the Strang splitting $\varphi_{h/2}^{[1]} \circ \varphi_h^{[2]} \circ \varphi_{h/2}^{[1]}$, we have to apply the symmetric BCH formula (A.41) to the differential operator

$$\Psi(h) = \mathrm{e}^{\frac{h}{2}F_1}\, \mathrm{e}^{hF_2}\, \mathrm{e}^{\frac{h}{2}F_1}, \tag{5.18}$$

leading to $\Psi(h) = \exp(Z(h))$ with

$$Z(h) = h(F_1 + F_2) - h^3 \left(\frac{1}{24} F_{112} + \frac{1}{12} F_{212} \right) + \mathcal{O}(h^5)$$

and the modified equation reads

$$f_h(x) = f_1(x) + f_2(x) - h^2 \Big(\frac{1}{24}(f_1, (f_1, f_2))(x) + \frac{1}{12}(f_2, (f_1, f_2))(x) \Big) + \mathcal{O}(h^4).$$

Notice that only even powers of h appear in the expression of $f_h(x)$ due to the time symmetry of the numerical scheme.

This procedure can be generalized to splitting methods of any order r. In fact, it is similar to the approach carried out in section 3.3 to obtain the order conditions, but now higher orders in h in the power series expansion of $F(h) = \log(\Psi(h))$ have to be computed. Thus, for the operator

$$\Psi(h) = \mathrm{e}^{a_1 hF_1}\, \mathrm{e}^{b_1 hF_2}\, \mathrm{e}^{a_2 hF_1} \cdots \mathrm{e}^{a_s hF_1}\, \mathrm{e}^{b_s hF_2}\, \mathrm{e}^{a_{s+1} hF_1}$$

associated with scheme (3.40), the modified vector field now reads

$$Z(h) = h(F_1 + F_2) + \sum_{i=1}^{\infty} h^{r+i} \sum_{j=1}^{c_{r+i}} v_{i,j}\, E_{r+i,j}, \tag{5.19}$$

where $E_{r+i,j}$ denotes the element j of a basis of the subspace $\mathcal{L}_{r+i}(A,B)$ and $v_{i,j}$ are fixed real numbers determined by the actual coefficients of the method.

The same considerations apply to composition methods, but this time one has to work with the product of exponentials of vector fields (3.47) and obtain the expression (3.48) up to the desired order.

5.3.1 Hamiltonian systems

One particular example deserves special attention, namely Hamiltonian systems of the form $H(q,p) = T(p) + V(q)$. In accordance with the treatment of Appendix A.3.2, in that case the operators F_1 and F_2 are the Hamiltonian vector fields corresponding to the kinetic and potential energy, respectively, $F_1 = L_{X_T}$, $F_2 = L_{X_V}$, so that the modified operator (5.19) associated with the splitting method (3.40) is itself Hamiltonian, and it is straightforward to get the expression of the corresponding modified Hamiltonian. In particular, for the Störmer–Verlet method (3.28),

$$S_h^{[2]} = \varphi_{h/2}^{[T]} \circ \varphi_h^{[V]} \circ \varphi_{h/2}^{[T]},$$

we have

$$Z(h) = h(L_{X_T} + L_{X_V}) - h^3 \left(\frac{1}{24}[L_{X_T},[L_{X_T},L_{X_V}]] + \frac{1}{12}[L_{X_V},[L_{X_T},L_{X_V}]] \right) + \mathcal{O}(h^5).$$

Due to the close connection between the Lie bracket $[L_{X_T}, L_{X_V}]$ and the Poisson bracket $\{T,V\}$, equation (A.30),

$$[L_{X_T}, L_{X_V}] = -L_{X_{\{T,V\}}},$$

it is clear that $Z(h) = hL_{X_{\tilde{H}}}$, where the modified Hamiltonian is given by

$$\tilde{H} = T + V - h^2 \left(\frac{1}{24}\{T,\{T,V\}\} + \frac{1}{12}\{V,\{T,V\}\} \right) + \mathcal{O}(h^4)$$

and we recover the expression (5.15).

5.3.2 Some physical insight

Consider now, for simplicity, the Hamiltonian $H = \frac{1}{2}p^T p + V(q)$, and the numerical approximation furnished by the symplectic Euler-TV scheme (3.26).

Since the associated differential operator is $\Psi(h) = \exp(hF_1)\exp(hF_2)$, the modified vector field $Z(h)$ is given by (5.17), i.e.,

$$Z(h) = h(L_{X_T} + L_{X_V}) + \frac{h^2}{2}[L_{X_T}, L_{X_V}] + \mathcal{O}(h^3),$$

so that the corresponding modified Hamiltonian is

$$\tilde{H}(q,p) = T + V - \frac{h}{2}\{T, V\} + \mathcal{O}(h^2) = \frac{1}{2}p^T p + V(q) + \frac{h}{2}p^T V_q(q) + \mathcal{O}(h^2).$$

Neglecting higher-order terms in h, $\tilde{H}$ can be written as

$$\tilde{H}(q,p) = \frac{1}{2}\left(p + \frac{h}{2}V_q\right)^T\left(p + \frac{h}{2}V_q\right) + \left(V(q) - \frac{h^2}{8}V_q^T V_q\right).$$

Given the near-identity canonical transformation

$$Q = q, \qquad P = p + \frac{h}{2}V_q(q),$$

the Hamiltonian in the new variables (Q, P) is just

$$\tilde{H}(Q,P) = \tilde{H}(q(Q,P), p(Q,P)) = \frac{1}{2}P^T P + \tilde{V}(Q), \tag{5.20}$$

where

$$\tilde{V}(Q) = V(Q) - \frac{h^2}{8}V_Q(Q)^T V_Q(Q).$$

In conclusion, the numerical solution provided by the symplectic Euler-TV scheme can be regarded, for sufficiently small values of h, as the exact solution of the Hamiltonian (5.20) involving a modified potential depending on h.

Let us analyze from this perspective three simple examples used several times within the book:

- The harmonic oscillator, with $V(q) = \frac{1}{2}q^2$. Then, clearly,

$$\tilde{V}(Q) = \frac{1}{2}\left(1 - \frac{h^2}{4}\right)Q^2,$$

which corresponds to an oscillator for $h < 2$ with frequency $\tilde{w} = \sqrt{1 - h^2/4}$. The case $h > 2$ corresponds to an inverted parabola with an unbounded solution. In consequence, if $h > 2$, the numerical scheme is unstable, as we already know (see subsection 3.3.4).

- The pendulum, with $V(q) = 1 - \cos(q)$. In that case

$$\tilde{V}(Q) = 1 - \cos(Q) - \frac{h^2}{8}\sin(Q)^2,$$

which corresponds to another pendulum without friction and whose length depends on the angle.

- The Kepler problem, with $V(q) = -1/r$, $r = \sqrt{q^T q}$. Then one has

$$\tilde{V}(Q) = -\frac{1}{R} - \frac{h^2}{8}\frac{1}{R^4}, \qquad R = \sqrt{Q^T Q}.$$

For sufficiently small values of h and for initial conditions leading to bounded trajectories of the Kepler problem, the numerical scheme provides solutions close to Keplerian trajectories but with a slight precession effect. This is similar to the precession caused by relativistic effects or due to the attraction of an oblate planet [246].

As we can see, this interpretation helps to get more insight into the numerical results observed for these examples in Chapter 1.

5.4 Estimates over long-time intervals

Convergence of the series (5.5) defining the modified equation, apart from the linear case analyzed at the beginning of the chapter, is the exception rather than the general rule. In consequence, an alternative has to be pursued to get rigorous estimates concerning the long-time behavior of the numerical solutions. Specifically, we first give bounds on the coefficient functions $f^{[j]}(x)$ of the modified equation, then determine an optimal truncation index, and finally estimate the difference between the numerical solution x_n and the exact solution $\tilde{x}(t_n)$ of the truncated modified equation. We only summarize here the main results and refer the reader to [170, 253] and references therein for a more comprehensive treatment.

Our basic assumption is the analyticity of $f(x)$ in the differential equation $\dot{x} = f(x)$ and the expression of the numerical method $\psi_h(x)$. More specifically, we assume that $f(x)$ is analytic in a complex neighborhood of x_0, verifying $\|f(x)\| \leq K$ for all $x \in B_{2\rho}(x_0)$, where

$$B_\rho(x_0) = \{x \in \mathbb{C}^d : \|x - x_0\| \leq \rho\}.$$

We also assume analyticity for the functions $f^{[j]}(x)$ of the modified equation on $B_{\rho/2}(x_0)$ and the functions $d_j(x)$ in the series (5.6) associated with the numerical method ψ_h. If a suitable truncation index for the formal series (5.5) is selected, so that one has

$$\dot{\tilde{x}} = f(\tilde{x}) + hf^{[2]}(\tilde{x}) + h^2 f^{[3]}(\tilde{x}) + \cdots + h^{N-1} f^{[N]}(\tilde{x}), \tag{5.21}$$

with $\tilde{x}(0) = x_0$ and exact flow $\tilde{\varphi}_t^{[N]}$, then there exist constants h_0 with $h \leq h_0/4$ and $\gamma > 0$ such that

$$\|\psi_h(x_0) - \tilde{\varphi}_h^{[N]}(x_0)\| \leq h\gamma K \mathrm{e}^{-h_0/h}. \tag{5.22}$$

In other words, the difference between the numerical solution $\psi_h(x_0)$ and the exact solution $\tilde{\varphi}_h^{[N]}(x_0)$ of the truncated modified equation (5.21) is exponentially small.

This result has important consequences regarding the long-time behavior of numerical schemes. Thus, in particular, suppose our numerical scheme is a symplectic integrator of order r applied with step size h to a Hamiltonian system. Then the corresponding modified equation is itself Hamiltonian with (truncated) Hamiltonian

$$\tilde{H}(x) = H(x) + h^r H_{r+1}(x) + \cdots + h^{N-1} H_N(x),$$

where now $x = (q, p)$. Denoting as before by $\tilde{\varphi}_t^{[N]}$ the flow of the truncated modified equation, it is clear that

$$\tilde{H}\big(\tilde{\varphi}_t^{[N]}(x_0)\big) = \tilde{H}(x_0) \qquad \text{for all } t.$$

Taking into account (5.22) and the bounds on the functions appearing in the modified equation (derivatives of the $\tilde{H}$ in this case), it follows that

$$\tilde{H}(x_{n+1}) - \tilde{H}\big(\tilde{\varphi}_h^{[N]}(x_n)\big) = \mathcal{O}\big(h\mathrm{e}^{-h_0/h}\big).$$

Next, from the telescopic sum

$$\tilde{H}(x_n) - \tilde{H}(x_0) = \sum_{j=1}^{n} \Big(\tilde{H}(x_j) - \tilde{H}(x_{j-1})\Big) = \sum_{j=1}^{n} \Big(\tilde{H}(x_j) - \tilde{H}\big(\tilde{\varphi}_h^{N}(x_{j-1})\big)\Big),$$

where the last identity follows from the conservation of $\tilde{H}$, one gets

$$\tilde{H}(x_n) - \tilde{H}(x_0) = \mathcal{O}\big(nh\mathrm{e}^{-h_0/h}\big).$$

In consequence, for $nh \leq \mathrm{e}^{h_0/2h}$, we arrive at

$$\tilde{H}(x_n) = \tilde{H}(x_0) + \mathcal{O}\big(\mathrm{e}^{-h_0/2h}\big).$$

If we assume in addition that the numerical solution stays in a compact set $\mathcal{K}$, then $H_{r+1}(x) + \cdots + h^{N-r-1} H_N(x)$ is uniformly bounded on $\mathcal{K}$ independently of h and N [170, p. 367], and finally

$$H(x_n) = H(x_0) + \mathcal{O}(h^r).$$

Equivalently, the error in the energy corresponding to the numerical solution is of order r over exponentially long-time intervals when a symplectic method is applied with constant step size in a compact region of the phase space [254]. This result provides a sound theoretical explanation for the phenomena reported in the numerical experiments of Chapters 1 and 2, in particular the good energy preservation observed for the symplectic Euler and the Störmer–Verlet schemes, as well as the midpoint rule and the Runge–Kutta–Gauss–Legendre methods.

Concerning symplectic methods defined by B-series, it has been shown that the only symplectic method (as B-series) that preserves the Hamiltonian for arbitrary Hamiltonian functions is the exact flow of the differential equation [104]. This result should be compared with that obtained by Ge and Marsden [156] on general symplectic methods (not necessarily defined as B-series) applied to a Hamiltonian system that does not have other integrals of motion than H: in that case, the only symplectic method preserving H constitutes a time re-parametrization of the exact flow. Basically, and in a simple way, a symplectic integrator can be considered as the exact solution of a modified Hamiltonian $\tilde{H}(q,p)$. If the method is energy-preserving, then $\tilde{H}(q,p)$ must be a function of $H(q,p)$, i.e., $\tilde{H}(q,p) = f(H(q,p))$ with Hamiltonian equations

$$\dot{x} = J\nabla_x \tilde{H}(x) = \alpha J \nabla_x H(x)$$

with $\alpha = f'(H(q,p)) = f'(H(q_0,p_0)) = const$, which is just equation (1.33) with a time re-parametrization.

With respect to the behavior of the error in position, as shown in [81, 170], if the Hamiltonian system is integrable and certain conditions on the frequencies at the initial point are satisfied, then

$$\|(q_n,p_n) - (q(t),p(t))\| \leq Cth^r, \qquad \text{for} \quad t = nh \leq h^{-r}, \qquad C = \text{const.},$$

that is, the global error grows at most linearly in time, whereas first integrals $I(q,p)$ that only depend on the action variables are well preserved on exponentially long-time intervals,

$$\|I(q_n,p_n) - I(q_0,p_0)\| \leq Ch^r \quad \text{for} \quad t = nh \leq h^{-r}.$$

In contrast, for a non-symplectic method of the same order r we can only assure that

$$H(x_{n+1}) - H(\varphi_h(x_n)) = \mathcal{O}(h^{r+1})$$

from a standard local error estimate and the bound obtained for the functions $f^{[j]}$ in the modified equation. Since H is constant along exact solutions (i.e., $H\big(\varphi_h(x_n)\big) = H(x_n)$), then

$$H(x_{n+1}) - H(x_n) = \mathcal{O}(h^{r+1})$$

and

$$H(x_n) - H(x_0) = \sum_{j=1}^{n} (H(x_j) - H(x_{j-1})) = \mathcal{O}(nh^{r+1}) = \mathcal{O}(th^r),$$

that is, the error of the energy grows linearly, whereas the global error in the solution typically increases quadratically with time, and the error in the first integrals drifts linearly from the correct value.

It is important to remark that the modified differential equation of a numerical scheme depends explicitly on the step size used, so that if h is changed, then one has a different modified equation. This fact helps to explain the poor long time behavior observed in practice when a symplectic scheme is implemented directly with a standard variable step-size strategy (see e.g., [84]).

5.5 Exercises

1. Verify that the modified Hamiltonian corresponding to the symplectic Euler-TV scheme (1.11) applied to the harmonic oscillator with $k = m = 1$ is indeed

$$\tilde{H}(q,p,h) = \frac{2\arcsin(h/2)}{h\sqrt{4-h^2}}(p^2 + q^2 + h\,p\,q).$$

2. Construct the modified equation corresponding to the explicit Euler method, the midpoint rule, and the leapfrog applied to the Lotka–Volterra equations (3.91). Comment on the results.

3. Apply the midpoint method to the pendulum problem with initial conditions $(q_0, p_0) = (1, 0)$ for the time interval $t \in [0, 10]$, and compare with the exact solution of (5.12) for different values of h. What is the order of the error in h?

4. Suppose the differential equation $\dot{x} = f(x)$ is such that $\mathrm{div} f(x) = 0$ and apply a volume-preserving integrator ψ_h. Show that the corresponding modified equation $\dot{\tilde{x}} = f_h(\tilde{x})$ is such that $\mathrm{div} f_h = 0$.

5. Obtain the first terms of the modified Hamiltonian corresponding to the Störmer–Verlet scheme $S_h^{[2]} = \varphi_{h/2}^{[V]} \circ \varphi_h^{[T]} \circ \varphi_{h/2}^{[V]}$ applied to the Hamiltonian $H(q,p) = T(p) + V(q)$.

6. Construct explicitly the modified vector field corresponding to the application of the second-order composition $S_h^{[2]} = \chi_{h/2} \circ \chi_{h/2}^*$ to equation $\dot{x} = f(x)$.

7. Solve the equations of motion corresponding to the Hamiltonian function $H = \alpha(p^2+q^2)^2$ with initial conditions $(q_0, p_0) = (\sqrt{2}, 0)$ when $\alpha = \frac{1}{16}, \frac{1}{8}$ and $\frac{1}{4}$ (take into account that H remains constant along the solution). Compare with the solution of the harmonic oscillator. Consider now the Hamiltonian $H = \log(p^2 + q^2)$ and repeat the computations.

Chapter 6

Time-splitting methods for PDEs of evolution

6.1 Introduction

Splitting methods can also be applied to partial differential equations (PDEs) of the form

$$u_t(x,t) \equiv \frac{\partial u}{\partial t}(x,t) = f(x, u(x,t)), \qquad u(x,0) = u_0(x), \tag{6.1}$$

where the unknown $u(x,t)$ is defined in a certain function space, the spatial variable $x \in \mathbb{R}^d$, and the operator $f(x,u)$ contains contributions coming from very different physical sources, so that one may decompose it into two (or more) parts, $f = f_1 + f_2$ and apply different schemes to solve approximately each sub-problem. For instance, in a reaction-diffusion system

$$u_t(x,t) = \nabla \cdot (D\,\nabla u) + g(u),$$

where D and g may also depend on x, it makes sense to split the diffusion from the reaction terms, i.e.,

$$f(u) = f_1(u) + f_2(u), \qquad \text{with} \qquad f_1(u) = \nabla \cdot (D\,\nabla u), \qquad f_2(u) = g(u),$$

where we have omitted the (possibly explicit) dependence on x. Once we specify appropriate boundary conditions and the way each sub-problem $u_t = f_i(u)$, $i = 1, 2$ is solved in practice, then the Lie–Trotter scheme can also be used in this setting. Thus, applying the implicit Euler scheme in each sub-problem leads to the so-called *Marchuk–Yanenko* operator-splitting scheme

$$\begin{aligned} u_{n+\frac{1}{2}} &= u_n + hf_1(u_{n+\frac{1}{2}}) \\ u_{n+1} &= u_{n+\frac{1}{2}} + hf_2(u_{n+1}), \qquad n = 0, 1, 2, \ldots \end{aligned} \tag{6.2}$$

where $h \equiv \Delta t$, the time step size. In spite of its low order of consistency and the fact that the intermediate stage $u_{n+\frac{1}{2}}$ is not a consistent approximation to the exact solution, it has been used in complicated problems, even involving non-smooth operators [157].

DOI: 10.1201/9781003527626-6

If the second-order implicit trapezoidal rule is used instead to integrate each sub-problem, the Lie–Trotter scheme results in Yanenko's Crank–Nicolson method [190, 232]:

$$\begin{aligned} u_{n+\frac{1}{2}} &= u_n + \frac{h}{2} f_1(u_n) + \frac{h}{2} f_1(u_{n+\frac{1}{2}}) \\ u_{n+1} &= u_{n+\frac{1}{2}} + \frac{h}{2} f_2(u_{n+\frac{1}{2}}) + \frac{h}{2} f_2(u_{n+1}), \qquad n = 0, 1, 2, \ldots, \end{aligned} \tag{6.3}$$

also of first order of consistency.

The *Peaceman–Rachford* scheme

$$\begin{aligned} u_{n+\frac{1}{2}} &= u_n + \frac{h}{2} f_1(u_{n+\frac{1}{2}}) + \frac{h}{2} f_2(u_n) \\ u_{n+1} &= u_{n+\frac{1}{2}} + \frac{h}{2} f_1(u_{n+\frac{1}{2}}) + \frac{h}{2} f_2(u_{n+1}), \qquad n = 0, 1, 2, \ldots, \end{aligned} \tag{6.4}$$

which is of second order of consistency, can also be interpreted in this way. Notice that, in contrast with methods (6.2) and (6.3), both f_1 and f_2 appear in each of the two stages, and thus the intermediate value $u_{n+\frac{1}{2}}$ provides a consistent approximation at $t = t_{n+\frac{1}{2}}$.

It is illustrative to analyze the particular case when f in (6.1) is a linear spatial differential operator. Assuming that an appropriate semidiscretization of (6.1) in the space variable x has been carried out, one ends up with system

$$\frac{dU}{dt} = F_1 U + F_2 U, \tag{6.5}$$

where $F_1, F_2 \in \mathbb{C}^{M\times M}$, $F_1F_2 \neq F_2F_1$ in general, and $U \in \mathbb{C}^M$ approximates u on the space grid points $x_1, \ldots, x_M$. Then, a step of the Marchuk–Yanenko scheme (6.2) reads

$$U_{n+1} = (I - hF_2)^{-1} (I - hF_1)^{-1} U_n, \tag{6.6}$$

where $U_n \approx (u(x_1, t_n), \ldots, u(x_M, t_n))^T$. Clearly, (6.6) can be obtained by applying the [0/1] Padé approximant to the exponentials in the Lie–Trotter scheme $U_{n+1} = \mathrm{e}^{hF_2}\, \mathrm{e}^{hF_1} U_n$, whence the first order of the approximation is obtained at once. On the other hand, applying (6.3) results in

$$U_{n+1} = \left(I - \frac{h}{2}F_2\right)^{-1} \left(I + \frac{h}{2}F_2\right) \left(I - \frac{h}{2}F_1\right)^{-1} \left(I + \frac{h}{2}F_1\right) U_n, \tag{6.7}$$

which corresponds to replacing each exponential in the Lie–Trotter scheme with the [1/1] Padé approximant. Although a second-order approximation of the exponentials is considered, their combination (6.7) is of first order, since in the end we are using a variant of the Lie–Trotter scheme.

Proceeding analogously, one can show that the Peaceman–Rachford scheme (6.4) applied to the linear equation (6.5) can be written as

$$U_{n+1} = \left(I - \frac{h}{2}F_2\right)^{-1} \left(I + \frac{h}{2}F_1\right) \left(I - \frac{h}{2}F_1\right)^{-1} \left(I + \frac{h}{2}F_2\right) U_n, \tag{6.8}$$

and one can show that both methods (6.7) and (6.8) are in fact conjugate.

All these algorithms can be formulated by applying properly chosen compositions of the implicit and explicit Euler methods. This observation leads eventually to the construction of methods in the same family but of higher order and/or with an improved behavior [52].

When applying splitting methods to the PDE (6.1), boundary conditions deserve a special discussion. Notice that they are defined (based very often on physical grounds) for the whole operator f, and they do not necessarily hold for the subproblems defined by each part f_1 and f_2. Therefore, we cannot expect the numerical solution obtained by a splitting method to belong to the domain of f. This results in severe order reductions in reaction-diffusion problems when Dirichlet or Neumann boundary conditions are considered [178, 190]. In particular, the order reduction for the Strang splitting is one in the infinity norm. Similar order reductions for advection-reaction problems have also been reported [189]. Although several procedures have been considered in the literature to avoid the order reduction in the case of reaction-diffusion problems [133, 134, 132, 7, 8], this is still the subject of active research.

On the other hand, even if the solution is sufficiently smooth and periodic boundary conditions are considered, applying splitting methods of order higher than two is not possible for certain problems involving the Laplacian operator [178]. This can be seen by considering the simple heat equation $u_t = \Delta u$ on $0 < x < 1$ with Dirichlet boundary conditions: the kth Fourier mode of the solution is $c_k \, \mathrm{e}^{-(k\pi)^2 t}$, which is generally not well-defined for $t < 0$.

There are, however, relevant problems where high-order splitting methods can be and have been safely used. An example in point is the time integration of the time-dependent Schrödinger equation with periodic boundary conditions. In this case, the combination of the Strang splitting in time and the Fourier collocation procedure in space is quite popular in chemical physics. These schemes have appealing structure-preserving properties, such as unitarity, symplecticity, and time-symmetry [228].

As a general treatment of splitting methods for PDEs goes well beyond the purpose of the present book, we will restrict ourselves to some relevant equations where preservation of qualitative properties is of interest and refer the reader to the classical references [331, 232], the multi-author book [157], and the review [52] for a more comprehensive coverage of this topic. Specifically, in this chapter we first review the application of splitting methods to the numerical time integration of the Schrödinger equation. In particular, we will see that, once a space discretization has been carried out and a discretized Hamiltonian H has been determined, one can build unitary integration methods providing high accuracy when the problem has sufficient spatial regularity. Next, we will show that the order-2 barrier for diffusion equations can be circumvented by considering methods with complex coefficients with positive real parts. These schemes possess some distinctive features that will be analyzed in detail.

6.2 Splitting methods for the time-dependent Schrödinger equation

The basic object to study the time evolution of a system in quantum mechanics is the (time-dependent) Schrödinger equation

$$i\frac{\partial\psi}{\partial t}(x,t) = \hat{H}\psi(x,t), \qquad \psi(x,0) = \psi_0(x), \tag{6.9}$$

where for convenience we have chosen units such that the reduced Planck constant $\hbar = 1$. Here $\hat{H}$ is the Hamiltonian operator, and the *wave function* $\psi : \Omega \subset \mathbb{R}^d \times \mathbb{R} \longrightarrow \mathbb{C}$ represents the state of the system. In the case of one particle, $d = 3$, and the quantity $|\psi(x,t)|^2$ represents a probability density for the position of the particle, in the sense that the probability of the particle being located in $S \subset \Omega$ at time t is $\int_S |\psi(x,t)|^2 dx$. The equation is defined in the Hilbert space $L^2(\Omega, \mathbb{C})$.

Very often the Hamiltonian adopts the form

$$\hat{H} = \hat{T}(\hat{P}) + \hat{V}(\hat{X}) \equiv \frac{1}{2\mu}\hat{P}^2 + \hat{V}(\hat{X}) \tag{6.10}$$

($\hat{P}^2 \equiv \sum_{j=1}^d \hat{P}_j^2$) in terms of the self-adjoint operators $\hat{X}$ (position) and $\hat{P}$ (momentum). They act on $\psi(x,t) \in L^2(\Omega, \mathbb{C})$ as

$$\hat{X}_j\psi(x,t) = x_j\,\psi(x,t), \qquad \hat{P}_j\psi(x,t) = -i\,\frac{\partial}{\partial x_j}\psi(x,t),$$

whereas μ represents the reduced mass of the system. The commutator of the operators is defined by their action on $\psi(x,t)$ as follows:

$$[\hat{X}_j, \hat{P}_k]\,\psi = \left(\hat{X}_j\hat{P}_k - \hat{P}_k\hat{X}_j\right)\psi = -ix_j\frac{\partial\psi}{\partial x_k} + i\frac{\partial}{\partial x_k}(x_j\psi) = i\,\delta_{jk}\,\psi,$$

so that $[\hat{X}_j, \hat{P}_k] = i\,\delta_{jk}$ plays here the role of the Poisson bracket in classical mechanical systems (c.f. section 1.4.1). Similarly, a simple calculation shows that

$$[\hat{T},\hat{V}]\,\psi = (\hat{T}\hat{V} - \hat{V}\hat{T})\,\psi = -\frac{1}{2\mu}\Delta(\hat{V}\,\psi) + \frac{1}{2\mu}\hat{V}\Delta\,\psi = -\frac{1}{2\mu}(\Delta\hat{V} + 2\nabla\hat{V}^T\nabla)\,\psi,$$

and

$$[\hat{V},[\hat{T},\hat{V}]]\,\psi = \frac{1}{\mu}\nabla\hat{V}^T\nabla\hat{V}\,\psi, \tag{6.11}$$

$$[\hat{V},[\hat{V},[\hat{T},\hat{V}]]]\,\psi = 0. \tag{6.12}$$

Notice the close similarity with classical mechanical Hamiltonian systems with quadratic kinetic energy.

Given a wave function $\psi(x,t)$, the quantities

$$N_{\psi}^2 = \int_{\Omega} \overline{\psi}(x,t)\psi(x,t)\, dx, \tag{6.13}$$

$$E_{\psi} = \int_{\Omega} \overline{\psi}(x,t)\hat{H}\psi(x,t)\, dx, \tag{6.14}$$

are the *norm* and *energy*, respectively. According to the previous comments, N_{ψ} is a constant of motion, whereas the energy E_{ψ} is also an invariant if the self-adjoint Hamiltonian operator is time independent.

The solution of (6.9) provides all the dynamical information on the physical system at any time and can be expressed as

$$\psi(x,t) = \hat{U}(t)\, \psi_0(x) \tag{6.15}$$

in terms of the (linear) *evolution operator* $\hat{U}$, which satisfies the equation [247]

$$i\, \frac{d\hat{U}(t)}{dt} = \hat{H}\hat{U}(t), \qquad \hat{U}(0) = I,$$

and is therefore unitary. When the Hamiltonian is explicitly time independent, the evolution operator is given formally by

$$\hat{U}(t) = \mathrm{e}^{-it\hat{H}}. \tag{6.16}$$

Except for academic problems, however, it is not possible to get a closed expression for the solution $\psi(x,t)$ or, alternatively, for the unitary operator $\hat{U}(t)$, and so one very often turns to numerical methods to construct reliable approximations. This process typically involves two stages. In the first a discrete representation of the initial wave function $\psi_0(x)$ and the operator $\hat{H}$ on an appropriate space grid is constructed. In the second, this finite representation of $\psi_0(x)$ is propagated in time until the desired final time with a numerical integrator in the usual step-by-step way.

6.2.1 Space discretization

There are many possible ways to discretize the Schrödinger equation in space: finite differences, spectral schemes based either on Galerkin methods with a basis of Hermite polynomials or with trigonometric polynomials, etc. [228]. Among them, collocation spectral methods have several appealing features: it is possible to get a faithful representation of the wave function with a relatively small grid size, they are simple to implement (especially in one-dimensional problems), and they lead to extremely high accuracy if the solution of the problem is sufficiently smooth [151, 162]. Moreover, when long-time integrations are required, the spatial discretization obtained by spectral methods does not cause a deterioration of the phase error as the integration in time goes on [74, 182].

To keep the treatment as simple as possible, we focus here on the one-dimensional case, although the same methods can be applied to more dimensions by taking tensor products of one-dimensional expansions [89, 228]. We assume that the wave function is negligible outside an interval $[\alpha, \beta]$ and that the problem can be formulated on this finite interval with periodic boundary conditions. Under such conditions it is possible to use the Fourier collocation approach. After rescaling, one may assume without loss of generality that the space interval is $[0, 2\pi]$ and $\mu = 1$, so that the problem reduces to solving

$$i \frac{\partial \psi}{\partial t}(x,t) = -\frac{1}{2} \frac{\partial^2 \psi}{\partial x^2}(x,t) + \hat{V}(x)\psi(x,t), \qquad 0 \leq x < 2\pi, \tag{6.17}$$

with $\psi(0,t) = \psi(2\pi, t)$ for all t. The goal is then to construct approximations to the solution based on the equidistant interpolation grid

$$x_j = \frac{2\pi}{M} j, \qquad j = 0, \ldots, M-1,$$

where M is even (although the formalism can also be adapted to an odd number of points) as

$$\psi_M(x,t) = \sum_{-M/2 \leq n < M/2} c_n(t)\, \mathrm{e}^{inx}, \qquad x \in [0, 2\pi). \tag{6.18}$$

The coefficients $c_n(t)$ are related to the grid values $\psi_M(x_j, t)$ through a discrete Fourier transform of length M, $\mathcal{F}_M$ [317], whose computation can be accomplished by the *Fast Fourier Transform* (FFT) algorithm with $\mathcal{O}(M \log M)$ floating point operations.

Next, the grid values $\psi_M(x_j, t)$ are determined by requiring that the approximation (6.18) satisfy the Schrödinger equation (6.17) precisely at the grid points x_j [228] (hence the name of *collocation*). This results in a system of M ordinary differential equations for the M point values $\psi_M(x_j, t)$,

$$i \frac{du}{dt} = \mathcal{F}_M^{-1} D_M \mathcal{F}_M\, u + V_M u \equiv H u, \qquad u = (u_0, u_1, \ldots, u_{M-1}), \tag{6.19}$$

where $u_j(t) = \psi_M(x_j, t) \approx \psi(x_j, t)$ and

$$D_M = -\frac{1}{2} \operatorname{diag}(n^2), \qquad V_M = \operatorname{diag}(\hat{V}(x_j)) \tag{6.20}$$

for $n = -M/2, \ldots, M/2 - 1$ and $j = 0, \ldots, M-1$. An important qualitative feature of this space discretization procedure is that it replaces the original Hilbert space $\mathcal{H} \equiv \mathcal{L}^2(0, 2\pi)$ defined by the quantum mechanical problem by a discrete one in which the action of operators is approximated by $M \times M$ (Hermitian) matrices. These matrices obey the same quantum mechanical commutation relations [206] in the limit of very large values of M. In this sense, the matrix H in (6.19) constitutes the discrete analogue of the Hamiltonian operator $\hat{H}$ in (6.9).

From a quantitative point of view, if the function ψ is sufficiently smooth and periodic, then the coefficients c_n exhibit a rapid decay, so that the value of M in the expansion (6.18) must not be very large to accurately represent the solution. Specifically, in [228] the following result is proved.

Theorem 4 *Suppose that the exact solution $\psi(x,t)$ of (6.17) is such that, for some $s \geq 1$, $\partial_x^{s+2}\psi(\cdot,t) \in \mathcal{H}$ for every $t \geq 0$. Then the error due to the approximation $\psi_M(x,t)$ defined by (6.18) in the collocation approach is bounded by*

$$\|\psi_M(\cdot,t) - \psi(\cdot,t)\| \leq C\, M^{-s}(1+t) \max_{0 \leq \tau \leq t} \left\|\partial_x^{s+2}\psi(\cdot,\tau)\right\|,$$

where C depends only on s.

When the problem is not periodic, the use of a truncated Fourier series introduces errors in the computation. In that case several techniques have been proposed to minimize its effects (see [24, 74] and references therein).

6.2.2 Splitting methods for time propagation

After a space discretization has been applied to eq. (6.17), one is left with a linear system of ODEs of the form

$$i\frac{du(t)}{dt} = Hu(t), \qquad u(0) = u_0 \in \mathbb{C}^M, \tag{6.21}$$

where H is a $M \times M$ Hermitian matrix. Since typically $u_j(t) \simeq \sigma\sqrt{\Delta x}\,\psi(x_j,t)$, the discrete version of the norm (6.13) of the wave function is defined as

$$N_u^2 = \|u\|^2 = \overline{u}^T u \simeq \sigma^2 \Delta x \sum_j \overline{\psi}(x_j,t)\psi(x_j,t) \simeq N_\psi^2.$$

Here $\Delta x = 2\pi/M$ denotes the size of the space grid, and σ is an arbitrary constant so that $\|u\| = 1$. In a similar way, we define the discrete energy by

$$E_u = \overline{u}^T H u.$$

Again, it is easy to show that both N_u and E_u are constants of motion, as in the continuous case: since $\overline{H}^T = H$, then

$$\frac{dE_u}{dt} = \frac{d\overline{u}^T}{dt} Hu + \overline{u}^T H \frac{du}{dt} = i\,\overline{u}^T \overline{H}^T Hu - i\,\overline{u}^T HHu = 0.$$

The grid size chosen for the spatial discretization has a direct consequence on the time propagation of the (discrete) wave function $u(t)$, since the matrix H representing the Hamiltonian has a discrete spectrum which depends on the scheme. This discrete representation, in addition, restricts the energy range of the problem [215].

The exact solution of eq. (6.21) is given by

$$u(t) = \mathrm{e}^{-itH}\, u_0, \tag{6.22}$$

but to compute the exponential of the $M \times M$ complex and full matrix $-itH$ (typically also of large norm) by diagonalizing H is not always the best procedure, especially for large values of M. In such cases one turns to time-stepping methods, advancing the approximate solution from time t_n to $t_{n+1} = t_n + h$. The goal is then to construct an approximation $u_{n+1} \approx u(t_{n+1}) = \mathrm{e}^{-ih\,H} u(t_n)$ as a map $u_{n+1} = \psi_h\, u_n$.

Notice that the discrete Hamiltonian H in (6.19) has the form

$$H = \mathcal{F}_M^{-1} D_M \mathcal{F}_M + V_M \equiv T + V,$$

where $V \equiv V_M$ is just a diagonal matrix (associated to the potential $\hat{V}$) and T, related to the kinetic energy $\hat{T}$, is also diagonal in a new representation defined by $\mathcal{F}_M$ (efficiently computable with the FFT algorithm). It turns out that the products $\mathrm{e}^{-ihV} u_n$ and $\mathrm{e}^{-ihT} u_n = \mathcal{F}_M^{-1} \mathrm{e}^{-ihD_M} \mathcal{F}_M u_n$ can be easily and efficiently computed, so that one may consider the Strang splitting scheme, which in this setting reads

$$u_{n+1} = S_h^{[2]} u_n \equiv \mathrm{e}^{-i\frac{h}{2}V}\, \mathrm{e}^{-ihT}\, \mathrm{e}^{-i\frac{h}{2}V}\, u_n. \tag{6.23}$$

When the Strang splitting in this way is combined with the Fourier collocation approach in space, the resulting method is often called the *split-step Fourier method* in the literature. It was first used for the nonlinear Schrödinger equation in [179], for the Fresnel equation in linear optics in [148], and for the (linear) Schrödinger equation in [140]. Algorithmically, the method proceeds from t_n to $t_{n+1} = t_n + h$ as follows:

$$\begin{array}{ll}
u \leftarrow \mathrm{e}^{-i\frac{h}{2}V} u & \\
u \leftarrow \mathcal{F}_M u & \text{(by the FFT algorithm)} \\
u \leftarrow \mathrm{e}^{-ihD_M} u & \\
u \leftarrow \mathcal{F}_M^{-1} u & \text{(by the inverse FFT algorithm)} \\
u_j \leftarrow \mathrm{e}^{-i\frac{h}{2}V} u. &
\end{array}$$

Error bounds are actually derived in [197], and they indicate that the method is indeed of order two. Specifically, assume that the potential $\hat{V}(x)$ is $\mathcal{C}^5$-smooth, periodic and bounded, $\|\hat{V}\psi\| \leq \beta\|\psi\|$, $\beta > 0$. Then, if $u(x,t)$ denotes the trigonometric interpolation polynomial of the pseudo-spectral method and $u_n(x)$ is the corresponding trigonometric interpolation polynomial built from the numerical approximations obtained from the Strang splitting (6.23) at time $t = t_n = nh$, one gets for the local and global errors the bounds

$$\begin{aligned}
\|u_1 - u(\cdot, h)\|_{L^2} &\leq C_1 h^3 \|u_0\|_{H^2}, \\
\|u_n - u(\cdot, t_n)\|_{L^2} &\leq C_2 h^2 \|u_0\|_{H^2},
\end{aligned} \tag{6.24}$$

respectively. Here, $\|\cdot\|_{H^2}$ denotes the usual Sobolev norm, and the constants C_1, C_2 are independent of the initial data u_0 and the discretization parameters M, n, and h, with $0 \leq t_n \leq t_\mathrm{f}$ for some finite t_f. The case when $\hat{V}$ is time-dependent and bounded for any t has been treated in [126], where new schemes are proposed and analyzed.

These results for the Strang splitting have been extended in [313] to splitting methods of the general form

$$u_{n+1} = \prod_{j=1}^{s} \mathrm{e}^{-ib_jhV}\,\mathrm{e}^{-ia_jhT}\,u_n = \mathrm{e}^{-ib_shV}\,\mathrm{e}^{-ia_shT}\cdots\mathrm{e}^{-ib_1hV}\,\mathrm{e}^{-ia_1hT}\,u_n \tag{6.25}$$

whose coefficients a_j, b_j satisfy the order conditions up to order r. In that case

$$\|u_n - u(\cdot,t_n)\|_{L^2} \leq C\|u(\cdot,0) - u_0\|_{L^2} + Ch^r\|u(\cdot,0)\|_{H^r}, \qquad 0 \leq t_n \leq t_\mathrm{f} \tag{6.26}$$

is valid with some constant C depending on t_f, but not on n and h. This error bound implies, in particular, that the splitting methods collected in section 3.7.1 retain their order of convergence when applied to the Schrödinger equation with periodic boundary conditions, provided that the initial conditions are sufficiently differentiable (see also [177]). Otherwise, an order reduction may occur.

One can also take advantage of the properties (6.11)-(6.12) and include the commutator $[V,[T,V]]$ in the composition (6.25), as in RKN splitting methods, so that one ends up with

$$u_{n+1} = \prod_{j=1}^{s} \mathrm{e}^{-i(b_jhV - c_jh^3[V,[T,V]])}\,\mathrm{e}^{-ia_jhT}\,u_n. \tag{6.27}$$

Even in this case the resulting schemes retain their order of convergence if the solution is sufficiently regular, as shown in [201]. It is then possible to apply also the RKN schemes presented in section 3.7.1.

Example 6.1. Let us consider the Schrödinger equation with a quadratic potential,

$$i\frac{\partial\psi}{\partial t}(x,t) = -\frac{1}{2}\frac{\partial^2\psi}{\partial x^2}(x,t) + \frac{1}{2}x^2\,\psi(x,t), \qquad -10 \leq x < 10, \tag{6.28}$$

and initial condition $\psi(x,0) = \psi_0(x) = \sigma\cos(x)\mathrm{e}^{-\frac{1}{2}(x-1)^2}$, where σ is a normalizing constant such that $N_{\psi_0} = 1$. Discretize in space taking $\Delta x = 20/M$ with $M = 128$. Integrate the discretized equation for $t \in [0,10]$ using the Strang splitting and the triple jump composition (3.31) with time steps $h = 1/2^k$, $k = 1,2,\ldots,10$, and measure the error in energy at the final time. Plot the accuracy versus the number of FFT calls (and its inverse). Plot also

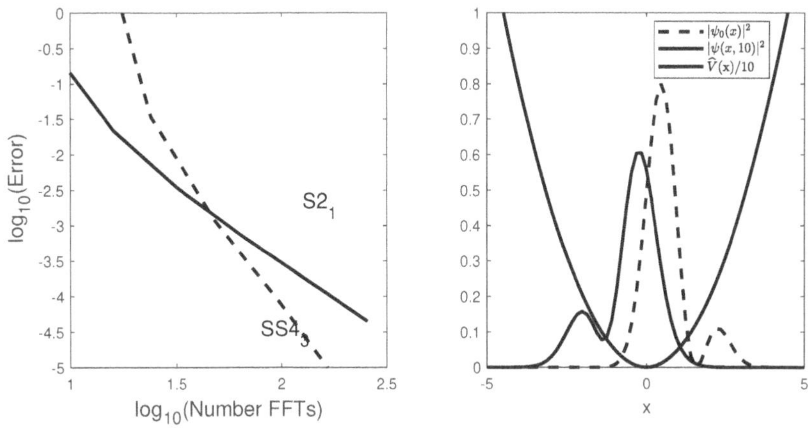

FIGURE 6.1: (Left) Error in energy versus number of FFT calls for the second-order Strang scheme $S2_1$ (6.23) and the triple jump composition $SS4_3$. (Right) Absolute value of the initial and final solutions at $t_f = 10$ and the potential (scaled by a factor of 10) for the Schrödinger equation (6.28).

the initial condition, the final solution, and the potential: $|\psi_0(x)|^2$, $|\psi(x,10)|^2$, and $\hat{V}(x)$.
Solution. Figure 6.1 shows the results obtained. We observe that if medium to high accuracy is desired, high-order splitting methods show the best performance. Notice also that the presence of negative coefficients in the fourth-order scheme does not affect the performance of the algorithm. □

Example 6.2. For the problem given in *Example 6.1* with the same initial condition and spatial discretization, integrate the discretized equation in $t \in [0, 100]$ using the Strang splitting with $h = 5/2^5$, and measure the error in norm and in energy along the time integration interval.
Solution. Figure 6.2 shows the results obtained with this unitary method. We clearly observe that the error in norm is due to round-off accuracy and the error in energy oscillates with no secular growth, in a similar way as the error in energy for symplectic methods when applied to classical Hamiltonian systems. □

Separating H into kinetic and potential energy is not the only possible splitting, however. If H in eq. (6.19) is a real symmetric matrix (as is often the case), then e^{-itH} is not only unitary but also symplectic with canonical coordinates $q = \mathrm{Re}(u) \in \mathbb{R}^M$ and momenta $p = \mathrm{Im}(u) \in \mathbb{R}^M$. In consequence, equation (6.21) is equivalent to [163, 164]

$$\dot{q} = Hp, \qquad \dot{p} = -Hq, \tag{6.29}$$

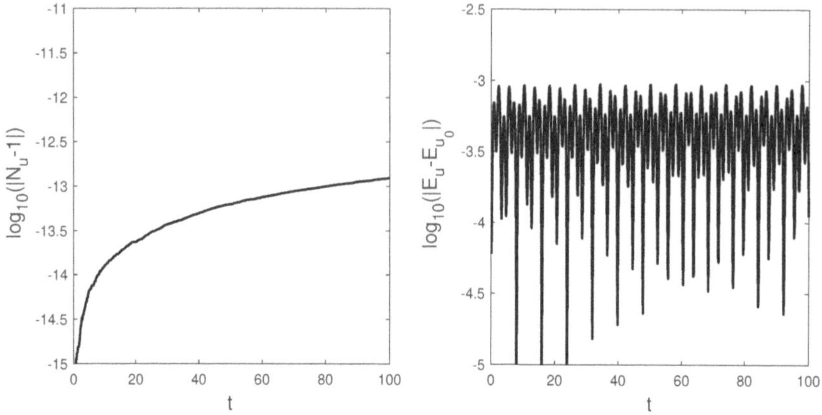

FIGURE 6.2: Error in norm (left) and in energy (right) for the Strang splitting (6.23) in the interval $t \in [0, 100]$ with time step size $h = 5/2^5$ for the Schrödinger equation with quadratic potential (6.28).

whose solution is given by

$$\begin{pmatrix} q(t) \\ p(t) \end{pmatrix} = O(tH) \begin{pmatrix} q_0 \\ p_0 \end{pmatrix}, \tag{6.30}$$

where

$$O(tH) = \begin{pmatrix} \cos(tH) & \sin(tH) \\ -\sin(tH) & \cos(tH) \end{pmatrix} \tag{6.31}$$

is an orthogonal and symplectic $2M \times 2M$ matrix. Computing $O(tH)$ exactly by diagonalizing the matrix H presents the same problems as its complex representation e^{-itH}, so that one proceeds in the usual step-by-step way by dividing the whole time interval into subintervals of length h and then approximating $O(hH)$ acting on the initial condition at each step $(q_n, p_n)^T$. In this respect, notice that if we introduce the nilpotent matrices A and B,

$$A = \begin{pmatrix} 0 & H \\ 0 & 0 \end{pmatrix}, \qquad B = \begin{pmatrix} 0 & 0 \\ -H & 0 \end{pmatrix}, \tag{6.32}$$

then

$$e^{ha_k A} = \begin{pmatrix} I & a_k hH \\ 0 & I \end{pmatrix}, \qquad e^{hb_k B} = \begin{pmatrix} I & 0 \\ -b_k hH & I \end{pmatrix},$$

and splitting methods of the form

$$\begin{pmatrix} q_{n+1} \\ p_{n+1} \end{pmatrix} = e^{ha_{s+1}A}\, e^{hb_s B}\, e^{ha_s A} \cdots e^{hb_1 B}\, e^{ha_1 A} \begin{pmatrix} q_n \\ p_n \end{pmatrix} \tag{6.33}$$

constitute a natural choice to construct high-order approximations. In fact, for a given k, there is only one independent condition to increase the order from $r = 2k - 1$ to $r = 2k$, and only two to increase the order from $r = 2k$ to $r = 2k + 1$ [48]. In this way, one can achieve order r ($r = 2, 4, 6, 8, 10$ and 12) with $s = r$ exponentials $e^{ha_j A}$ and $e^{hb_j B}$ [163]. Notice that the evaluation of the exponentials of A and B only requires computing the products Hp and Hq, and this can be done with the FFT algorithm for real vectors (instead of complex vectors). Although these methods are neither unitary nor unconditionally stable, they are symplectic and conjugate to unitary schemes for sufficiently small values of h. In consequence, neither the average error in energy nor the norm of the solution increases with time. In other words, the error in norm and in energy are both approximately preserved along the evolution, since the committed error is only local and does not propagate with time [50].

Although initially motivated by the time integration of the Schrödinger equation, methods of the form (6.33) can be generalized in several ways, and in particular, to construct an algorithm to approximate $e^{-itH}v$ for any real symmetric matrix H and any complex vector v by only carrying out matrix-vector products of the form Hv [51].

The processing technique has also been used to construct splitting schemes in this setting with two different goals in mind: to attain maximal stability and maximal accuracy. They have the general structure

$$P^{-1}(hH) \cdot K(hH) \cdot P(hH),$$

where K (the kernel) is built as a composition (6.33) with a large number of stages s and P (the processor) is taken as a polynomial. Thus, in [47, 48] kernels with up to 19, 32 and 38 stages are proposed, either to construct methods of orders $r = 10$, 16 and 20, or to bring highly accurate second-order methods with a much enlarged stability domain.

This approach to approximate $e^{-ihH}u$ is closely related to other polynomial approximations of the form

$$e^{-ihH}u \approx P_m(hH)u, \tag{6.34}$$

where $P_m(y)$ is a polynomial in y approximating the exponential e^{-iy}. There are different choices for such a polynomial. For instance, one may consider a truncated Chebyshev series expansion of e^{-iy} for an appropriate real interval of y or the Lanczos method, where the polynomial is determined by a Galerkin approximation on the Krylov space spanned by $u_n, Hu_n, \ldots, H^{m-1}u_n$ (and thus a different polynomial is implicitly selected in every time step) [228]. Both approximations have quite similar error bounds (see [228]), and their relative performance depends on the particular problem considered.

In contrast with splitting methods based on the separation of T and V, eq. (6.25), which preserve unitarity by construction and are thus unconditionally stable, the previous polynomial approximations suffer from a step size restriction. Given h (the time step size), Δx (the space grid size), and m (the degree of the polynomial), these methods must satisfy the restriction

$$\frac{h}{m\Delta x^2} \leq C,$$

for a certain constant C depending on the actual method. In this way, the time interval that one can advance per matrix-vector product is proportional to Δx^2 or, equivalently, the number of matrix-vector products to reach the final time is inversely proportional to Δx^2.

In the previous treatment, we have assumed that the potential in (6.10) does not depend explicitly on time or, if it does, only varies slowly with time, so that in each sub-interval $[t_n, t_{n+1}]$ the corresponding matrix V is obtained from the average of $\hat{V}(x,t)$ on this interval. In general, however, one has to deal with situations in which this approximation is no longer valid. In that case, and in contrast with other approaches based on Chebyshev or Lanczos approximations, splitting methods can still be applied with some appropriate modifications [52]. Magnus integrators and CFQM methods also constitute a standard choice in this case.

6.2.3 Splitting methods for non-linear Schrödinger equations

Time-splitting methods, including extensions involving modified potentials, have also been applied to the time integration of the nonlinear Schrödinger equation [20, 314, 315, 44]. Their long-term behavior has been examined by deriving the associated modified equation and analyzing its analytical characteristics [124, 139, 155].

As an illustrative example, we consider next the important case of the Gross–Pitaevskii equation in one dimension,

$$i\frac{\partial\psi}{\partial t}(x,t) = -\frac{1}{2\mu}\frac{\partial^2\psi}{\partial x^2}(x,t) + \big(\hat{V}(x) + g|\psi(x,t)|^2\big)\psi(x,t) \tag{6.35}$$

with asymptotic boundary conditions $\psi(x,t) \to 0$ as $|x| \to \infty$. It is used as a model of the Bose–Einstein condensate of an atomic species trapped in the external potential $\hat{V}(x)$, with the parameter g originating from the mean-field interaction between the particles: repulsive forces lead to $g > 0$, whereas $g < 0$ represents attractive forces.

Equation (6.35) possesses strong conservation properties. In particular, the norm

$$N_\psi^2 = \int_\Omega |\psi(x,t)|^2 dx$$

and the associated energy

$$E_\psi = \int_\Omega \left(\frac{1}{2\mu}\left|\frac{\partial\psi}{\partial x}(x,t)\right|^2 + \hat{V}(x)\,|\psi(x,t)|^2 + \frac{g}{2}\,|\psi(x,t)|^4 \right) dx \tag{6.36}$$

are both preserved for all times where the solution is defined. The Laplacian term can be solved as in the linear case, whereas for the nonlinear part we may use the following result.

Given a real-valued function F, the solution of

$$i\frac{\partial\psi}{\partial t}(x,t) = F(x,|\psi(x,t)|)\psi(x,t) \tag{6.37}$$

verifies $|\psi(x,t)| = |\psi(x,0)|$, since

$$\frac{d}{dt}|\psi(x,t)|^2 = \frac{d\overline{\psi}(x,t)}{dt}\psi(x,t) + \overline{\psi}(x,t)\frac{d\psi(x,t)}{dt} = 0$$

due to the fact that F is real. In consequence, the solution of (6.37) is

$$\psi(x,t) = \mathrm{e}^{-itF(x,|\psi(x,0)|)}\psi(x,0), \tag{6.38}$$

and the initial value problem

$$i\frac{\partial\psi}{\partial t}(x,t) = \left(\hat{V}(x) + g|\psi(x,t)|^2\right)\psi(x,t), \qquad \psi(x,0) = \psi_0(x) \tag{6.39}$$

is also solvable, so that time-splitting methods can be used also here after space discretization in a similar way as in the linear case (which corresponds to $g = 0$ in (6.35)).

Example 6.3. Take the Gross–Pitaevskii equation (6.35) with $\mu = 1$, a quadratic potential $\hat{V}(x) = \frac{1}{2}x^2$, and coupling constant $g = -1$. For the same initial condition, discretization parameters, time interval, and time-splitting methods as in *Example 6.1*, plot the error in energy versus the number of FFT calls (and its inverse). Plot also the initial condition $|\psi_0(x)|^2$, the final solution $|\psi(x,10)|^2$ and the potential $\hat{V}(x)$. Use the solution of (6.39) as given by (6.38).

Solution. Figure 6.3 shows the results obtained. As in the linear case (see Figure 6.1), high-order splitting methods are more efficient for medium to high accuracy. □

6.3 Splitting methods for parabolic evolution equations

Linear parabolic partial differential equations can be properly formulated as an initial value problem in a particular Banach space (i.e., a complete normed space) of functions $\mathcal{X}$,

$$\frac{\partial u}{\partial t}(x,t) = L\,u(x,t), \qquad u(x,0) = u_0(x), \tag{6.40}$$

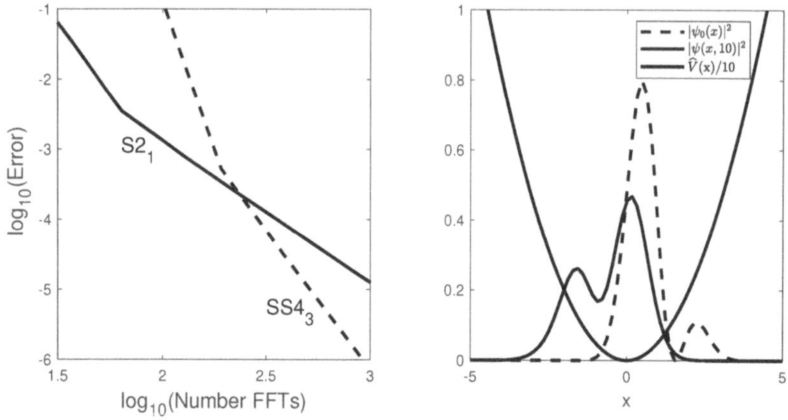

FIGURE 6.3: (Left) Error in energy versus number of FFT calls for the second-order Strang scheme $S2_1$ (6.23) and the triple jump composition $SS4_3$. (Right) Absolute value of the initial and final solutions at $t_f = 10$ and the potential $\hat{V}(x) = \frac{1}{2}x^2$ (scaled by a factor of 10) for the Gross–Pitaevskii equation (6.35) with $\mu = 1$ and $g = -1$.

so that the abstract theory of semigroups and groups of linear operators developed in the functional analysis literature [135, 273, 333] can be readily applied, assuring the existence, uniqueness and regularity of the solution u of (6.40). This requires previously showing that the differential operator L is indeed the infinitesimal generator of a C_0 semigroup $T(t)$ on $\mathcal{X}$. In that case, if $u_0 \in \mathcal{D}(L)$, the domain of L (which is dense in $\mathcal{X}$), then $u(t) = T(t)u_0$ is a classical solution of (6.40) [272]. Since in the special case of a bounded linear operator L the solution is given by the familiar expression $u(t) = \mathrm{e}^{tL}u_0$ [135], the semigroup $T(t)$ is also denoted by the symbol e^{tL}; see [135, 273, 333] for an introduction to the theory of C_0 semigroups.

The problem we are mainly interested in is when the linear operator L is the sum of two linear, possibly unbounded, operators A and B, so that the differential equation (6.40) reads

$$u_t(x,t) = L\,u(x,t) = (A+B)\,u(x,t), \qquad u(x,0) = u_0(x) \in \mathcal{X} \tag{6.41}$$

and the operators A, B and L are the infinitesimal generators of C_0 semigroups on $\mathcal{X}$, denoted by e^{tA}, e^{tB} and e^{tL}, respectively. Moreover, we assume that the C_0 semigroups e^{tA} and e^{tB} can be obtained in a simpler way than $\mathrm{e}^{t(A+B)}$.

A prototypical example is the linear heat equation with potential

$$u_t(x,t) = \frac{1}{2}\Delta u(x,t) - \hat{V}(x)u(x,t). \tag{6.42}$$

Here $\hat{V}(x) \geq 0$, $t \geq 0$ and $x \in \mathbb{R}^d$ (or $x \in \mathbb{T}^d$). In that case $(Au)(x) = \frac{1}{2}\Delta u(x)$, $(Bu)(x) = -\hat{V}(x)u(x)$ and A generates only a C_0 semigroup [135].

In this setting, one is naturally inclined to consider similar strategies as in ordinary differential equations defined in $\mathbb{R}^d$ for obtaining approximate solutions of (6.41) on a given time interval $t \in [0, t_f]$: consider N subdivisions of size $h \equiv \Delta t$ and apply splitting methods of the form

$$\Psi(h) = \prod_{j=1}^{s} \mathrm{e}^{b_j hB}\, \mathrm{e}^{a_j hA} \tag{6.43}$$

so that the $\Psi(h)u_0$ approximates the exact solution $u(h) = \mathrm{e}^{hL}u_0$ of the problem. Here, for clarity, we have omitted the dependence on x of the solution u.

As we know, in the finite-dimensional case, the coefficients a_i, b_i are chosen in such a way that the Taylor expansions in h of $\Psi(h)$ and e^{hL} agree up to terms h^r, in which case method (6.43) is said to be of (classical) order r. A natural question then arises: under which conditions (if any) does a splitting method (6.43) of classical order r provide a *convergent* approximation to the classical solution of (6.41) of the same order?

This problem has been analyzed in [177], establishing the following result. Assuming that $\|\mathrm{e}^{tA}\| \le \mathrm{e}^{\omega t}$, $\|\mathrm{e}^{tB}\| \le \mathrm{e}^{\omega t}$ for the same value of $\omega \ge 0$ and all $t \ge 0$, and that for any operator E_{r+1} obtained as the product of exactly $r+1$ factors chosen amongst A and B, there is a constant $\tilde{C} > 0$ such that

$$\|E_{r+1}\mathrm{e}^{t(A+B)}u_0\| \le \tilde{C},$$

then the splitting method (6.43) with all $a_j \ge 0$ and $b_j \ge 0$ and of classical order r retains its order when applied to (6.41):

$$\|(\Psi(h)^n - \mathrm{e}^{nhL})u_0\| \le Ch^r, \qquad \text{for} \qquad n\,h \le t_f, \tag{6.44}$$

where the constant C is independent of n and h on the bounded time interval $[0, t_f]$.

In practice, however, the positivity requirement on the coefficients restricts the splitting method to be of order two at most. If, in addition, $[B, [A, B]]$ is a bounded operator and $\|\mathrm{e}^{t[B,[A,B]]}\| \le \mathrm{e}^{\omega t}$, then the same result (6.44) also holds for splitting methods involving double commutators,

$$\Psi(h) = \prod_{j=1}^{s} \mathrm{e}^{b_j hB - c_j h^3[B,[A,B]]}\, \mathrm{e}^{a_j hA}$$

and positive a_j, b_j [201]. Efficient schemes within this class up to order 4, specially tailored to the problem at hand, have been proposed in [43].

An important application of splitting methods for this type of problem arises in connection with the imaginary-time propagation method. This procedure can be used to compute the ground state (and its corresponding eigenvalue) of a quantum system with Hamiltonian $H = -\frac{1}{2}\Delta + \hat{V}$ [18, 216, 21].

Essentially, under the time transformation $t = -is$, the time-dependent Schrödinger equation (6.17) is transformed into

$$\frac{\partial \psi}{\partial s}(x,s) = \frac{1}{2}\frac{\partial^2 \psi}{\partial x^2}(x,s) - \hat{V}(x)\psi(x,s), \qquad \psi(x,0) = \psi_0(x), \tag{6.45}$$

that is, a linear heat equation of the form (6.42). If we denote the (real) eigenvalues of H as E_j, with $E_0 < E_1 < \cdots$, and the corresponding eigenfunctions as ϕ_j, $j = 0, 1, 2, \ldots$, the initial wave function $\psi_0(x)$ can be expanded in the orthonormal basis $\{\phi_j\}$,

$$\psi_0(x) = \sum_{j\geq 0} c_j\, \phi_j(x), \qquad c_j = \langle \phi_j \,|\, \psi(\cdot, 0)\rangle,$$

where $\langle \cdot \,|\, \cdot \rangle$ is the usual L^2-scalar product. Then the solution of (6.45) can be written as

$$\psi(x,s) = \mathrm{e}^{-sH}\psi(x,0) = \sum_{j\geq 0} \mathrm{e}^{-sE_j}\, c_j\, \phi_j(x). \tag{6.46}$$

Notice that, for sufficiently large s, one gets $\psi(x,s) \to \mathrm{e}^{-sE_0} c_0 \phi_0$, since the other exponentials decay more rapidly. In other words, any given wave function at $s = 0$ for which $c_0 \neq 0$ converges towards the ground state solution when $s \to \infty$. Once an accurate approximation to ϕ_0 is obtained, the associated eigenvalue E_0 is easily obtained by computing $E_0 = \langle \phi_0 \,|\, H\phi_0 \rangle$. Other functions ϕ_j can also be approximated, e.g., by propagating different wave functions simultaneously in time [5].

Example 6.4. To illustrate this technique, we next consider the same problem as in *Example 6.1*, but now integrating in imaginary time in the interval $s \in [0, 20]$. Equivalently, we integrate numerically the diffusion equation

$$\frac{\partial \psi}{\partial s}(x,s) = \frac{1}{2}\frac{\partial^2 \psi}{\partial x^2}(x,s) - \frac{1}{2}x^2\, \psi(x,s), \qquad -10 \leq x < 10 \tag{6.47}$$

using the Strang splitting and the triple jump composition (3.31) with time steps $\Delta s = 1/2^k$, $k = 1, 2, \ldots, 10$. The numerical solution converges (for nearly all initial conditions) to the function $\phi_0(x) = \sigma \mathrm{e}^{-\frac{1}{2}x^2}$ describing the ground state. Here, σ is a normalizing constant. We take as initial conditions $\psi_0(x) = \sigma \cos(x)\mathrm{e}^{-\frac{1}{2}(x-1)^2}$ and compute the numerical solution and norm of the wave function after each time step, i.e., $\tilde{\psi}(x, s_n + \Delta s) = \frac{1}{\|\psi\|}\psi(x, s_n + \Delta s)$. We take $\tilde{\psi}(x, s_n + \Delta s)$ as the initial condition at the next step and repeat the process. At the final time $s = 20$, we compute the error

$$E_g = \|\hat{\phi}_0(x) - \tilde{\psi}(x, 20)\|$$

and plot $\psi_0(x), \psi(x, 20)$ and $\hat{V}(x)$. Here $\hat{\phi}_0(x)$ stands for the function $\phi_0(x)$ computed at the grid points. The experiment is carried out with two different space grids, corresponding to $M = 128$ and $M = 256$.

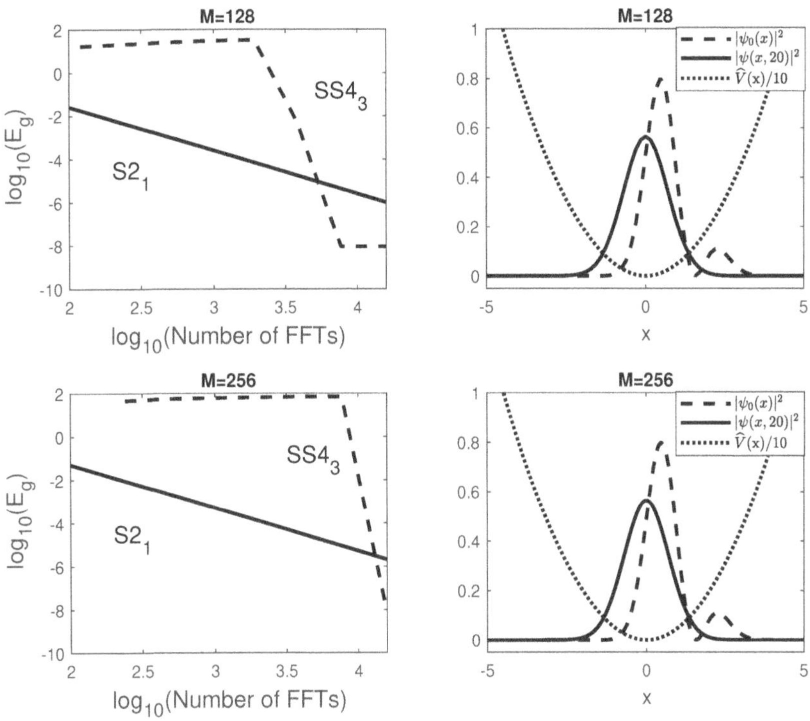

FIGURE 6.4: Numerical integration of the Schrödinger equation in imaginary time with a quadratic potential, eq. (6.47) using a mesh $\Delta x = 20/M$ with $M = 128$ (top figures) and $M = 256$ (bottom figures). (Left) Error in the solution describing the ground state of the system versus number of FFT calls for the second-order Strang scheme $S2_1$ (6.23) and the triple jump composition $SS4_3$. (Right) Absolute value of the initial and final solutions at $s_f = 20$ and the potential (scaled by a factor of 10).

Solution. Figure 6.4 shows the corresponding results. The numerical solution converges to the ground state as observed in the right panels and does so with the order of accuracy of the corresponding method. Notice the effect of the negative coefficient in scheme (3.31): unless it is used with a tiny step size, the numerical result becomes unstable [21], and this effect becomes more pronounced with increasing spatial mesh resolution. □

One possible way to circumvent this order barrier for splitting methods when applied to problem (6.41) consists in taking schemes with *complex* coefficients a_i, b_i having *positive* real parts. As shown in [96, 178], any splitting method $\Psi(h) = \prod_{j=1}^{s} \mathrm{e}^{b_j h B}\, \mathrm{e}^{a_j h A}$ within this class still retains its classical order if the previous assumptions are conveniently modified. Specifically, one has to extend the notion of a C_0 semigroup $T(t)$ to the sector Σ_θ in the complex plane for some angle $0 < \theta < \pi/2$,

$$\Sigma_\theta = \{t \in \mathbb{C} : |\arg(t)| < \theta\},$$

so that $T(t)$ is analytic in t for all $t \in \Sigma_\theta$ [135, 273]. Thus, if

- L, A and B generate analytic semigroups on $\mathcal{X}$ (now a complex Banach space) in the sector Σ_θ, $0 < \theta < \pi/2$, with

$$\|e^{tA}\| \leq e^{\omega|t|}, \qquad \|e^{tB}\| \leq e^{\omega|t|}$$

  for $\omega \geq 0$ and all $t \in \Sigma_\theta$;

- $\|E_{r+1} e^{t(A+B)} u_0\| \leq C$, where E_{r+1} is a composition of the operators A and B that consist of exactly $r+1$ factors,

then the splitting method of classical order r with coefficients $a_i, b_i \in \Sigma_\theta \subset \mathbb{C}$ satisfies

$$\|(\Psi(h)^n - e^{nhL})u_0\| \leq Ch^r, \qquad 0 \leq nh < t_f. \tag{6.48}$$

Since $\Psi(h)u_0$ is complex valued, this approach cannot be applied in principle when the operators A and B are real, i.e., for problems defined in a real Banach space $\mathcal{X}$. The most straightforward remedy consists of projecting the numerical solution after each time step on the real axis, i.e., computing the approximations $u_n \approx u(t_n)$, as $u_n = \mathrm{Re}(\Psi(h)u_{n-1})$. In that case one still has the bound (6.48) for the resulting integration scheme [178].

Although this is only valid in the linear case, similar results are observed in practice in the nonlinear heat equation $\partial_t u = \Delta u + F(u)$ with periodic boundary conditions, at least when F is analytic [96, 39], and so they constitute a strong motivation to study splitting methods with complex coefficients in general. This is precisely the subject of the following section.

6.4 Splitting and composition methods with complex coefficients

It has been known for a long time that, besides real solutions, the order conditions arising from splitting and composition methods (see section 3.3) also admit complex solutions [26, 310, 311, 312]. In fact, some of the resulting methods were explored in the context of Hamiltonian [101] and quantum mechanics [25, 276] before being used more recently in parabolic partial differential equations [96, 178] as a strategy to develop effective high-order methods.

We have already seen at the end of the previous section that a strategy to develop effective methods of order higher than two for systems evolving in a semigroup, such as the heat equation[1], consists precisely in applying splitting schemes with complex coefficients having a positive real part. Moreover, the large number of complex solutions for the order conditions might offer more flexibility in the final choice of coefficients and perhaps lead to schemes with smaller truncation errors and new symmetries.

For problems formulated in the real line, however, one should be aware that the use of complex arithmetic introduces an additional computational cost with respect to methods with purely real coefficients. From a theoretical point of view, the vector field appearing in the differential equation has to be analytic at least in a domain containing the path where the actual integration is carried out in the complex plane. Otherwise, order reductions are to be expected unless the implementation is conveniently adapted.

In this section we review several issues involved in the construction and analysis of splitting methods with complex coefficients and also summarize some of their most salient properties with regard to preservation of qualitative properties, both in classical Hamiltonian and quantum systems.

For brevity, we denote in the sequel a complex number a with a positive real part as $a \in \mathbb{C}_+$, so that we are interested in methods whose coefficients $a_j, b_j \in \mathbb{C}_+$.

6.4.1 Compositions

Most of the existing splitting methods with complex coefficients have been constructed by applying the composition technique of subsection 3.2. Here, for completeness, we review the process, starting with the Lie–Trotter scheme $\chi_h = \varphi_h^{[1]} \circ \varphi_h^{[2]}$ as the basic method and composing it with different weights. In the simplest case, the scheme

$$\phi_h^{[2]} \equiv \chi_{\gamma_{1,2}h} \circ \chi_{\gamma_{1,1}h}, \tag{6.49}$$

is of order 2 if the coefficients satisfy equation (3.37), that is,

$$\gamma_{1,1} = \frac{1}{2} \pm \frac{1}{2}i, \qquad \gamma_{1,2} = \overline{\gamma}_{1,1} = \frac{1}{2} \mp \frac{1}{2}i.$$

If we try to construct a method of order three, a composition of at least four maps χ_h is needed. Although there are eight solutions (+cc), only four have a positive real part. In fact, two of these solutions result from composing $\phi_h^{[2]}$, namely

$$\phi_h^{[3]} = \phi_{\gamma_{2,2}h}^{[2]} \circ \phi_{\gamma_{2,1}h}^{[2]} \tag{6.50}$$

by requiring that $\gamma_{2,1} + \gamma_{2,2} = 1$, $\gamma_{2,1}^3 + \gamma_{2,2}^3 = 0$. Notice that in (6.50) $\gamma_{2,2} = \overline{\gamma}_{2,1}$. This procedure can in principle be repeated by considering the recurrence

$$\phi_h^{[r+1]} = \phi_{\gamma_{r,2}h}^{[r]} \circ \phi_{\gamma_{r,1}h}^{[r]}, \qquad r = 1, 2, \ldots \tag{6.51}$$

[1]This constitutes in fact Problem 10 in the list of open problems posed in [243].

with $\gamma_{r,1} + \gamma_{r,2} = 1$, $\gamma_{r,1}^{r+1} + \gamma_{r,2}^{r+1} = 0$, to construct higher-order methods with *any* first-order integrator $\phi_h^{[1]}$ (not necessarily the Lie–Trotter scheme). It turns out, however, that $\phi_h^{[4]}$ and higher-order schemes obtained from (6.51) have at least one coefficient with a negative real part [39]. Thus, if one is interested in schemes having only coefficients in $\mathbb{C}_+$, $r = 3$ constitutes indeed an order barrier for this type of composition [178].

Let us now consider the Strang splitting as the basic method. Then, composition (3.29) with $s = 2$ already provides a method of order three,

$$S_h^{[3]} = S_{\gamma_2 h}^{[2]} \circ S_{\gamma_1 h}^{[2]}, \qquad \text{with} \qquad \gamma_1 = \frac{1}{2} \pm i\frac{\sqrt{3}}{6}, \qquad \gamma_2 = \overline{\gamma}_1. \tag{6.52}$$

Again, higher-order methods can be obtained by applying the recursive procedure

$$S_h^{[r+1]} = S_{\gamma_{r,2} h}^{[r]} \circ S_{\gamma_{r,1} h}^{[r]}, \qquad r = 2, 3, \ldots, \tag{6.53}$$

again with $\gamma_{r,1} + \gamma_{r,2} = 1$, $\gamma_{r,1}^{r+1} + \gamma_{r,2}^{r+1} = 0$. The solution with the smallest phase is given by

$$\gamma_{k,1} = \frac{1}{2} + \frac{i}{2} \tan\left(\frac{\pi}{2(r+1)}\right), \qquad \gamma_{r,2} = \overline{\gamma}_{r,1}, \tag{6.54}$$

but only schemes up to order 6 with coefficients in $\mathbb{C}_+$ are possible with this approach [178].

As we have already seen, the triple jump (3.32) allows one to raise the order by two. Thus, starting from a time-symmetric second-order method $S_h^{[2]}$, the composition

$$S_h^{[4]} = S_{\gamma_3 h}^{[2]} \circ S_{\gamma_2 h}^{[2]} \circ S_{\gamma_1 h}^{[2]}, \tag{6.55}$$

has, apart from the real solution (3.31),

$$\gamma_1 = \gamma_3 = \frac{1}{2 - 2^{1/3} e^{2i\ell\pi/3}}, \qquad \gamma_2 = 1 - 2\alpha_1, \qquad \ell = 1, 2, \tag{6.56}$$

corresponding to a time-symmetric composition, and

$$\gamma_1 = \overline{\gamma}_3 = \frac{1}{4} \pm i\frac{1}{4}\sqrt{\frac{5}{3}}, \qquad \gamma_2 = \frac{1}{2}, \tag{6.57}$$

possessing the same symmetry as (6.50) and (6.52). Palindromic methods up to order 8 with coefficients having a positive real part are possible by applying the triple jump composition, whereas if one considers instead the *quadruple jump*,

$$S_{\gamma_{k,1} h}^{[2k]} \circ S_{\gamma_{k,2} h}^{[2k]} \circ S_{\gamma_{k,2} h}^{[2k]} \circ S_{\gamma_{k,1} h}^{[2k]},$$

one can achieve order 14 with all coefficients in $\mathbb{C}_+$. These order barriers have been rigorously proved in [39].

6.4.2 Palindromic and symmetric-conjugate splitting methods

Compositions (6.52) and (6.55) can be reformulated as splitting methods when $S_h^{[2]}$ is the Strang splitting. Thus,

$$S_h^{[3]} = \varphi_{\bar{a}_1 h}^{[1]} \circ \varphi_{\bar{b}_1 h}^{[2]} \circ \varphi_{a_2 h}^{[1]} \circ \varphi_{b_1 h}^{[2]} \circ \varphi_{a_1 h}^{[1]}, \tag{6.58}$$

with $a_1 = \frac{1}{4} + i\frac{\sqrt{3}}{12}$, $b_1 = \frac{1}{2} + i\frac{\sqrt{3}}{6}$, $a_2 = \frac{1}{2}$, whereas the palindromic version of (6.55) reads

$$\psi_{h,p}^{[4]} = \varphi_{a_1 h}^{[1]} \circ \varphi_{b_1 h}^{[2]} \circ \varphi_{a_2 h}^{[1]} \circ \varphi_{b_2 h}^{[2]} \circ \varphi_{a_2 h}^{[1]} \circ \varphi_{b_1 h}^{[2]} \circ \varphi_{a_1 h}^{[1]}, \tag{6.59}$$

with $a_1 = \frac{\gamma_1}{2}$, $b_1 = \gamma_1$, $a_2 = \frac{1}{2} - a_1$, $b_2 = 1 - 2b_1$. Finally, solution (6.57) leads to

$$\psi_{h,sc}^{[4]} = \varphi_{\bar{a}_1 h}^{[1]} \circ \varphi_{\bar{b}_1 h}^{[2]} \circ \varphi_{\bar{a}_2 h}^{[1]} \circ \varphi_{b_2 h}^{[2]} \circ \varphi_{a_2 h}^{[1]} \circ \varphi_{b_1 h}^{[2]} \circ \varphi_{a_1 h}^{[1]}, \tag{6.60}$$

with

$$a_1 = \frac{b_1}{2}, \qquad b_1 = \frac{1}{4} \pm i\frac{1}{4}\sqrt{\frac{5}{3}}, \qquad a_2 = \frac{1}{4}(2b_1 + 1), \qquad b_2 = \frac{1}{2}.$$

Notice the different pattern in the distribution of the coefficients: whereas $\psi_{h,p}^{[4]}$ is a *palindromic* (and hence time-symmetric or time-reversible) composition, coefficients in $S_h^{[3]}$ and $\psi_{h,sc}^{[4]}$ are symmetric in the real part and antisymmetric in the imaginary part.

Example 6.5. Given the simple two-level quantum system described by [247]

$$i\frac{dU}{dt} = HU = (\sigma_1 + \sigma_2)U, \quad \text{with} \quad \sigma_1 = \begin{pmatrix} 0 & 1 \\ 1 & 0 \end{pmatrix}, \ \sigma_2 = \begin{pmatrix} 0 & -i \\ i & 0 \end{pmatrix} \tag{6.61}$$

and $U(0) = I$, we test both the preservation of unitarity in the numerical approximations and the computational efficiency of the previous third- and fourth-order schemes. To this end, we take $\varphi_h^{[j]} = \mathrm{e}^{-ih\sigma_j}$, $j = 1, 2$, and compare with the exact solution $\mathrm{e}^{-ih(\sigma_1+\sigma_2)}$. Notice that, for problems formulated in complex variables, the use of complex coefficients does not increase in general the overall cost of the algorithm.

We fix $t_{\mathrm{f}} = 1000$ as the final time, and adjust h (and therefore the number of iterations n) so that all the methods require the same computational effort. Specifically, $n = 6000$ ($h = 1/6$) for scheme (6.58) and $n = 4000$ ($h = 1/4$) for the fourth-order methods (6.59) and (6.60). Finally, we compute $|\|U_{\mathrm{app}}(jh)\| - 1|$, $j = 1, 2, \ldots, n$, where $U_{\mathrm{app}}(jh)$ denotes the approximate solution after j steps, which is shown in Figure 6.5 (left). Dashed (red) line corresponds to scheme (6.59), the solid (black) line is produced by (6.60), and finally the dashed-dotted line is the result obtained with the third-order scheme (6.58).

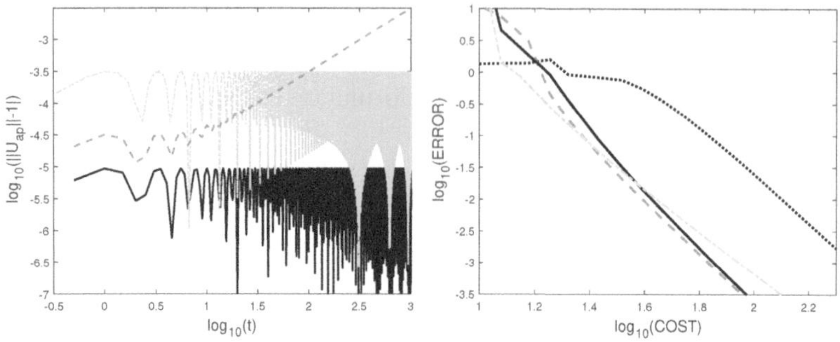

FIGURE 6.5: Left: evolution of error in unitarity for system (6.61) integrated with third- and fourth-order splitting methods with complex coefficients. Right: efficiency diagram computed for a final time $t_f = 10$. In addition to the methods depicted in the left graph, we also include for comparison the triple jump with real coefficients (dotted line). Different patterns in the coefficients lead to a different qualitative behavior. In particular, the palindromic scheme (6.59) exhibits an error in unitarity that increases over time.

The different behavior of the integrators is apparent: whereas the error in the preservation of unitarity does not grow with time for methods (6.60) and (6.58), it certainly does for the palindromic scheme (6.59).

Figure 6.5 (right) shows an efficiency diagram of these integrators, together with the classical triple jump with real coefficients (dotted line). It is produced by integrating eq. (6.61) with different values of h and computing the error of the approximation (in the 2-norm) at the final time $t_f = 10$ as a function of the computational cost (estimated as the number of exponentials involved in the whole integration). We observe the surprisingly good performance of scheme (6.58) and the superior efficiency of the fourth-order methods with complex coefficients in comparison with their real counterparts. This is related to the difference in size of the main error term in the expansion of the modified operator associated with each fourth-order method, which is typically associated with the coefficient multiplying $\mathcal{E}_5 = |\sum_{j=1}^{3} \gamma_j^5|$. Thus, $\mathcal{E}_5$ is about 200 times smaller for the schemes (6.56) and (6.57) than for the triple jump with real coefficients [40]. □

The previous example shows that the time-symmetric character of a method with complex coefficients does not necessarily guarantee good preservation properties, whereas the situation is different for schemes (6.60) and (6.58). They belong to the general class of methods

$$\psi_h = \varphi_{a_{s+1}h}^{[1]} \circ \varphi_{b_s h}^{[2]} \circ \varphi_{a_s h}^{[1]} \circ \cdots \circ \varphi_{a_2 h}^{[1]} \circ \varphi_{b_1 h}^{[2]} \circ \varphi_{a_1 h}^{[1]} \tag{6.62}$$

whose coefficients verify

$$a_{s+2-j} = \overline{a}_j, \qquad b_{s+1-j} = \overline{b}_j, \qquad j = 1, 2, \ldots \tag{6.63}$$

and are properly called *symmetric-conjugate* [38]. They possess the following distinctive feature: assume our differential equation $\dot{x} = f(x)$ is reversible with respect to complex conjugation,

$$\overline{f(x)} = -f(\overline{x}) \qquad \text{for all } x \in \mathbb{C},$$

and the same holds for each part f_j in $f = f_1 + f_2$. Then scheme (6.62)-(6.63) verifies

$$\overline{\psi}_h = \psi_h^{-1}, \tag{6.64}$$

so that the map ψ_h is also reversible with respect to complex conjugation [170, section V.1]. Property (6.64), *not* satisfied by palindromic compositions involving complex coefficients, has some major consequences. Suppose we are dealing with the linear problem

$$i\frac{dU}{dt} = HU, \qquad U(0) = I, \tag{6.65}$$

where H is a $M \times M$ real matrix of the form $H = A + B$, with A, B also real, so that the symmetric-conjugate method (6.62)-(6.63) reads

$$\psi_h = \mathrm{e}^{-ih\overline{a}_1 A}\, \mathrm{e}^{-ih\overline{b}_1 B}\, \mathrm{e}^{-ih\overline{a}_2 A} \cdots \mathrm{e}^{-iha_2 A}\, \mathrm{e}^{-ihb_1 B}\, \mathrm{e}^{-iha_1 A}. \tag{6.66}$$

It is proved in [30] that if all the eigenvalues of H are real and simple, then, for sufficiently small h, there exist real matrices D_h (diagonal) and P_h (invertible) such that

$$\psi_h^n = P_h\, \mathrm{e}^{-inhD_h} P_h^{-1},$$

all the eigenvalues of the linear map ψ_h have modulus 1 and the norm $|u_n|^2$, and energy $\overline{u}_n^T H u_n$ are almost preserved for large n, with $u_n = \psi_h^n u_0$. In other words, any symmetric-conjugate splitting method ψ_h applied to (6.65) is similar to a unitary method for sufficiently small values of the step size h.

It is worth recalling that the time-dependent Schödinger equation, once discretized in space, leads to an equation of the form (6.65), with both A and B real and symmetric matrices: $A = T$ is associated with the second-order periodic spectral differentiation matrix, and $B = V$ is the (diagonal) matrix corresponding to the potential evaluated at the grid points. It is then admissible to use symmetric-conjugate methods in this setting, since they also guarantee preservation of invariants for sufficiently small values of h. We should notice, however, that both positive and negative imaginary parts are present in the scheme if $a_j \in \mathbb{C}$ (since the method is consistent), and this may lead to severe instabilities due to the unboundedness of the Laplace operator [96, 178]. It seems then appropriate to consider in this setting only symmetric-conjugate splitting methods with $0 < a_j < 1$ and $b_j \in \mathbb{C}_+$. Different schemes of orders 3 up to 6 within this family have been proposed and tested in [30], showing also promising results in terms of efficiency.

6.4.3 Alternating-conjugate splitting methods

We next consider the schemes

$$\widetilde{S}_h^{[3]} = \varphi_{a_1 h}^{[1]} \circ \varphi_{b_1 h}^{[2]} \circ \varphi_{a_2 h}^{[1]} \circ \varphi_{\overline{b}_1 h}^{[2]} \circ \varphi_{\overline{a}_1 h}^{[1]},$$

$$\widetilde{\psi}_{h,p}^{[4]} = \varphi_{\overline{a}_1 h}^{[1]} \circ \varphi_{\overline{b}_1 h}^{[2]} \circ \varphi_{\overline{a}_2 h}^{[1]} \circ \varphi_{\overline{b}_2 h}^{[2]} \circ \varphi_{\overline{a}_2 h}^{[1]} \circ \varphi_{\overline{b}_1 h}^{[2]} \circ \varphi_{\overline{a}_1 h}^{[1]},$$

$$\widetilde{\psi}_{h,sc}^{[4]} = \varphi_{a_1 h}^{[1]} \circ \varphi_{b_1 h}^{[2]} \circ \varphi_{a_2 h}^{[1]} \circ \varphi_{b_2 h}^{[2]} \circ \varphi_{\overline{a}_2 h}^{[1]} \circ \varphi_{\overline{b}_1 h}^{[2]} \circ \varphi_{\overline{a}_1 h}^{[1]},$$

obtained from (6.58)-(6.60) by reversing the sign of the imaginary part of the respective coefficients, and form the fourth-order compositions

$$\mathcal{C}_h^{(1)} \equiv S_{h/2}^{[3]} \circ \widetilde{S}_{h/2}^{[3]}, \qquad \mathcal{C}_h^{(2)} \equiv \psi_{h/2,p}^{[4]} \circ \widetilde{\psi}_{h/2,p}^{[4]}, \qquad \mathcal{C}_h^{(3)} \equiv \psi_{h/2,sc}^{[4]} \circ \widetilde{\psi}_{h/2,sc}^{[4]}. \tag{6.67}$$

Notice that $\mathcal{C}_h^{(2)}$ is now symmetric-conjugate, whereas both $\mathcal{C}_h^{(1)}$ and $\mathcal{C}_h^{(3)}$ are time-symmetric, but all of them are formed by alternating a given composition with the one obtained by reversing the sign of the imaginary part of the coefficients. They are called *alternating-conjugate* splitting methods [31].

Example 6.6. We take problem (6.65) with $M = 10$, generate a random Hermitian matrix $H \in \mathbb{C}^{M\times M}$ with multiple eigenvalues, and decompose H as a sum of two (complex) Hermitian matrices A, B. Then we compute the matrices generated by the previous schemes (6.58)–(6.60) for different values of h, determine their eigenvalues ω_j and compute

$$D_h = \max_{1\le j\le M} |(|\omega_j| - 1)|. \tag{6.68}$$

This quantity is finally plotted as a function of h. The results are shown in Figure 6.6 (left): $D_h > 0$ for any $h > 0$ for both the palindromic (6.59) and the symmetric-conjugate fourth-order splitting method (6.60), whereas $S_h^{[3]}$ behaves as a unitary method for values of h in an interval $(0, h^*)$ for some h^* depending on the particular H taken ($|\omega_j| = 1$ for all j, except round-off). If, on the other hand, the alternating-conjugate compositions (6.67) are considered, then the resulting approximations have all their eigenvalues $|\omega_j| = 1$ for small enough h (Figure 6.6, right). We thus may conclude that concatenating a given method with the same composition with complex conjugate coefficients leads to a new scheme with improved preservation properties, at least for unitary flows. One may wonder why scheme $S_h^{[3]}$ still behaves as a unitary map here: the reason is that it is itself an alternating-conjugate method, as (6.52) clearly shows. □

The most general form of an alternating-conjugate (AC) splitting method $\mathcal{C}_h$ is as follows. Given $m \ge 1$ operators $A_1, \ldots, A_m$ and certain coefficients $\alpha_1, \cdots, \alpha_m \in \mathbb{C}$, we take the compositions

$$\psi_h = \varphi_{h\alpha_1}^{[A_1]} \circ \cdots \circ \varphi_{h\alpha_m}^{[A_m]}, \qquad \widetilde{\psi}_h = \varphi_{h\overline{\alpha}_1}^{[A_1]} \circ \cdots \circ \varphi_{h\overline{\alpha}_m}^{[A_m]}$$

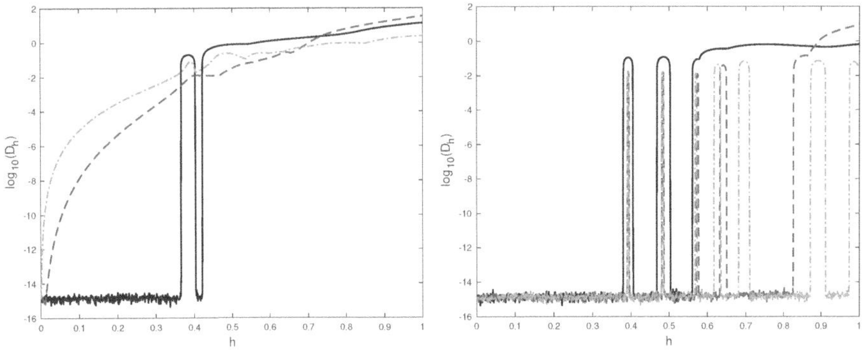

FIGURE 6.6: Left panel: function D_h defined by (6.68) vs. the step size h obtained by schemes (6.58)-(6.60) applied to eq. (6.65) when H is a 10×10 complex Hermitian matrix with multiple eigenvalues. The schemes depicted are $S_h^{[3]}$ (black), $\psi_{h,sc}^{[4]}$ (blue) and $\psi_{h,p}^{[4]}$ (red). Right panel corresponds to the alternating-conjugate version of the previous methods (6.67). Only alternating-conjugate methods behave as unitary integrators in this case.

and then we concatenate them, resulting in

$$\mathcal{C}_h := \psi_h \circ \widetilde{\psi}_h = \varphi_{h\alpha_1}^{[A_1]} \circ \cdots \circ \varphi_{h\alpha_m}^{[A_m]} \circ \varphi_{h\overline{\alpha}_1}^{[A_1]} \circ \cdots \circ \varphi_{h\overline{\alpha}_m}^{[A_m]}. \tag{6.69}$$

Notice that if ψ_h is symmetric-conjugate, $\alpha_{m-j+1} = \overline{\alpha}_j$ and $A_{m-j+1} = A_j$, then $\widetilde{\psi}_h$ is also symmetric-conjugate, and the full method $\mathcal{C}_h$ is palindromic, so that it is also time-symmetric. Particular examples are $\mathcal{C}_h^{(1)}$ and $\mathcal{C}_h^{(3)}$ in (6.67).

The linear case, where each flow corresponds to the exponential of the corresponding matrix, $\varphi_{h\alpha_j}^{[A_j]} = \mathrm{e}^{h\alpha_j A_j}$, has been treated in [31], extending the previous results for symmetric-conjugate methods to linear unitary and Hermitian flows, even when H has multiple eigenvalues. What is more striking, analogous conclusions can be drawn for linear Hamiltonian systems of the form

$$\dot{x} = JSx, \tag{6.70}$$

where $x = (q,p)^T \in \mathbb{R}^{2d}$, J is the basic canonical matrix (1.34), and S is a real symmetric $2d \times 2d$ matrix. The corresponding Hamiltonian function

$$H(x) = \frac{1}{2} x^T S x \tag{6.71}$$

as well as the actions

$$I_j = \frac{1}{2}(q_j^2 + p_j^2), \qquad j = 1, \ldots, d, \tag{6.72}$$

are integrals of the motion, and the dynamics are globally bounded. It turns out that the numerical flow generated by an AC method is also bounded, the numerical solution remains almost real for all times, and the actions and the Hamiltonian function of the original system are almost preserved.

On the other hand, if one applies an alternating-conjugate method of order r to a nonlinear problem defined in a Lie group $\mathcal{G}$, then it is conjugate to a method that preserves the Lie group structure up to order $2r+1$. This is so up to order $2r+2$ if ψ_h is symmetric-conjugate and the order r is even, whereas if ψ_h is symmetric-conjugate and its order r is odd, then the resulting AC method is time-symmetric and of order $r+1$ [31].

6.5 Exercises

1. Show that Yanenko's Crank–Nicolson and the Peaceman–Rachford methods, when applied to the linear system (6.5), i.e., schemes (6.7) and (6.8), are conjugate to each other.

2. Verify identities (6.11) and (6.12).

3. Prove that the norm of a wave function, defined in (6.13), is a constant of motion for the Schrödinger equation and, if the Hamiltonian is autonomous, the energy as defined in (6.14) is also a constant of motion.

4. Verify that if H is a real symmetric $M \times M$ matrix, then $\exp(-itH)$ is symplectic with canonical coordinates $q = \mathrm{Re}(u) \in \mathbb{R}^M$ and momenta $p = \mathrm{Im}(u) \in \mathbb{R}^M$, and that equation $i\dot{u} = Hu$ is equivalent to the pair $\dot{q} = Hp$, $\dot{p} = -Hq$.

5. Repeat the numerical experiments carried out in *Example 6.1* for the one-dimensional Schrödinger equation (6.17) with the Pöschl–Teller potential given by
$$V(x) = -\frac{a^2}{2\mu}\frac{\lambda(\lambda-1)}{\cosh^2(ax)},$$
and take $\mu = 1745$, $a = 2$, $\lambda = 24.5$, and the same initial conditions on the interval $x \in [-10, 10]$.

6. The energy for the ground state of the Pöschl–Teller potential is given by
$$E_k = -\frac{a^2}{2\mu}(\lambda-1)^2.$$
Repeat *Example 6.4* for this problem, but measure the error in the energy for the ground state versus the number of FFTs.

7. Repeat *Example 6.4* with $M = 64$ and 256 and analyze the results.

8. Repeat the experiments leading to Figures 6.1 and 6.4, but include the results obtained using the fourth-order splitting method with modified potentials (3.62), taking into account the double commutator (6.11).

9. Consider the Gross–Pitaevskii equation (6.35) studied in *Example 6.3*. Apply the RKN splitting method $NB4_6$ (see section 3.7.1) to solve this problem and obtain the plot shown in Figure 6.3, including the new results.

10. Show that the coefficients of the composition method (6.53) with the smallest phase are given by (6.54).

11. Solve the order conditions to be satisfied by the coefficients of the triple jump composition (6.55) and obtain all the (real and complex) solutions.

12. Given the harmonic oscillator with Hamiltonian $H = \frac{1}{2}(p^2 + q^2)$ and initial conditions $q_0 = p_0 = 1$, integrate the corresponding equations of motion until the final time $t_f = 20000\pi$ with the composition methods (6.52) (step size $h = \pi/7$), (6.56), and (6.57) with step size $h = 2\pi/9$. Compute the error in position and in energy along the integration. Compare with the corresponding alternating-conjugate methods obtained from them. Comment on the results.

13. The purpose of this exercise is to illustrate the behavior of splitting methods with complex coefficients to certain semi-linear reaction-diffusion equations. To this end, let us consider the scalar 1D-equation

$$u_t = u_{xx} + u(1-u), \qquad u(x,0) = u_0(x) = \sin(2\pi x)$$

with periodic boundary conditions in $x \in [0,1]$ and the splitting into linear and nonlinear parts. Discretize in space with second-order finite differences ($M = 100$) and integrate in time until $t = 1$ with the second-order Strang splitting and the composition methods (6.52), (6.56) and (6.57). Obtain the corresponding efficiency diagrams. Notice that, although the theoretical framework developed in this chapter does not cover this nonlinear problem, the results achieved are largely similar to the linear case.

Chapter 7

Geometric integrators in action: other relevant applications

7.1 The gravitational N-body problem

The important problem in Classical Mechanics of a planetary system modeled as N bodies (a massive star and $N-1$ planets) under mutual gravitational Newtonian interaction constitutes an excellent test bench to illustrate the behavior and performance of geometric numerical integrators. In fact, symplectic integrators have been used in this setting both for theoretical purposes and for carrying out practical computations.

An early example of the first aspect corresponds to Newton's proof of Kepler's second law of motion in the 2-body problem, that is, the conservation of areal velocity (the area swept out by the radius vector per unit time). In fact, the symplectic Euler method is used in the original proof in an inverse way: the properties of the numerical solution are pulled back to the continuous problem to establish a general property. This is essentially the content of Proposition 1 in the *Philosophiae Naturalis Principia Mathematica*, as noticed by Wanner [322] and Chin [112].

On the other hand, the long-time numerical integrations of the whole Solar System carried out in [309, 328], suggesting its chaotic nature, represented a major boost for the development of new and efficient symplectic algorithms. The integrator used in these simulations, the so-called Wisdom–Holman map, is just a variant of the Strang splitting conveniently adapted to the structure of the problem.

The gravitational N-body problem can be modeled as a Hamiltonian system with Hamiltonian function

$$H(q,p) = \sum_{i=0}^{N-1} \frac{1}{2m_i} p_i^T p_i - G \sum_{i=1}^{N-1} \sum_{j=0}^{i-1} \frac{m_i m_j}{\|q_i - q_j\|}. \tag{7.1}$$

Here, (q,p) denotes the "supervectors" composed by the positions $q_i \in \mathbb{R}^3$ and momenta $p_i \in \mathbb{R}^3$ of the "Sun" ($i = 0$) and the $N-1$ planets ($i = 1, \ldots, N-1$) in some Cartesian coordinate system: $q = (q_0, q_1, \ldots, q_{N-1})^T$, $p = (p_0, p_1, \ldots, p_{N-1})^T$. In (7.1), m_i is the mass of the i-th body, and G is the universal gravitational constant.

DOI: 10.1201/9781003527626-7

The equations of motion (1.32) read

$$
\begin{aligned}
\frac{dq_i}{dt} &= \nabla_{p_i} H = \frac{1}{m_i} p_i, \qquad\qquad i = 0, \ldots, N-1, \\
\frac{dp_i}{dt} &= -\nabla_{q_i} H = -G \sum_{\substack{j=0 \\ j \neq i}}^{N-1} \frac{m_i m_j}{\|q_i - q_j\|^3} (q_i - q_j), \qquad i = 0, \ldots, N-1.
\end{aligned} \tag{7.2}
$$

Since the kinetic energy $T(p)$ and the potential energy $V(q)$ are in this case

$$
T(p) = \sum_{i=0}^{N-1} \frac{1}{2m_i} p_i^T p_i, \qquad\qquad V(q) = -G \sum_{i=1}^{N-1} \sum_{j=0}^{i-1} \frac{m_i m_j}{\|q_i - q_j\|}, \tag{7.3}
$$

it also makes sense to separate the Hamiltonian (7.1) as $H(q,p) = T(p) + V(q)$, and apply the splitting methods designed for problems with this structure: schemes of type $\mathrm{S}r_s$ or $\mathrm{NB}r_s$, as given in section 3.7.1. However, this choice is suboptimal for planetary systems, where planets describe near-Keplerian orbits around the central star. An alternative procedure taking advantage of the hierarchical nature of the motion of the planets around the central massive body consists in introducing a new set of coordinates and conjugate momenta $\hat{q}, \hat{p}$ so that the transformed H can be written as

$$
\hat{H}(\hat{q}, \hat{p}) = H_1(\hat{q}, \hat{p}) + H_2(\hat{q}), \tag{7.4}
$$

with an integrable part H_1 (corresponding to the Keplerian motion of the planets) and a small perturbation H_2 (that accounts for the gravitational interaction of the planets among themselves), and then applying the second-order scheme Strang splitting (3.21) to this new Hamiltonian. This is the approach carried out in [328].

Specifically, $(\hat{q}_i, \hat{p}_i)$, $i = 0, 1, \ldots, N-1$ correspond to the so-called Jacobi coordinates, in which $\hat{q}_0$ is the position of the center of mass of the system, $\hat{q}_1$ is the relative position of the first planet with respect to the central star, and for $i = 2, \ldots, N-1$, $\hat{q}_i$ is the position of the i-th planet relative to the center of mass of the central star and the planets with lower indices. In other words,

$$
\hat{q}_0 = \frac{1}{M_{N-1}} \sum_{j=0}^{N-1} m_j q_j, \qquad \hat{q}_i = q_i - \frac{1}{M_i} \sum_{j=0}^{i-1} m_j q_j, \quad i = 1, \ldots, N-1, \tag{7.5}
$$

where $M_i = \sum_{j=0}^{i} m_j$ for $i = 0, \ldots, N-1$. This can be written in more compact way as $\hat{q} = Aq$, where A is a $3N \times 3N$ invertible matrix with mass-dependent entries, $\hat{q} = (\hat{q}_0, \ldots, \hat{q}_{N-1})^T$ and $q = (q_0, \ldots, q_{N-1})^T$.

The conjugate momenta $\hat{p}_i$, $i = 0, 1, \ldots, N-1$, are uniquely determined by the requirement that the linear change of variables be canonical (i.e., $p = A^T \hat{p}$), so that the transformed Hamiltonian function is obtained by rewriting (7.1) in the new variables: $\hat{H}(\hat{q}, \hat{p}) \equiv H(A^{-1}\hat{q}, A^T \hat{p})$.

It is not difficult to check that the kinetic energy, expressed in terms of $\hat{p}$, has the same diagonal structure as in p:

$$T = \sum_{i=0}^{N-1} \frac{1}{2m_i} p_i^T p_i = \sum_{i=0}^{N-1} \frac{1}{2\hat{m}_i} \hat{p}_i^T \hat{p}_i,$$

where

$$\hat{m}_0 = M_{N-1}, \qquad \hat{m}_i = \frac{M_{i-1}}{M_i} m_i, \qquad i = 1, \dots, N-1.$$

Then, the Hamiltonian (7.1) in the new variables $\hat{H}(\hat{q}, \hat{p})$ can be split as (7.4), where

$$\begin{aligned} H_1(\hat{q}, \hat{p}) &= \frac{1}{2\hat{m}_0} \hat{p}_0^T \hat{p}_0 + \sum_{i=1}^{N-1} \left(\frac{1}{2\hat{m}_i} \hat{p}_i^T \hat{p}_i - G \frac{m_0 m_i}{\|\hat{q}_i\|} \right), \\ H_2(\hat{q}) &= V(q) + G \sum_{i=1}^{N-1} \frac{m_0 m_i}{\|\hat{q}_i\|} \\ &= G m_0 \sum_{i=1}^{N-1} m_i \left(\frac{1}{\|\hat{q}_i\|} - \frac{1}{\|q_i - q_0\|} \right) - G \sum_{i=2}^{N-1} \sum_{j=1}^{i-1} \frac{m_i m_j}{\|q_i - q_j\|} \end{aligned} \tag{7.6}$$

and q has to be expressed in terms of $\hat{q}$ according to $q = A^{-1}\hat{q}$.

Observe that the potential energy does not depend on $\hat{q}_0$, so $\hat{p}_0$ is constant (in fact, it is the linear momentum of the system), and therefore we can remove it from $\hat{H}$ if we assume that the center of the mass is at rest. We have also that, for fixed $\hat{q}$ and varying mass ratios,

$$\hat{q}_i = q_i - q_0 + \mathcal{O}(\varepsilon), \qquad H_2(\hat{q}) = \mathcal{O}(\varepsilon) \quad \text{as} \quad \varepsilon \equiv \max_{1 \le i \le N-1} \left(\frac{m_i}{m_0} \right) \to 0. \tag{7.7}$$

Hamiltonian H_1 in eq. (7.6) can then be seen as a collection of $N-1$ two-body problems, and H_2 as a perturbation. It turns out that the flow $\varphi_h^{[H_1]}$ can be computed with the MATLAB function

```
[q p] = phiKepler(q0,p0,t,mu)
```

introduced in Chapter 1 to solve each two-body problem and using the algorithm proposed in [122, p. 165] for the (dimensionless) Hamiltonians

$$H_{1,i} = \frac{1}{\hat{m}_i} \left(\frac{1}{2} \hat{p}_i^T \hat{p}_i - (G m_0 m_i \hat{m}_i) \frac{1}{\|\hat{q}_i\|} \right)$$

with $\mu = G m_0 m_i \hat{m}_i$. In this respect, it is important to remark that propagating the system under the scaled Hamiltonian αH for a time interval h is equivalent to propagating under the original Hamiltonian H for a rescaled time interval αh.

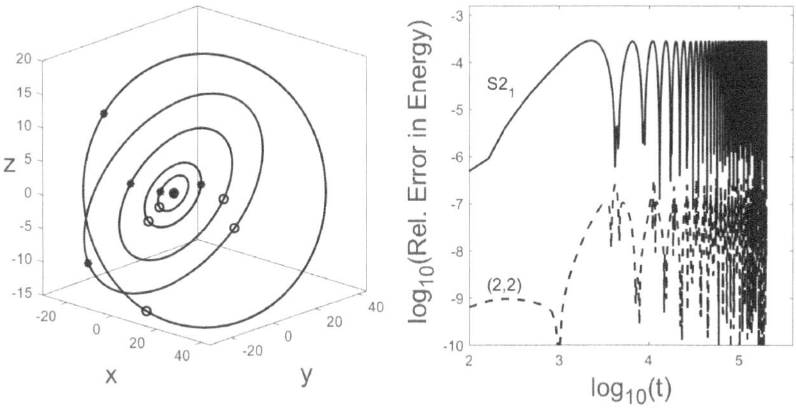

FIGURE 7.1: Left: Trajectories of the six-body system modeling the outer Solar System. Right: Relative error in energy of schemes (2,2) (for perturbed systems) and $S2_1$ (for the separation into kinetic and potential energy) as a function of time for an interval of 200 000 days using 2500 steps.

On the other hand, H_2 only depends on $\hat{q}$, and thus its flow $\varphi_h^{[H_2]}$ can be explicitly evaluated in an efficient way. Notice that the number of terms in H_1 grows linearly with the number of bodies N, whereas the number of terms in H_2 grows quadratically.

In what follows, we consider the outer Solar System, modeled as a six-body system with the inner Solar System ($i = 0$), the four giant planets ($i = 1, 2, 3, 4$), and Pluto ($i = 5$), all considered as point masses. The initial conditions for each planet are taken at Julian time (TDB) 2440400.5 (28 June 1969), obtained from the DE430 ephemerides [150] and normalized so that the center of mass of the system is at rest.

The trajectories for 200 000 days are shown in Figure 7.1 (left panel), with the initial and final positions of each body indicated by stars and circles, respectively.

For this model of the Solar System, the parameter ε introduced in (7.7) corresponds to the ratio of the mass of Jupiter to the mass of the inner Solar system, that is, $\varepsilon \approx 10^{-3}$. In consequence, the splitting methods for near-integrable systems presented in subsection 3.4.3 are particularly appropriate. In fact, the Wisdom–Holman map, in the terminology of Subsection 3.4.3, corresponds to a method of generalized order $(2, 2)$.

To illustrate this feature, we show in the right panel of Figure 7.1 the relative error in energy committed by the Strang splitting $S2_1$ when the system is separated into kinetic and potential energy, eq. (7.1), and when it is considered as a near-integrable problem, eq. (7.5), leading to a $(2, 2)$ scheme. In both cases we use 2500 steps of a length of 80 days per step to fulfill 2×10^5 days (a relatively short time integration of approximately 46 periods of Jupiter and 2 periods of Pluto). The error is computed after each step.

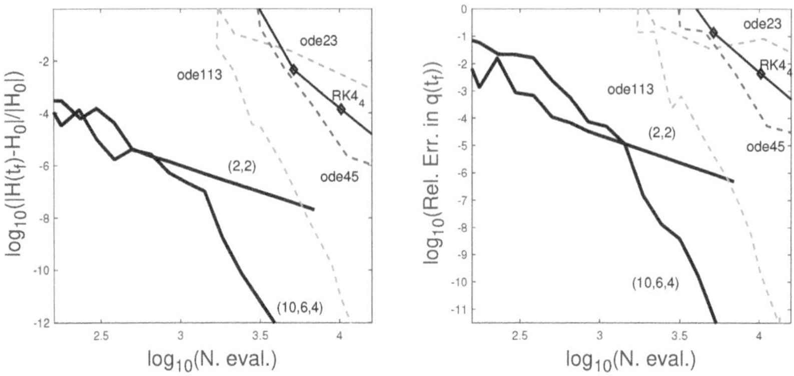

FIGURE 7.2: Efficiency diagrams showing the relative error in energy (left) and in positions (right) at the final time $t_{\rm f} = 2 \times 10^5$ days vs. the number of force evaluations, obtained with different schemes: MATLAB routines `ode23`, `ode45` and `ode113`, the standard 4-stage fourth-order RK methods, RK4_4, and splitting methods of generalized order (2,2) and (10,6,4).

We observe that the (2,2) scheme achieves an accuracy approximately three orders of magnitude higher than that of $S2_1$, corresponding to a factor of $\varepsilon \approx 10^{-3}$ for this problem, while both methods involve comparable computational cost. This improvement becomes even more pronounced when evaluating the error in positions and momenta, as the (2,2) scheme is conjugate to a method with a leading error term of order $\varepsilon^2 h^3$, thereby introducing an additional factor of ε in the overall error.

While the Wisdom–Holman map (i.e., the (2,2) scheme applied to the Hamiltonian (7.4)) has been the method of choice for exploring some qualitative features of the Solar System, such as whether it exhibits regular or chaotic behavior [309, 328], splitting methods of higher generalized order are clearly preferred when high accuracy is essential, even for short-time integrations.

This aspect is illustrated in Figure 7.2, where we depict efficiency diagrams for the error in energy (left) and in positions (right) at the final time $t_{\rm f} = 2 \times 10^5$ days obtained with different integration methods. On the one hand, the previous $(2, 2)$ scheme and the fourth-order $\text{NIA}(10, 6, 4)_8$ splitting method for near-integrable problems collected in subsection 3.7.1. On the other hand, as representatives of conventional integrators, the explicit variable-step and variable-order methods from MATLAB `ode23` (with embedded RK methods of order two and three), `ode45` (with embedded RK methods of order four and five), and `ode113` (with multistep methods of orders from 1 to 13). For comparison, we also include the standard explicit 4-stage fourth-order RK method, RK4_4. The exact solution is determined here by numerical computation with a very high accuracy.

Owing to the smoothness of the underlying problem and the low per-step computational cost of multistep methods, the `ode113` integrator delivers the most accurate results among the conventional schemes tested[1]. Nonetheless, these standard integrators are markedly outperformed across all accuracy requirements by the specially tailored splitting methods for this specific problem. This advantage is expected to widen for longer integrations as a result of the fundamentally different error propagation mechanisms characteristic of the respective methods.

7.2 Hamiltonian Monte Carlo

In certain applications it is often required to evaluate integrals in a space of large dimension. Examples include the posterior in Bayesian inference or the average energy in statistical physics. Given a certain function $f : \mathbb{R}^d \longrightarrow \mathbb{R}$, the problem consists in computing

$$I(f) = \int_{\mathbb{R}^d} f(x)\,\mu(x)\,dx, \tag{7.8}$$

where $\mu(x)$ is a given probability distribution defined in $\mathbb{R}^d$ (the *target distribution*). It turns out, however, that $I(f)$ cannot be determined analytically in general, and even if d is not very large, conventional quadrature rules might not be feasible. In those circumstances, the integral $I(f)$ can still be numerically estimated with Monte Carlo methods. The idea is to generate a set of independent and identically distributed (i.i.d.) samples $\{x^{(i)}\}_{i=1}^N$ from the target density $\mu(x)$, evaluate the function f at these samples, and then use the estimator [72]

$$I_N(f) = \frac{1}{N}\sum_{i=1}^{N} f(x^{(i)}). \tag{7.9}$$

Then, assuming that $I(f) < \infty$, the estimate $I_N(f)$ is unbiased and, by the law of large numbers,

$$\lim_{N\to\infty} I_N(f) = I(f)$$

almost surely. If the standard deviation

$$\sigma_f^2 = \int_{\mathbb{R}^d} \left(f(x) - I(f)\right)^2 \mu(x)dx$$

is finite, then the central limit theorem guarantees that [11]

$$N^{1/2}(I_N(f) - I(f)) \Longrightarrow \mathcal{N}(0, \sigma_f^2) \qquad \text{as } N \to \infty,$$

[1] See also [168] for further numerical evidence when the system (7.1) is separated as kinetic energy + potential energy.

where $\Longrightarrow$ denotes convergence in distribution [282] and $\mathcal{N}$ stands for the normal distribution. In other words, the standard deviation of the error $I_N(f) - I(f)$ decreases as the inverse square root of the number of samples. Thus, reducing this *Monte Carlo error* by a factor of 2 requires quadrupling the number of i.i.d. samples.

An advantage of Monte Carlo integration over other procedures is that it places the integration grid points (samples) in regions of high probability. It happens, however, that unless $\mu(x)$ has standard form, e.g., Gaussian, we cannot directly generate i.i.d. samples $\{x^{(i)}\}$, and thus we have to use other techniques such as rejection sampling, importance sampling or Markov chain Monte Carlo (MCMC) methods [11].

Discussing MCMC methods in detail is well beyond the scope of this work, and the interested reader is referred to the collective book [76] for a comprehensive treatment of this important topic. Briefly, MCMC is a strategy for generating samples $x^{(i)}$ while exploring the state space using a Markov chain mechanism. It is constructed in such a way that the chain spends more time in regions with higher probability, and the generated samples $x^{(i)}$ mimic samples drawn from the target distribution $\mu(x)$.

We recall that a Markov chain with state space $\mathbb{R}^d$ is a sequence of random d-vectors $\{x^{(i)}\}_{i\in\mathbb{N}}$ verifying

$$\mathbb{P}(x^{(i+1)} | \, x^{(i)}, \ldots, x^{(1)}) = \mathbb{P}(x^{(i+1)} | \, x^{(i)}),$$

that is, the only information required for generating the new state $x^{(i+1)}$ of the chain is the current state $x^{(i)}$. The chain is homogeneous if $\mathbb{P}(x^{(i+1)} | \, x^{(i)})$ remains invariant for all i, in which case the evolution of the chain depends solely on the current state and a fixed transition matrix. For any starting point, under certain conditions (irreducibility and aperiodicity [11]), the chain converges to an invariant (or stationary) distribution, and the purpose of MCMC methods is to generate a Markov chain that has the target $\mu(x)$ as its invariant distribution. Then, $I(f)$ is estimated by (7.9).

A well known MCMC is the Metropolis–Hastings (MH) algorithm [248, 180]. It involves transforming a given proposal chain (for which $\mu(x)$ is not invariant) into another chain which leaves $\mu(x)$ invariant. This is done with an accept/reject mechanism. Thus, an MH step with proposal distribution $q(x^*|x)$ involves, given the current state $x^{(i)}$, sampling a candidate value x^* according to $q(x^*|x^{(i)})$. The Markov chain then takes $x^{(i+1)} = x^*$ with acceptance probability

$$\mathcal{A}(x^{(i)}, x^*) = \min \left\{ 1, \frac{\mu(x^*) \, q(x^{(i)}|x^*)}{\mu(x^{(i)}) \, q(x^*|x^{(i)})} \right\};$$

otherwise it remains at the same state, $x^{(i+1)} = x^{(i)}$. The algorithm requires a careful design of the proposal distribution $q(x^*|x)$ to be efficient. In particular, if $q(x^*|x^{(i)}) = q(x^{(i)}|x^*)$, i.e., the proposal is reversible, then the acceptance simplifies to

$$\mathcal{A}(x^{(i)}, x^*) = \min\left\{1, \frac{\mu(x^*)}{\mu(x^{(i)})}\right\}, \tag{7.10}$$

and so the normalizing constant of the target distribution $\mu(x)$ is not required.

Sometimes it is easier to sample from an augmented distribution $\mu(x, u)$, where u is an auxiliary variable, than directly from $\mu(x)$. Then, it is possible to get marginal samples $x^{(i)}$ by sampling $(x^{(i)}, u^{(i)})$ according to $\mu(x, u)$ and, subsequently discarding the samples $u^{(i)}$. This is precisely the idea behind the *Hamiltonian* (or hybrid) *Monte Carlo (HMC) method*, a technique first proposed in the context of lattice quantum chromodynamics [128], and then used in data science and statistics [223, 265].

At this point, the reader might wonder how geometric numerical integrators, designed for *deterministic* problems, can be used in this *probabilistic* setting. To make clear the connection, from now on we denote $q \equiv x$ and $p \equiv u$ for the auxiliary variable.

Suppose we have target distribution $\mu(q)$ in $\mathbb{R}^d$. If $V : \mathbb{R}^d \longrightarrow \mathbb{R}$ denotes the negative logarithm of the (not necessarily normalized) probability density function, we then may write

$$\mu(q) = \frac{1}{Z_q} \mathrm{e}^{-V(q)}, \qquad \text{where } Z_q = \int_{\mathbb{R}^d} \mathrm{e}^{-V(q)} dq.$$

In HMC, V can be seen as the *potential energy* of a mechanical system with coordinates q. Then the auxiliary momenta $p \in \mathbb{R}^d$ and the kinetic energy $T(p) = (1/2)p^T M^{-1} p$ are introduced to form the Hamiltonian function

$$H(q, p) = \frac{1}{2} p^T M^{-1} p + V(q), \tag{7.11}$$

where M is a positive-definite symmetric matrix chosen by the user. In statistical physics, the Boltzmann–Gibbs distribution corresponding to this function H is given by [210]

$$\mu(q, p) = (2\pi)^{-d/2} \,|\det M|^{-1/2}\, \mathrm{e}^{-T(p)}\, Z_q^{-1}\, \mathrm{e}^{-V(q)} \tag{7.12}$$

(with inverse temperature equal to 1). It is important to remark that the Hamilton equations of motion preserve $\mu(q, p)$ exactly [72], whereas the p-marginal is Gaussian with zero mean and covariance matrix M. Thus, the Markov chain $\{(q^{(i)}, p^{(i)})\}_{i \in \mathbb{N}}$ produced from the initial state $(q^{(0)}, p^{(0)})$ with $p^{(i)} \sim \mathcal{N}(0, M)$ by integrating Hamilton's equations corresponding to H with initial condition $(q^{(i)}, p^{(i)})$ has $\mu(q, p)$ as an invariant distribution.

In most cases of practical interest, however, the exact flow φ_t corresponding to this Hamiltonian system cannot be obtained. We have then to resort to numerical approximations, and it is here where structure-preserving methods come into play. Specifically, the numerical integrator has to be both reversible (time-symmetric) and volume-preserving. In addition, an accept/reject mechanism has to be introduced to cope with the fact that the volume in phase

TABLE 7.1: HMC algorithm. The function $H = (1/2)p^T M^{-1} p + V(q)$ is the Hamiltonian. The algorithm generates a Markov chain $q^{(0)} \mapsto q^{(1)} \mapsto \cdots \mapsto q^{(i_{\max})}$ reversible with respect to the target probability distribution $\mu(q) \propto \exp(-V(q))$.

Given $q^{(0)} \in \mathbb{R}^d$, $i_{\max} \geq 1$, set $i = 0$.

1. (Momentum refreshment.) Draw $p^{(i)} \sim \mathcal{N}(0, M)$.
2. (Integration leg.) Compute (q^*, p^*) (q^* is the proposal) by integrating, with a reversible, volume-preserving integrator with step size h, the equations of motion derived from the Hamiltonian (7.11) over an interval $0 \leq t \leq Lh$ with initial condition $(q^{(i)}, p^{(i)})$.
3. (Accept/reject.) Calculate

$$\mathcal{A}^{(i)} = \min\left(1, \exp(H(q^{(i)}, p^{(i)}) - H(q^*, p^*))\right)$$

   and draw $u^{(i)} \sim \mathcal{U}(0,1)$. If $\mathcal{A}^{(i)} > u^{(i)}$, set $q^{(i+1)} = q^*$ (acceptance); otherwise, set $q^{(i+1)} = q^{(i)}$ (rejection).
4. Set $i = i + 1$. If $i = i_{\max}$ stop; otherwise, go to step 1.

---

space *and* the energy (and therefore the Boltzmann–Gibbs distribution) cannot be simultaneously preserved by any numerical scheme.

The resulting basic HMC algorithm is described in Table 7.1, although more elaborate possibilities exist [72]. Notice that the accept/reject mechanism here is a particular case of the rule (7.10).

One of the most salient features of the HMC algorithm is that it is able to generate proposal moves that, while being far away from the current state of the Markov chain, have a high probability of acceptance, thus reducing the correlation between samples. Since low acceptance rates are not desired and the acceptance probability depends on the energy error $\Delta H = H(q^{(i)}, p^{(i)}) - H(q^*, p^*)$, it is important to perform the integration so as to have small energy errors. In this connection we recall the results of Chapter 5 relative to symplectic integrators: they preserve energy with errors of order $\mathcal{O}(h^r)$ over exponentially long time intervals, r being the order of the method. Moreover, due to symplecticity, the schemes are also volume-preserving. If in addition we take into account the special structure of H, it is clear that palindromic splitting methods constitute the natural choice for determining the proposal (q^*, p^*).

As the main contribution to the computational cost of HMC resides in the numerical integration of the equations of motion, it is of paramount interest to apply splitting methods requiring as few evaluations of the force $-\nabla_q V(q)$ as possible, with small energy errors and able to use large step sizes h. All these features make the Störmer–Verlet method the most obvious choice. Indeed, the energy errors committed by this method are $\mathcal{O}(h^4)$, so that acceptance rates are likely to be high even if the step size is not small [58].

Example 7.2. Suppose we have to generate samples from the (non-normalized) target probability distribution

$$\mu(q) \propto \mathrm{e}^{-V(q)}, \qquad \text{with } V(q) = \frac{1}{5}q - q^2 + \frac{1}{4}q^4.$$

To do that, we apply the HMC algorithm of Table 7.1 with the velocity Verlet scheme (1.49) with $M = 1$, step size $h = 0.05$, $L = 20$ steps, and $i_{\max} = 50000$. Figure 7.3 shows the histogram we have obtained, together with the normalized probability density. With this choice of parameters, there are 920 rejections. □

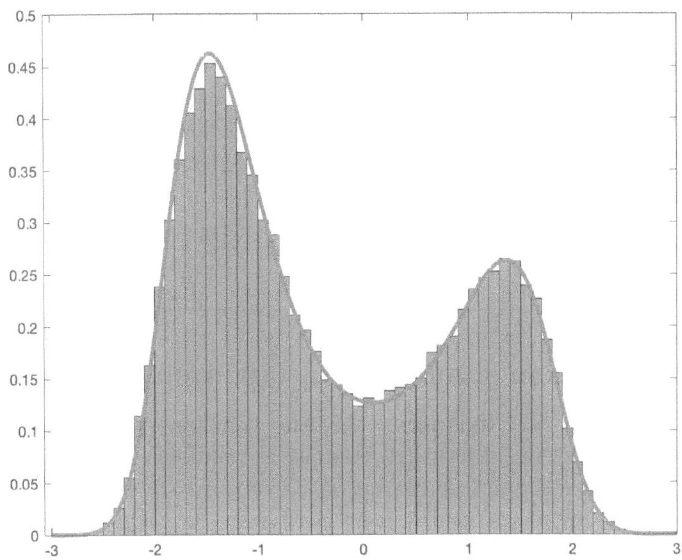

FIGURE 7.3: Normalized probability density, $\mathrm{e}^{-V(q)}$, with $V(q) = q/5 - q^2 + q^4/4$ and the corresponding histogram obtained by the HMC algorithm with the velocity Verlet method.

When the dimension $d = 1$, the velocity version of Störmer–Verlet is indeed the best choice, as discussed in [72]. However, as d increases, the energy error for the overall system will likely be large unless the step size h is very small. This is what happens when the procedure is applied to a multivariate Gaussian model with d large. In those circumstances, it is possible to design splitting methods with two and three stages more efficient than Störmer–Verlet with respect to acceptance ratios. In particular, the 3-stage method

$$\psi_h = \varphi_{a_1 h}^{[T]} \circ \varphi_{b_1 h}^{[V]} \circ \varphi_{a_2 h}^{[T]} \circ \varphi_{b_2 h}^{[V]} \circ \varphi_{a_2 h}^{[T]} \circ \varphi_{b_1 h}^{[V]} \circ \varphi_{a_1 h}^{[T]} \qquad (7.13)$$

with

$$a_1 = 0.11888010966548, \quad b_1 = 0.29619504261126, \quad a_2 = \frac{1}{2} - a_1, \quad b_2 = 1 - 2b_1,$$

provides better results at the same computational cost as the standard Störmer–Verlet method [58]. The rationale behind this and other multi-stage splitting methods presented in [58] is that the coefficients a_j, b_j are chosen to minimize the energy error in the proposal for relatively large values of h (and not, as in the usual numerical integration domain, to increase the accuracy in the limit $h \to 0$). An extension of this idea is the so-called Adaptive Integration Approach (AIA) [146, 6]: here the user chooses the value of h to be employed, and then the AIA algorithm itself finds the coefficients of the method within a given family of s-stage splitting integrators providing the best acceptance rate. The efficiency can be further enhanced by using a conveniently modified version of processing so that time-reversible kernels provide time-reversible integrations [35].

In any case, the HMC algorithm requires careful examination to select suitable values for the step size h (too large a step size increases rejection rates, while too small a step size increases computational cost) and the number of steps of integration, L, which together determine the length of the trajectory in fictitious time hL.

7.3 Hamiltonian simulation in quantum computing

It is a well known fact that simulating the full time evolution of an arbitrary quantum system on classical computers requires exponential amounts of computational resources. In particular, to record the state of the system is already cumbersome, since the associated wave function belongs to a vector space whose dimension grows exponentially with the size of the system. This and other reasons led Feynman to speculate on the possibility of using a specific quantum system (a quantum computer) to simulate the behavior of arbitrary quantum systems whose dynamics are determined by local interactions [147]. The conjecture was later shown to be correct [224], at least in theory. Quantum computers nowadays can simulate a variety of systems arising in quantum chemistry, quantum field theory, condensed matter physics, etc. (see e.g., [32, 106] and the review [251]).

Since simulating the time evolution of a quantum system with (a time-independent) Hamiltonian H requires approximating the unitary operator e^{-itH} ($\hbar = 1$), it makes sense to consider the application of the geometric integration techniques developed in the previous chapters. Thus, if H is the sum over many local interactions, an obvious approach is to apply splitting methods. Suppose that for a system composed of n variables, H can be decomposed as

$$H = \sum_{j=1}^{\ell} H_j, \tag{7.14}$$

where each H_j acts on a space of dimension m_j encompassing at most k of the variables, and ℓ is a polynomial in n. This is the case, in particular, of the Hubbard, Ising and Heisenberg models [266]. The important point is that, whereas e^{-itH} is generally unfeasible to compute, each e^{-itH_j} acts on a much smaller subsystem and can be evaluated by quantum circuits. In fact, the explicit quantum simulation algorithm proposed in [224] is based on the Lie–Trotter method

$$\chi_t = \mathrm{e}^{-itH_\ell} \cdots \mathrm{e}^{-itH_2}\, \mathrm{e}^{-itH_1},$$

whereas other proposals include the Strang splitting

$$S_t^{[2]} = \mathrm{e}^{-i\frac{t}{2}H_1}\, \mathrm{e}^{-i\frac{t}{2}H_2} \cdots \mathrm{e}^{-itH_\ell} \cdots \mathrm{e}^{-i\frac{t}{2}H_2}\, \mathrm{e}^{-i\frac{t}{2}H_1},$$

and especially the quintuple jump recursion (3.33)

$$S_t^{[2k+2]} = \left(S_{\gamma_{2k}t}^{[2k]}\right)^2 \circ S_{(1-4\gamma_{2k})t}^{[2k]} \circ \left(S_{\gamma_{2k}t}^{[2k]}\right)^2 \tag{7.15}$$

with $\gamma_{2k} = 1/(4-4^{1/(2k+1)})$, due to its improved efficiency. These schemes are known in this setting as *product formulas* [105, 106] and the procedure is sometimes called *Trotterization.*

Product formulas of this type provide approximations U_{app} to the exact evolution e^{-itH} and the goal is, given a time t and a maximum simulation error ε, to find an algorithm (a quantum circuit) U_{app} such that $\|U_{\mathrm{app}} - \mathrm{e}^{-itH}\| < \varepsilon$. It is therefore important to analyze and eventually provide tight bounds for the error ε committed by this type of approximations when applied to Hamiltonian systems of the form (7.14). Thus, if $H = H_1 + H_2 + H_3$ and H_j are Hermitian, it is shown in [107] that

$$\|S_t^{[4]} - \mathrm{e}^{-itH}\| \le t^5 \sum_{i,j,k,l,m=1}^{3} c_{i,j,k,l,m} \|[H_i,[H_j,[H_k,[H_l,H_m]]]]\|,$$

where the coefficients $c_{i,j,k,l,m}$ are computed explicitly. In general,

$$\|S_t^{[2k]} - \mathrm{e}^{-itH}\| = \mathcal{O}\left(\tilde{\alpha}_{\mathrm{comm}} t^{2k+1}\right),$$

where $\tilde{\alpha}_{\mathrm{comm}}$ are bounds on the nested commutators involving $2k+1$ operators H_j, although tighter bounds are clearly desirable. It is said then that these product formulas exhibit an error with commutator scaling [107]. As usual, if t is large, then the whole interval is divided into N steps, and (7.15) is applied within each step. In that case,

$$\|\left(S_{t/N}^{[2k]}\right)^N - \mathrm{e}^{-itH}\| = \mathcal{O}(\varepsilon)$$

for a given $\varepsilon > 0$ provided that

$$N = \mathcal{O}\left(\frac{\tilde{\alpha}_{\text{comm}}^{1/(2k)}\, t^{1+1/(2k)}}{\varepsilon^{1/(2k)}}\right).$$

Multi-product expansions (4.66) are also used for Hamiltonian simulation [108, 335]. Their main advantage is that they only require a sum of polynomially many unitary operations to achieve the same accuracy as product formulas with exponentially many unitary operators would, although especially adapted algorithms to perform linear combinations of unitary operators on quantum computers are needed [108].

Several strategies based on structure-preserving methods can be applied when the Hamiltonian depends explicitly on time. The first, most obvious, is to adapt product formulas by following the approach we considered in subsection 3.6, and in particular by introducing the time t as two new coordinates. This technique is particularly advantageous if the time-dependent part is cheap to compute, for instance when $H(t) = a(t)H_1 + b(t)H_2$ [191].

For general time-dependent Hamiltonian simulation, a quantum algorithm based on the Magnus expansion that features both commutator scaling and logarithmic dependence on the derivatives is proposed in [10] and further analyzed in [138, 70] in the context of unbounded Hamiltonians, whereas [71] combines approximations rendered by truncating the Magnus series with product formulas to get efficient schemes. The resulting methods, however, require the implementation of exponentials of commutators. This difficulty can be overcome either by using specially designed product formulas [92] or commutator-free quasi-Magnus methods similar to schemes (4.48) and (4.49) [90].

7.4 Exercises

1. Given the change of coordinates $(q,p) \longmapsto (Q,P)$ defined by $Q = Aq$, $P = Bp$, with A, B non-singular matrices, find the relation between A and B so that the transformation is canonical (or symplectic).

2. Consider the gravitational interaction of the Sun and two planets, or alternatively, the Hamiltonian (7.1) with $N = 3$. Write explicitly the 3×3 matrix A associated to the Jacobi coordinates. Write explicitly the Hamiltonian (7.6) in this case, and the resulting equations of motion.

3. (See reference [52]). Consider again the pendulum described by equation (1.12) with parameter $k = 1$ and initial conditions $(q_0, p_0) = (\frac{1}{10}, 0)$, corresponding to small-amplitude oscillations. Decompose the Hamiltonian

into a harmonic oscillator and a perturbative remainder as follows:

$$H = \frac{1}{2}(p^2 + q^2) + \left(1 - \frac{1}{2}q^2 - \cos q\right) \equiv H_1(q,p) + H_2(q). \qquad (7.16)$$

Integrate the system until the final time $t_f = 500$ using a time step $h = \frac{5}{12}$. Apply the Strang splitting in two different configurations: first, by separating the Hamiltonian into kinetic and potential energy, and second, using the decomposition (7.16). In each case, compute the relative energy error along the trajectory as

$$\frac{|H(q_n, p_n) - H(q_0, p_0)|}{|H(q_0, p_0)|}.$$

Compare the results obtained with both splitting strategies with those presented in the right panel of Figure 7.1.

4. Determine the hyper-parameters h and L in Example 7.2 to be used by the HMC algorithm of Table 7.1 with the velocity Verlet scheme so that $hL = 1$ and the number of rejections is reduced to a minimum.

5. (See reference [58]). Consider the (non-normalized) target probability distribution

$$\mu(q) \propto \exp\left(\frac{1}{2}\sum_{j=1}^{d} j^2 q_{(j)}^2\right)$$

with $d = 1, 2, 4, \ldots, 1024$ and mass matrix $M = I$. Generate $i_{\max} = 5000$ samples $q^{(i)}$ with the velocity version of the Störmer–Verlet method and the 3-stage scheme (7.13), and plot the resulting acceptance probability as a function of d for both integrators. Use in all cases $L = 2d$ time steps for the integration. Take $q^{(0)} \sim \mathcal{U}(d, 1)$ and "randomize" the step-size h at the beginning of each Markov step by allowing $\pm 20\%$ variations around a mean value $h_0 = 3/d$:

$$h = (1 + u)h_0, \qquad u \sim \mathcal{U}(-0.2, 0.2).$$

Comment on the results.

6. Repeat *Example 7.2* with the scheme (7.13) and parameters $h = 0.15$, $L = 7$. Get the corresponding histogram and the number of rejections obtained.

Appendix A

Additional mathematical results

A.1 Differential equations, vector fields and flows

An *ordinary differential equation* of the form

$$\dot{x} = \frac{dx}{dt} = f(t, x), \tag{A.1}$$

with $x \in \mathbb{R}^d$ and $f : \mathbb{R} \times \mathbb{R}^d \longrightarrow \mathbb{R}^d$, admits infinitely many different solutions $x(t)$. From this set it is possible to single out certain solutions by introducing additional requisites. Thus, if one specifies the initial condition $x(t_0) = x_0$ for given t_0 and x_0, the resulting *initial value problem*

$$\frac{dx}{dt} = f(t, x), \qquad x(t_0 = 0) = x_0 \in \mathbb{R}^d \tag{A.2}$$

has one and only one solution, provided f satisfies some simple regularity conditions [114].

Boundary value problems also occur frequently in practice. Here the solution $x(t)$ of the differential equation (A.1) has to satisfy, in addition, a boundary condition, i.e., a functional relation linking the values of x and/or its derivatives at two different points.

In order to keep the presentation as simple as possible, let us assume that the problem (A.2) is autonomous; in other words, f does not depend explicitly on time t. With $\dot{x} = f(x)$, $x \in \mathbb{R}^d$, we associate a *vector field* f and the corresponding *flow* φ_t [16]. For each value of t, φ_t maps $\mathbb{R}^d$ in $\mathbb{R}^d$ so that $\varphi_t(\alpha)$ is the value at time t of the solution with initial condition α at time 0, whereas, for fixed x_0 and varying t, $\varphi_t(x_0)$ provides the solution of the initial value problem (A.2). The flow φ_t defines a one-parameter group of diffeomorphisms for which f is the velocity vector field.

In addition, for each infinitely differentiable map $G : \mathbb{R}^d \longrightarrow \mathbb{R}$, $G(\varphi_t(x))$ admits the representation

$$G(\varphi_t(x)) = \Phi^t[G](x), \tag{A.3}$$

where the operator Φ^t acts on differentiable functions [267]. To get some insight into Φ^t, we associate with the vector field f the first-order differential operator (also called *Lie derivative* or *Lie operator*) L_f. It maps the smooth

DOI: 10.1201/9781003527626-A

function $G : \mathbb{R}^d \longrightarrow \mathbb{R}$ into the real-valued function $L_f G : \mathbb{R}^d \longrightarrow \mathbb{R}$ such that, for $x \in \mathbb{R}^d$, (see [16, Chapter 8])

$$(L_f\, G)(x) = \left.\frac{d}{dt}\right|_{t=0} G(\varphi_t(x)),$$

that is

$$L_f G(x) = \sum_{i=1}^{d} f_i(x) \frac{\partial G}{\partial x_i} = f(x) \cdot \nabla G(x).$$

In coordinates, the operator L_f has the form

$$L_f = \sum_{i=1}^{d} f_i \frac{\partial}{\partial x_i}. \tag{A.4}$$

It follows from the chain rule that

$$\frac{d}{dt} G(\varphi_t(x_0)) = (L_f G)(\varphi_t(x_0)). \tag{A.5}$$

If one has instead a vector function $F : \mathbb{R}^d \longrightarrow \mathbb{R}^m$, then L_f acts on each component function F_i as before, so that $L_f F(x) = F'(x) f(x)$, where $F'(x)$ stands for the Jacobian matrix.

The Lie derivative L_f satisfies some remarkable properties. Given two functions ψ_1, ψ_2, it can be easily verified that

$$\begin{aligned} &L_f(\alpha_1 \psi_1 + \alpha_2 \psi_2) = \alpha_1 L_f \psi_1 + \alpha_2 L_f \psi_2, \qquad && \alpha_1, \alpha_2 \in \mathbb{R} \\ &L_f(\psi_1\, \psi_2) = (L_f \psi_1)\, \psi_2 + \psi_1\, L_f \psi_2 && \end{aligned}$$

and by induction we can prove the Leibniz rule

$$L_f^k(\psi_1\, \psi_2) = \sum_{i=0}^{k} \binom{k}{i} (L_f^i \psi_1) \left(L_f^{k-i} \psi_2\right),$$

with $L_f^i \psi = L_f \left(L_f^{i-1} \psi\right)$ and $L_f^0 \psi = \psi$, justifying the name of Lie derivative. In addition, the correspondences between the vector field f, the flow φ_t and the Lie derivative L_f are one to one.

Given two vector fields f and g, then

$$\alpha_1 L_f + \alpha_2 L_g = L_{\alpha_1 f + \alpha_2 g}, \qquad \alpha_1, \alpha_2 \in \mathbb{R}$$

but, in general, the operators L_f and L_g do not commute. Its commutator

$$[L_f, L_g] = L_f\, L_g - L_g\, L_f \tag{A.6}$$

is nevertheless a first-order linear differential operator [16]. An explicit calculation shows that

$$(L_f\, L_g - L_g\, L_f) G(x) = \sum_{i,j=1}^{d} \left(f_i \frac{\partial g_j}{\partial x_i} - g_i \frac{\partial f_j}{\partial x_i} \right) \frac{\partial G}{\partial x_j}.$$

Therefore, we can associate a new vector field to this differential operator, denoted by $w = (f, g)$ and called the *Lie bracket* of the vector fields f and g. Its components are

$$w_i = (f, g)_i = \sum_{j=1}^{d} \left(f_j \frac{\partial g_i}{\partial x_j} - g_j \frac{\partial f_i}{\partial x_j} \right) \tag{A.7}$$

and $L_w = L_{(f,g)} = [L_f, L_g]$. The Lie bracket just defined satisfies, in addition, the Jacobi identity: since

$$[[L_f, L_g], L_h] + [[L_g, L_h], L_f] + [[L_h, L_f], L_g] = 0,$$

one has analogously

$$((f, g), h) + ((g, h), f) + ((h, f), g) = 0.$$

From (A.5), if we apply the operator L_f iteratively to the function G, we get

$$\frac{d^k}{dt^k} G(\varphi_t(x_0)) = (L_f^k G)(\varphi_t(x_0)), \qquad k \geq 1.$$

Therefore, the Taylor series of $G(\varphi_t(x_0))$ at $t = 0$ is given by

$$G(\varphi_t(x_0)) = \sum_{k \geq 0} \frac{t^k}{k!} (L_f^k G)(x_0) \equiv \exp(tL_f)[G](x_0). \tag{A.8}$$

The operator $X \equiv \exp(tL_f)$ is called *Lie transformation* and can be seen as the formal solution of the operator equation

$$\frac{dX}{dt} = X\, L_f, \qquad X(0) = I, \tag{A.9}$$

for, on the one hand,

$$\frac{d}{dt} G\big(\varphi_t(x)\big) = \frac{d}{dt} \big(X(t)[G]\big)(x) = \left(\frac{dX(t)}{dt} [G] \right)(x),$$

and on the other hand

$$\frac{d}{dt} G\big(\varphi_t(x)\big) = (L_f G)\big(\varphi_t(x)\big) = X(t)[L_f G](x).$$

Moreover, the solution of the problem can be formally written as follows. If we replace $G(x)$ in (A.8) by $\mathrm{Id}(x) = x$, the identity map, we get the Taylor series of $\varphi_t(x_0)$:

$$\varphi_t(x_0) = \sum_{k \geq 0} \frac{t^k}{k!} (L_f^k\, \mathrm{Id})(x_0) \equiv \exp(tL_f)[\mathrm{Id}](x_0). \tag{A.10}$$

Comparing with (A.3), it is clear that formally

$$G(\varphi_t(x)) = \Phi^t[G](x) = \exp(tL_f)[G](x). \tag{A.11}$$

Example A.1. For the simple mathematical pendulum (1.12) with $k = 1$, $x = (q,p)^T$, the vector field $f(x) = (p, -\sin q)^T$ and therefore

$$L_f = p\frac{\partial}{\partial q} - \sin q \frac{\partial}{\partial p}.$$

Given the function $G(q,p) = q^2 - 3qp$, it is clear that $L_f G = p(2q - 3p) + 3q \sin q$. The expression (A.10) reads in this case

$$q(t) = \exp(tL_f)[\mathrm{Id}](q_0) = \sum_{k\geq 0} \frac{t^k}{k!}(L_f^k \mathrm{Id})(q_0), \qquad p(t) = \exp(tL_f)[\mathrm{Id}](p_0),$$

where

$$\begin{aligned} (L_f \mathrm{Id})(q_0) &= p_0, & (L_f^2 \mathrm{Id})(q_0) &= -\sin q_0, \\ (L_f \mathrm{Id})(p_0) &= -\sin q_0, & (L_f^2 \mathrm{Id})(p_0) &= -p_0 \cos q_0, \end{aligned}$$

so that we recover the Taylor expansion of the exact solution:

$$\begin{aligned} q(t) &= q_0 + tp_0 - \frac{t^2}{2}\sin q_0 + \cdots \\ p(t) &= p_0 - t\sin q_0 - \frac{t^2}{2} p_0 \cos q_0 + \cdots \end{aligned}$$

□

Suppose now that $\varphi_{t_1}^{[1]}$ and $\varphi_{t_2}^{[2]}$ are the flows corresponding to the differential equations $\dot{x} = f_1(x)$ and $\dot{x} = f_2(x)$, respectively. Then, by applying (A.10) to $\varphi_{t_2}^{[2]}$, we get

$$G(x) \equiv \varphi_{t_2}^{[2]}(x) = \exp(t_2 L_{f_2})[\mathrm{Id}](x). \tag{A.12}$$

If we consider now expression (A.8) for f_1, $t = t_1$ and $G(x)$ given by (A.12) we get finally for the composition of flows

$$\left(\varphi_{t_2}^{[2]} \circ \varphi_{t_1}^{[1]}\right)(x_0) = \exp(t_1 L_{f_1}) \exp(t_2 L_{f_2})[\mathrm{Id}](x_0). \tag{A.13}$$

It is important to notice how the indices 1 and 2 in (A.13) appear depending on whether we are dealing with maps or with exponentials of operators. The flows $\varphi_{t_1}^{[1]}$ and $\varphi_{t_2}^{[2]}$ commute everywhere for all sufficiently small t_1 and t_2 if and only if $[L_{f_1}, L_{f_2}] = 0$ [16].

Proceeding by induction, one has an analogous identity to (A.13) when more flows are involved:

$$\left(\varphi_{t_m}^{[m]} \circ \cdots \circ \varphi_{t_2}^{[2]} \circ \varphi_{t_1}^{[1]}\right)(x_0) = \exp(t_1 L_{f_1}) \exp(t_2 L_{f_2}) \cdots \exp(t_m L_{f_m})[\mathrm{Id}](x_0). \tag{A.14}$$

Example A.2. The scalar differential equation

$$\dot{x} = x^2 + \frac{1}{2x}, \qquad x(0) = x_0$$

has an analytical solution that is relatively involved. However, each one of the equations

$$\dot{x} = f_1(x) = x^2, \qquad \dot{x} = f_2(x) = \frac{1}{2x},$$

with the same initial condition $x(0) = x_0$, lead to the simple expressions

$$\varphi_s^{[1]}(x_0) = \exp(sL_{f_1})[\mathrm{Id}](x_0) = \sum_{n=0}^{\infty} s^n x_0^{n+1} = \frac{x_0}{1 - s x_0},$$

$$\varphi_t^{[2]}(x_0) = \exp(tL_{f_2})[\mathrm{Id}](x_0) = \sum_{n=0}^{\infty} \frac{(-1)^{n+1}(2(n-1)-1)t^n}{2^n x_0^{2n-3}} = \sqrt{t + x_0^2},$$

for their solution. Then, clearly,

$$\left(\varphi_s^{[1]} \circ \varphi_t^{[2]}\right)(x_0) = \frac{\sqrt{t + x_0^2}}{1 - s\sqrt{t + x_0^2}}. \tag{A.15}$$

The same result is obtained by applying the formalism of Lie transformations:

$$\exp(tL_{f_2}) \exp(sL_{f_1})[\mathrm{Id}](x_0) = \exp(tL_{f_2}) \sum_{n=0}^{\infty} s^n x_0^{n+1}$$

$$= \sum_{n=0}^{\infty} s^n \left(\exp(tL_{f_2})[\mathrm{Id}](x_0)\right)^{n+1} = \frac{\sqrt{t + x_0^2}}{1 - s\sqrt{t + x_0^2}}.$$

Analogously,

$$\varphi_t^{[2]} \circ \varphi_s^{[1]}(x_0) = \exp(sL_{f_1}) \exp(tL_{f_2})[\mathrm{Id}](x_0) = \sqrt{t + \left(\frac{x_0}{1 - s x_0}\right)^2}. \tag{A.16}$$

If $s = t = h$, with h the time step, then (A.15) and (A.16) correspond to both versions of the Lie–Trotter method, while if one takes $s = a_i h$, $t = b_i h$, then they correspond to one stage in a splitting method. □

Since L_{f_1} and L_{f_2} do not commute, then

$$\exp(tL_{f_1}) \exp(tL_{f_2})[\mathrm{Id}](x_0) \neq \exp(t(L_{f_1} + L_{f_2}))[\mathrm{Id}](x_0),$$

which is precisely the exact solution $\varphi_t(x_0)$ of $\dot{x} = f_1(x) + f_2(x)$. By using the Baker–Campbell–Hausdorff formula (see section A.4), we can obtain the difference between both operators. On the other hand, it is true that

$$\exp(\alpha_1 L_f) \exp(\alpha_2 L_f)[\mathrm{Id}](x_0) = \exp((\alpha_1 + \alpha_2) L_f)[\mathrm{Id}](x_0),$$
$$\exp(0 L_f)[\mathrm{Id}](x_0) = \mathrm{Id}(x_0) = x_0,$$

for all $\alpha_1, \alpha_2 \in \mathbb{R}$, so that $(\exp(L_f)[\mathrm{Id}](x_0))^{-1} = \exp(-L_f)[\mathrm{Id}](x_0)$.

A.2 Numerical integrators and series of differential operators

In the same way as for the exact flow φ_t of (A.1), we can associate an operator $X(h)$ to each integrator $\chi_h : \mathbb{R}^d \longrightarrow \mathbb{R}^d$. Specifically, $X(h)$ has the form

$$X(h) = I + \sum_{n \geq 1} h^n X_n, \tag{A.17}$$

where I denotes the identity operator and each linear differential operator X_n acts on smooth functions G as

$$X_n[G](x) = \frac{1}{n!} \frac{d^n}{dh^n} \bigg|_{h=0} G(\chi_h(x)), \tag{A.18}$$

so that $G(\chi_h(x)) = X(h)[G](x)$. In fact, it is possible to write $X(h)$ formally as the exponential of another operator $Y(h)$. This can be accomplished by introducing the series of vector fields

$$Y(h) = \sum_{n \geq 1} h^n Y_n = \log(X(h)) = \sum_{m \geq 1} \frac{(-1)^{m+1}}{m} \left(hX_1 + h^2 X_2 + \cdots \right)^m, \tag{A.19}$$

so that

$$g(\chi_h(x)) = X(h)[g](x) = \exp(Y(h))[g](x). \tag{A.20}$$

The numerical flow can be thus expressed, in analogy with (A.10), as

$$x_1 = \chi_h(x_0) = X(h)[\mathrm{Id}](x_0) = \exp(Y(h))[\mathrm{Id}](x_0).$$

The operators Y_n appearing in the series $Y(h)$ can be expressed in terms of X_k as [49]

$$Y_n = \sum_{m \geq 1}^{n} \frac{(-1)^{m+1}}{m} \sum_{j_1 + \cdots + j_m = n} X_{j_1} \cdots X_{j_m}.$$

Now, by comparing (A.20) with (A.11) for $t = h$, it is clear that the integrator χ_h is of order r if

$$Y_1 = L_f, \qquad Y_n = 0 \quad \text{for} \quad 2 \leq n \leq r.$$

For the adjoint method $\chi_h^* = \chi_{-h}^{-1}$ one has analogously

$$G(\chi_h^*(x)) = \exp(-Y(-h))[G](x).$$

In consequence, χ_h is time-symmetric (i.e., $\chi^* = \chi_h$) if and only if $Y(h) = hY_1 + h^3 Y_3 + \cdots$.

Example A.3. For the explicit Euler method $x_1 = x_0 + h\, f(x_0)$ one has

$$x_1 = X(h)[\mathrm{Id}](x_0) = (\mathrm{Id} + L_f \mathrm{Id})(x_0) = \exp(Y(h))[\mathrm{Id}](x_0)$$

so that, according to (A.19) ($X_1 = L_f$, $X_n = 0$, $n \geq 2$),

$$Y(h) = \sum_{n\geq 1} h^n Y_n = \sum_{n \geq 1} h^n \frac{(-1)^{n+1}}{n} L_f^n$$

and thus $Y_1 = L_f$, $Y_2 = L_f^2$, etc., corresponding to a first-order integrator. □

Suppose now we are dealing with a composition method

$$\psi_h = \chi_{\alpha_{2s}h} \circ \chi_{\alpha_{2s-1}h}^* \circ \cdots \circ \chi_{\alpha_2 h} \circ \chi_{\alpha_1 h}^* \tag{A.21}$$

of a basic scheme χ_h and its adjoint. Then we can proceed in the same way, thus leading to a series $\Psi(h) = I + h\Psi_1 + h^2\Psi_2 + \cdots$ of differential operators of the form

$$\Psi(h) = X(-\alpha_1 h)^{-1} X(\alpha_2 h) \cdots X(-\alpha_{2s-1}h)^{-1} X(\alpha_{2s}h), \tag{A.22}$$

where the series $X(h)$ is given by (A.17)–(A.18), and

$$X(h)^{-1} = I + \sum_{m\geq 1} (-1)^{m+1} \left(hX_1 + h^2 X_2 + \cdots\right)^m. \tag{A.23}$$

In this way $g(\psi_h(x)) = \Psi(h)[g](x)$ and, since $X(h) = \exp(Y(h))$, we have the formal identity

$$\Psi(h) = \mathrm{e}^{-Y(-h\alpha_1)}\, \mathrm{e}^{Y(h\alpha_2)} \cdots \mathrm{e}^{-Y(-h\alpha_{2s-1})}\, \mathrm{e}^{Y(h\alpha_{2s})}, \tag{A.24}$$

which can be expressed as

$$\Psi(h) = \exp(Z(h)) = \sum_{n\geq 1} h^n Z_n$$

by virtue of the Baker–Campbell–Hausdorff theorem (see section A.4). The order of the composition method can therefore be checked by comparing the series of differential operators $Z(h)$ with the series $\exp(hL_f)$ corresponding to the exact flow of the system. Specifically, the composition method is of order r if

$$Z_1 = L_f, \qquad Z_n = 0 \quad \text{for} \quad 2 \leq n \leq r. \tag{A.25}$$

When f in $\dot{x} = f(x)$ can be decomposed in two parts, $f(x) = f_1(x) + f_2(x)$, we can apply a splitting method of the form

$$\psi_h = \varphi^{[1]}_{a_{s+1}h} \circ \varphi^{[2]}_{b_s h} \circ \varphi^{[1]}_{a_s h} \circ \cdots \circ \varphi^{[1]}_{a_2 h} \circ \varphi^{[2]}_{b_1 h} \circ \varphi^{[1]}_{a_1 h}. \tag{A.26}$$

Then, the corresponding series $\Psi(h)$ of differential operators associated to ψ_h can be formally expressed as [49, 52]

$$\Psi(h) = \mathrm{e}^{a_1 h F_1}\, \mathrm{e}^{b_1 h F_2} \cdots \mathrm{e}^{a_s h F_1}\, \mathrm{e}^{b_s h F_2}\, \mathrm{e}^{a_{s+1} h F_1}, \tag{A.27}$$

where, for simplicity, we have denoted by F_1 and F_2 the Lie derivatives associated with f_1 and f_2, respectively, i.e.

$$F_1 \equiv f_1(x) \cdot \nabla, \qquad F_2 \equiv f_2(x) \cdot \nabla \tag{A.28}$$

so that

$$g(\varphi^{[1]}_t(x)) = \mathrm{e}^{tF_1} g(x), \qquad g(\varphi^{[2]}_t(x)) = \mathrm{e}^{tF_2} g(x).$$

Notice that the exponentials of Lie derivatives in (A.24) and (A.27) appear in a reversed order with respect to the maps in the integrators (A.21) and (A.26), respectively. This is a consequence of property (A.13).

A.3 Lie algebras and Lie groups

A.3.1 Lie algebras

A *Lie algebra* is a vector space $\mathfrak{g}$ together with a map $[\cdot,\cdot]$ from $\mathfrak{g} \times \mathfrak{g}$ into $\mathfrak{g}$ called Lie bracket, with the following properties:

1. $[\cdot,\cdot]$ is bilinear.
2. $[X, Y] = -[Y, X]$ for all $X, Y \in \mathfrak{g}$.
3. $[X, [Y, Z]] + [Y, [Z, X]] + [Z, [X, Y]] = 0$ for all $X, Y, Z \in \mathfrak{g}$.

Condition 2 is called *skew symmetry* and Condition 3 is the *Jacobi identity*. One should remark that $\mathfrak{g}$ can be any vector space and that the Lie bracket operation $[\cdot,\cdot]$ can be any bilinear, skew-symmetric map that satisfies the Jacobi identity. Thus, in particular, *the Lie bracket of vector fields (A.7) makes the vector space of vector fields on* $\mathbb{R}^d$ (in general, on a manifold) *into a Lie algebra* [16]. Also, the space of all $d \times d$ (real or complex) matrices is a Lie algebra with the Lie bracket defined as the commutator $[A, B] = AB - BA$.

Associated with any $X \in \mathfrak{g}$ we can define a linear map $\mathrm{ad}_X : \mathfrak{g} \longrightarrow \mathfrak{g}$ which acts according to

$$\mathrm{ad}_X Y = [X, Y], \qquad \mathrm{ad}_X^j Y = [X, \mathrm{ad}_X^{j-1} Y], \qquad \mathrm{ad}_X^0 Y = Y, \qquad j \in \mathbb{N} \tag{A.29}$$

for all $Y \in \mathfrak{g}$. The "ad" operator allows one to express nested Lie brackets in an easy way. Thus, for instance, $[X, [X, [X, Y]]]$ can be written as $\mathrm{ad}_X^3 Y$. Moreover, as a consequence of the Jacobi identity, we have the following properties:

1. $\mathrm{ad}_{[X,Y]} = \mathrm{ad}_X \mathrm{ad}_Y - \mathrm{ad}_Y \mathrm{ad}_X = [\mathrm{ad}_X, \mathrm{ad}_Y]$
2. $\mathrm{ad}_Z [X, Y] = [X, \mathrm{ad}_Z Y] + [\mathrm{ad}_Z X, Y]$.

It is said that ad_X is a *derivation* in the Lie algebra [204].

A.3.2 The Lie algebra of Hamiltonian functions

It is also clear that for a given Hamiltonian system with Hamiltonian function $H(q,p)$ defined on $D \subset \mathbb{R}^d \times \mathbb{R}^d$, the set of (sufficiently smooth) functions on D acquires the structure of a Lie algebra with the Poisson bracket (1.35). This is the *Lie algebra of Hamiltonian functions.*

In addition, we can associate with any such function $F : D \longrightarrow \mathbb{R}$ a vector field X_F defined as $X_F = J \nabla_x F(x)$ and called a *Hamiltonian vector field.* In particular, if equation (A.2) is derived from a Hamiltonian, then $f = X_H = (\nabla_p H, -\nabla_q H)^T$. The Lie derivative associated with X_F verifies, for any function $G : D \longrightarrow \mathbb{R}$,

$$L_{X_F} G = \sum_{i=1}^{2d} (J \nabla_x F)_i \frac{\partial G}{\partial x_i} = (J \nabla_x F)^T \nabla_x G = -(\nabla_x F)^T J \nabla_x G = -\{F, G\}.$$

Alternatively, $\{F, G\} = -L_{X_F} G = L_{X_G} F$. Moreover, by restating the Jacobi identity (A.40) satisfied by the Poisson bracket of any three functions F, G, H as $\{H, \{F, G\}\} = -\{\{H, G\}, F\} + \{\{H, F\}, G\}$, we have

$$L_{X_C} = -L_{X_F} L_{X_G} + L_{X_G} L_{X_F} = -[L_{X_F}, L_{X_G}], \qquad \text{where} \qquad C = \{F, G\}, \tag{A.30}$$

so that the Lie bracket of the Hamiltonian vector fields X_F, X_G is itself Hamiltonian with associated function C, the Poisson bracket of F and G. In particular, the set of Hamiltonian vector fields forms a subalgebra of all vector fields. These results generalize to any symplectic manifold [2].

Example A.4. Let $d = 1$ and consider the functions

$$U = \frac{1}{2} p^2, \qquad V = \frac{1}{2} q^2, \qquad W = -q\,p.$$

They verify

$$\{U, V\} = W, \qquad \{U, W\} = 2U, \qquad \{V, W\} = -2V.$$

Then

$$X_U = J\nabla U = (p,\, 0)^T, \qquad X_V = (0,\, -q)^T, \qquad X_W = (-q,\, p)^T; \quad \text{(A.31)}$$

the corresponding Lie derivatives are

$$L_{X_U} = p\frac{\partial}{\partial q}, \qquad L_{X_V} = -q\frac{\partial}{\partial p}, \qquad L_{X_W} = -q\frac{\partial}{\partial q} + p\frac{\partial}{\partial p}$$

and the identities (A.30) follow readily. □

A.3.3 Structure constants

If $\mathfrak{g}$ is any finite-dimensional Lie algebra, we can introduce a basis $\{V_1, V_2, \ldots, V_r\}$ of $\mathfrak{g}$ so that the Lie bracket of any two basis elements can be expressed again in terms of the basis. Thus there exist certain constants c_{ij}^k, $i, j, k = 1, \ldots, r$ called the *structure constants* of $\mathfrak{g}$ such that

$$[V_i, V_j] = \sum_{k=1}^{r} c_{ij}^k V_k, \qquad i, j = 1, \ldots, r. \quad \text{(A.32)}$$

Notice that the structure constants can be used to recover the Lie algebra from (A.32) and the bilinearity of the Lie bracket [267]. In *Example A.4*, the Hamiltonian vector fields X_U, X_V, X_W (or their corresponding Lie derivatives) form a Lie (sub)algebra of dimension 3 characterized by the basic Lie brackets

$$(X_U, X_V) = X_W, \qquad (X_U, X_W) = 2X_U, \qquad (X_V, X_W) = -2X_V.$$

In addition, this Lie algebra is isomorphic to the 2×2 matrix Lie algebra (with the usual commutator as Lie bracket) with basis

$$A = \begin{pmatrix} 0 & 1 \\ 0 & 0 \end{pmatrix}, \qquad B = \begin{pmatrix} 0 & 0 \\ -1 & 0 \end{pmatrix}, \qquad C = \begin{pmatrix} -1 & 0 \\ 0 & 1 \end{pmatrix} \quad \text{(A.33)}$$

since $[A, B] = C$, $[A, C] = 2A$, $[B, C] = -2B$. In fact, this is a particular illustration of Ado's theorem [267].

Theorem 5 *Let $\mathfrak{g}$ be a finite-dimensional Lie algebra. Then $\mathfrak{g}$ is isomorphic to a subalgebra of $\mathfrak{gl}(d)$ for some d, where $\mathfrak{gl}(d)$ denotes the Lie algebra of all $d \times d$ matrices with the Lie bracket being the matrix commutator.*

A.3.4 Lie groups

A Lie group is a differentiable manifold $\mathcal{G}$ which is also a group and such the group product $\mathcal{G} \times \mathcal{G} \longrightarrow \mathcal{G}$ and the inverse map $g \longmapsto g^{-1}$, $g \in \mathcal{G}$, are differentiable.

Familiar examples of Lie groups are *matrix Lie groups* [283]. The *general linear group* $\mathrm{GL}(d)$ is the group of all $d \times d$ invertible matrices and is itself a matrix Lie group. The corresponding Lie algebra is $\mathfrak{gl}(d)$. Other relevant examples of matrix Lie groups are the following.

- All elements of $\mathrm{GL}(d)$ with unit determinant form the *special linear group* $\mathrm{SL}(d)$. Its Lie algebra $\mathfrak{sl}(d)$ consists of all matrices with zero trace. Matrices (A.33) form a basis of $\mathfrak{sl}(2)$.

- The set of $d \times d$ real orthogonal matrices forms the *orthogonal group* $\mathrm{O}(d)$. Its Lie algebra $\mathfrak{so}(d)$ consists of $d \times d$ skew-symmetric matrices.

- The *special orthogonal group* $\mathrm{SO}(d) = \mathrm{SL}(d) \cap \mathrm{O}(d)$ consists of $d \times d$ real orthogonal matrices with unit determinant.

- The *symplectic group* $\mathrm{Sp}(d)$ is formed by all $2d \times 2d$ real matrices M verifying $M^T J M = J$, where J is the canonical matrix (1.34). The corresponding Lie algebra $\mathfrak{sp}(d)$ is the set of matrices B such that $JB + B^T J = 0$.

- The *special unitary group* $\mathrm{SU}(d)$ is made out of all complex unitary matrices with unit determinant.

Let $\mathcal{G}$ be a matrix Lie group. Then, the set of all matrices X such that $\mathrm{e}^{tX} \in \mathcal{G}$ for all real numbers t form a Lie algebra [283]. Thus, the matrix exponential establishes a connection between matrix Lie groups and Lie algebras. More generally, the tangent space $T\mathcal{G}_e$ to a Lie group at the identity e has a natural Lie algebra structure $\mathfrak{g}$ with the Lie bracket defined as

$$[X, Y] = \left.\frac{\partial^2}{\partial s \partial t}\right|_{s=t=0} \varphi_{sX} \circ \psi_{tY} \circ \varphi_{-sX} \circ \psi_{-tY}, \tag{A.34}$$

where φ_{sX} and ψ_{tY} are the flows corresponding to $X, Y \in T\mathcal{G}_e$, respectively [16]. This algebra is called the *Lie algebra of the Lie group* $\mathcal{G}$. It can be shown that the Lie bracket of left-invariant vector fields on a Lie group $\mathcal{G}$ is a left-invariant vector field whose value at $e \in \mathcal{G}$ is precisely the Lie bracket (A.34) of the values of the original vector fields at the identity e [275].

More specifically, if we denote by $L_a : \mathcal{G} \longrightarrow \mathcal{G}$, for any element $a \in \mathcal{G}$, the map such that $L_a(x) = ax$, then the smooth vector field X on $\mathcal{G}$ is *left-invariant* if and only if $X_a = (dL_a)_e X_e$, where $(dL_a)_e$ denotes the differential of L_a at the identity of $\mathcal{G}$. The left-invariant smooth vector fields forms a subalgebra $\mathfrak{g}$ of the Lie algebra of all smooth vector fields. The linear map $X \longmapsto X_e$ of $\mathfrak{g}$ onto $T\mathcal{G}_e$ is an isomorphism [204, 275].

The exponential mapping $\exp : \mathfrak{g} \longrightarrow \mathcal{G}$ is defined as $\exp(X) = \beta(1)$, where $\beta(t) \in \mathcal{G}$ is the one-parameter group solution of the differential equation

$$\frac{d\beta(t)}{dt} = X\beta(t), \qquad \beta(0) = e$$

and one writes $\exp(tX) = \beta(t)$. This exponential map coincides with the usual exponential matrix function if $\mathcal{G}$ is a matrix Lie group.

For any element $a \in \mathcal{G}$ consider the smooth isomorphism $\Phi_a(x) = axa^{-1}$, $x \in \mathcal{G}$, and its differential $(d\Phi_a)_e : \mathfrak{g} \longrightarrow \mathfrak{g}$, denoted by Ad_a. In the case of a linear group, then $\mathrm{Ad}_a X = aXa^{-1}$ (both a and X are now matrices). It is clear that the mapping $\mathrm{Ad} : a \longmapsto \mathrm{Ad}_a$ is a homomorphism of $\mathcal{G}$ onto the Lie group of all nonsingular linear operators of the vector space $\mathfrak{g}$. This is the *adjoint representation* of $\mathcal{G}$.

The differential $(d\,\mathrm{Ad})_e$ of the homomorphism Ad at the point e is precisely the linear "ad" operator introduced in (A.29) [204, 275]:

$$\mathrm{ad}_X(Y) = \left.\frac{d}{ds}\right|_{s=0} \mathrm{Ad}_{\sigma(s)} Y,$$

where $\sigma(s)$ is a smooth curve on $\mathcal{G}$ such that $\sigma(0) = I$ and $\dot{\sigma}(0) = X$. Consequently,

$$\mathrm{Ad}_{\exp X} = \mathrm{e}^{\mathrm{ad}_X}.$$

In the special case that $\mathcal{G}$ is a linear group, then this formula reads explicitly [283]

$$\mathrm{e}^X Y \mathrm{e}^{-X} = \mathrm{e}^{\mathrm{ad}_X} Y = \sum_{k=0}^{\infty} \frac{1}{k!} \mathrm{ad}_X^k Y.$$

Therefore

$$\mathrm{e}^X \mathrm{e}^Y \mathrm{e}^{-X} = \mathrm{e}^Z, \qquad \text{with} \qquad Z = \mathrm{e}^{\mathrm{ad}_X} Y. \tag{A.35}$$

A.3.5 Free Lie algebras

In the analysis of splitting methods in geometric numerical integration and also in the development of Lie-group methods, the concept of *free Lie algebra* plays a fundamental role. In both cases it is necessary to carry out computations in a Lie algebra, and so it is worthwhile to analyze the general case where no particular algebraic structure is assumed, except for what is common to all Lie algebras. This is the intuitive origin of a free Lie algebra. More specifically, given an arbitrary index set I (either finite or countably infinite), a Lie algebra $\mathfrak{g}$ is *free* over the set I if [260]

1. for every $i \in I$, there corresponds an element $X_i \in \mathfrak{g}$;

2. for any Lie algebra $\mathfrak{h}$ and any function $i \mapsto Y_i \in \mathfrak{h}$, there exists a unique Lie algebra homomorphism $\pi : \mathfrak{g} \to \mathfrak{h}$ satisfying $\pi(X_i) = Y_i$ for all $i \in I$.

If $\mathcal{T} = \{X_i : i \in I\} \subset \mathfrak{g}$, then the algebra $\mathfrak{g}$ can be viewed as the set of all Lie brackets of X_i. In this sense, we can say that $\mathfrak{g}$ is the free Lie algebra generated by $\mathcal{T}$ and we denote $\mathfrak{g} = \mathcal{L}(X_1, X_2, \ldots)$.

It is important to remark that $\mathfrak{g}$ is a universal object and that computations in $\mathfrak{g}$ can be applied in any particular Lie algebra $\mathfrak{h}$ via the homomorphism π [260], just by replacing each abstract element X_i with the corresponding Y_i.

In practical calculations, it is useful to represent a free Lie algebra by means of a basis (in the vector space sense). There are several systematic procedures to construct such a basis. One of them makes use of the so-called Hall sets, whose elements can be viewed as completed bracketed words formed by letters in some alphabet [73, 296]. Denoting such a set by H, we can equip it with a total order by means of length $\ell(\cdot)$ of the elements of H as follows. First it is assumed that $\mathcal{T} \subset H$, and $\ell(X) = 1$ if $X \in \mathcal{T}$. If $w \notin \mathcal{T}$ is a member of H, then it is of the form $w = [u, v]$, with $u, v \in H$, and we set $\ell(w) = \ell(u) + \ell(v)$. Then $u < v$ if $\ell(u) < \ell(v)$, whereas elements of the same length are ordered internally as we please (typically by following some lexicographical rule). Elements of length 2 in the Hall set are of the form $[X, Y]$, with $X, Y \in \mathcal{T}$ and $X < Y$. Elements of length greater than or equal to 3 are included if and only if they are of the form $[u, [v, w]]$, $u, v, w \in H$, $[v, w] \in H$, $v \leq u < [v, w]$.

For instance, suppose that the free Lie algebra is generated by just two elements, $\mathcal{T} = \{X, Y\}$, and thus it is denoted as $\mathcal{L}(X, Y)$. Then, the elements of the Hall basis with length ≤ 4 are the following:

$$
\begin{array}{lll}
X \quad Y & & \\
[X,Y] & & \\
[X,[X,Y]] & [Y,[X,Y]] & \\
[X,[X,[X,Y]]] & [Y,[X,[X,Y]]] & [Y,[Y,[X,Y]]].
\end{array}
$$

It is possible to compute the number of elements in the Hall basis with precisely n iterated brackets. This corresponds to the dimension of the linear subspace in the free Lie algebra generated by all the independent Lie brackets of order n, denoted by $\mathcal{L}_n(X, Y)$. Its dimension c_n is provided by the Magnus–Witt formula [207, 243]:

$$
c_n = \frac{1}{n} \sum_{d|n} \mu(d) s^{n/d}, \tag{A.36}
$$

where s is the number of generators, the sum is over all (positive) divisors d of the degree n, and $\mu(d)$ is the Möbius function, defined by the rule $\mu(1) = 1$, $\mu(d) = (-1)^k$ if d is the product of k distinct prime factors and $\mu(d) = 0$ otherwise [243].

The number c_n grows very fast indeed with n (asymptotically, one has $c_n = \mathcal{O}(2^n/n)$). To illustrate this fact, consider again the case of two generators, $\mathcal{L}(X, Y)$. In that case it is customary to denote by $\mathcal{L}_n(X, Y)$ the linear subspace generated by the independent Lie brackets of order n (i.e., of nested commutators involving n elements X and Y), so that $c_n = \dim \mathcal{L}_n(A, B)$. For $n \leq 12$ we have the following result:

n	1	2	3	4	5	6	7	8	9	10	11	12
c_n	1	1	2	3	6	9	18	30	56	99	186	335

A.4 The Baker–Campbell–Hausdorff formula

Another powerful result in the theory of Lie algebras and Lie groups is the Baker–Campbell–Hausdorff (BCH) formula (or theorem). In particular, it allows one to explicitly write the operation of multiplication in a Lie group in terms of the Lie bracket operation in its Lie algebra and also prove the existence of a local Lie group with a given Lie algebra [127, 161].

Let X and Y be non-commutative operators. By introducing the formal series for the exponential function

$$\mathrm{e}^X \, \mathrm{e}^Y = \sum_{p,q=0}^{\infty} \frac{1}{p! \, q!} X^p \, Y^q$$

and substituting this series in the formal series defining the logarithm function, one gets

$$\Phi(X, Y) = \log(\mathrm{e}^X \, \mathrm{e}^Y) = \sum_{k=1}^{\infty} \frac{(-1)^{k-1}}{k} \sum \frac{X^{p_1} Y^{q_1} \ldots X^{p_k} Y^{q_k}}{p_1! \, q_1! \ldots p_k! \, q_k!},$$

where the inner summation extends over all non-negative integers p_1, q_1, $\ldots$, p_k, q_k for which $p_i + q_i > 0$ ($i = 1, 2, \ldots, k$). Gathering together the terms for which $p_1 + q_1 + p_2 + q_2 + \cdots + p_k + q_k = m$ we can write

$$\Phi(X, Y) = X + Y + \sum_{m \geq 2} \Phi_m(X, Y), \tag{A.37}$$

where $\Phi_m(X, Y)$ is a homogeneous polynomial of degree m in the non-commuting variables X and Y.

The Baker–Campbell–Hausdorff (BCH) theorem asserts that every polynomial $\Phi_m(X, Y)$ in (A.37) is indeed a homogeneous Lie polynomial, that is, a linear combination of commutators of the form $[V_1, [V_2, \ldots, [V_{m-1}, V_m] \cdots]]$ with $V_i \in \{X, Y\}$ for $1 \leq i \leq m$, the coefficients being rational constants. As usual, the commutator $[X, Y]$ is defined as $XY - YX$. This theorem proves to be very useful in various fields of mathematics (see [68] for a comprehensive treatment).

If $\mathbb{K}$ is any field of characteristic zero, let us denote by $\mathbb{K}\langle X,Y\rangle$ the associative algebra of polynomials in the non-commuting variables X and Y [275]. With the operation $X,Y \longmapsto [X,Y]$ one can introduce a commutator Lie algebra $[\mathbb{K}\langle X,Y\rangle]$ in a natural way. Then, in $[\mathbb{K}\langle X,Y\rangle]$, the set of all Lie polynomials in X and Y (i.e., all possible expressions obtained from X, Y by addition, multiplication by numbers and the Lie operation $[a,b] = ab - ba$) forms a subalgebra $\mathcal{L}(X,Y)$, which in fact is a free Lie algebra with generators X, Y [275]. With this notation, the BCH theorem can be formulated as four statements, each one more stringent than the preceding [323]. Specifically,

(A) The equation $\mathrm{e}^X\mathrm{e}^Y = \mathrm{e}^\Phi$ has a solution Φ in $\mathbb{K}\langle X,Y\rangle$.

(B) The solution Φ lies in $\mathcal{L}(X,Y)$.

(C) The exponent Φ is an analytic function of X and Y.

(D) The exponent Φ can be computed by a series

$$\Phi(X,Y) = X + Y + \Phi_2(X,Y) + \Phi_3(X,Y)\cdots \tag{A.38}$$

where every polynomial $\Phi_n(X,Y) \in \mathcal{L}_n(X,Y)$.

In [323] the question of the global validity of the BCH theorem is analyzed in detail. It is shown that, whereas statement (A) is always true, statements (B), (C) and (D) are only globally valid in a free Lie algebra. Thus, in particular, if X and Y are elements of a normed algebra (for instance $d \times d$ matrices), the resulting series of normed elements is not guaranteed to converge out of a neighborhood of zero and therefore cannot be used to compute Φ unless the norm of X and Y are sufficiently small.

Although the BCH theorem establishes the precise algebraic structure of the exponent Φ in $\mathrm{e}^X\,\mathrm{e}^Y = \mathrm{e}^\Phi$, it does not provide simple ways to compute explicitly the series (A.38). As a matter of fact, the problem of effectively computing the BCH series up to arbitrary degree has a long history, and different procedures have been proposed over the years (see [94] for a review). Most of the procedures lead to expressions where not all the iterated commutators are linearly independent, due to the Jacobi identity and other identities appearing at higher degrees. Equivalently, the resulting expressions are not formulated directly in terms of a basis of the free Lie algebra $\mathcal{L}(X,Y)$.

To illustrate this point, let us consider the explicit expression for Φ_m obtained by Dynkin [130, 129] in the form

$$\Phi_m(X,Y) = \sum_{p_i,q_i} \frac{(-1)^{m-1}}{m} \frac{[X^{p_1}Y^{q_1}\ldots X^{p_m}Y^{q_m}]}{(\sum_{i=1}^m(p_i+q_i))\,p_1!\,q_1!\ldots p_m!\,q_m!}. \tag{A.39}$$

Here the summation is taken over all non-negative integers $p_1, q_1, \ldots, p_m, q_m$ such that $p_1+q_1>0, \ldots, p_m+q_m>0$ and $[X^{p_1}Y^{q_1}\ldots X^{p_m}Y^{q_m}]$ denotes the right-nested commutator based on the *word* $X^{p_1}Y^{q_1}\ldots X^{p_m}Y^{q_m}$. Thus, for instance,

$$[XY^2X^2Y] \equiv [XYYXXY] \equiv [X,[Y,[Y,[X,[X,Y]]]]].$$

Expression (A.39) can be used in principle to compute Φ_n in the BCH series up to any desired order. One should notice, however, that not all the terms are independent, due to the many existing redundancies. Thus, in particular, $[X^3Y^1] = [X^1Y^0X^2Y^1] = [X,[X,[X,Y]]]$. An additional source of redundancies arises from the Jacobi identity [320]:

$$[X_1,[X_2,X_3]] + [X_2,[X_3,X_1]] + [X_3,[X_1,X_2]] = 0, \qquad \text{(A.40)}$$

for any three variables X_1, X_2, X_3 and other identities obtained from it.

An efficient algorithm to construct the BCH series up to an arbitrary degree in terms of commutators is proposed in [94], based on the results in [263]. In particular, explicit expressions of Φ_m up to $m=20$ in terms of the classical Hall and Lyndon bases of $\mathcal{L}(X,Y)$ can be obtained with relatively modest computer requirements. They can be found at the website

`https://www.gicas.uji.es/Research/bch.html`.

For illustration, the first terms in the classical Hall basis read

$$\begin{aligned}
\Phi_2 &= \frac{1}{2}[X,Y], \qquad \Phi_3 = \frac{1}{12}\big([X,[X,Y]] - [Y,[X,Y]]\big),\\
\Phi_4 &= -\frac{1}{24}[Y,[X,[X,Y]]],\\
\Phi_5 &= -\frac{1}{720}[X,[X,[X,[X,Y]]]] - \frac{1}{180}[Y,[X,[X,[X,Y]]]]\\
&\quad + \frac{1}{180}[Y,[Y,[X,[X,Y]]]] + \frac{1}{720}[Y,[Y,[Y,[X,Y]]]]\\
&\quad - \frac{1}{120}[[X,Y],[X,[X,Y]]] - \frac{1}{360}[[X,Y],[Y,[X,Y]]],\\
\Phi_6 &= \frac{1}{1440}[Y,[X,[X,[X,[X,Y]]]]] + \frac{1}{360}[Y,[Y,[X,[X,[X,Y]]]]]\\
&\quad + \frac{1}{1440}[Y,[Y,[Y,[X,[X,Y]]]]] + \frac{1}{240}[[X,Y],[Y,[X,[X,Y]]]]\\
&\quad + \frac{1}{720}[[X,Y],[Y,[Y,[X,Y]]]] - \frac{1}{240}[[X,[X,Y]],[Y,[X,Y]]].
\end{aligned}$$

When obtaining the order conditions of time-symmetric splitting methods, it is helpful to compute the operator W defined by

$$\exp\left(\frac{1}{2}X\right)\exp(Y)\exp\left(\frac{1}{2}X\right) = \exp(W). \qquad \text{(A.41)}$$

This is the so-called symmetric BCH formula. Two applications of the usual BCH formula give the expression of $W(X,Y)$. More efficient procedures exist, however, that allow one to construct explicitly the series $\sum_{n\geq 1} W_n(X,Y)$ defining W in terms of independent commutators involving X and Y up to an arbitrarily high degree. In general, $W_{2m} = 0$ for $m \geq 1$, whereas terms W_{2m+1} up to W_{19} in both Hall and Lyndon bases can again be found at the website

`https://www.gicas.uji.es/Research/bch.html`.

For the first terms one has

$$
\begin{aligned}
W_1(X,Y) &= X + Y, \\
W_3(X,Y) &= -\frac{1}{24}[X,[X,Y]] - \frac{1}{12}[Y,[X,Y]], \\
W_5(X,Y) &= \frac{7}{5760}[X,[X,[X,[X,Y]]]] + \frac{7}{1440}[Y,[X,[X,[X,Y]]]] \\
&\quad + \frac{1}{180}[Y,[Y,[X,[X,Y]]]] + \frac{1}{720}[Y,[Y,[Y,[X,Y]]]] \\
&\quad + \frac{1}{480}[[X,Y],[X,[X,Y]]] - \frac{1}{360}[[X,Y],[Y,[X,Y]]], \\
W_7(X,Y) &= -\frac{31}{967680}[X,[X,[X,[X,[X,[X,Y]]]]]] \\
&\quad - \frac{31}{161280}[Y,[X,[X,[X,[X,[X,Y]]]]]] \\
&\quad - \frac{13}{30240}[Y,[Y,[X,[X,[X,[X,Y]]]]]] - \frac{53}{120960}[Y,[Y,[Y,[X,[X,[X,Y]]]]]] \\
&\quad - \frac{1}{5040}[Y,[Y,[Y,[Y,[X,[X,Y]]]]]] - \frac{1}{30240}[Y,[Y,[Y,[Y,[Y,[X,Y]]]]]] \\
&\quad - \frac{53}{161280}[[X,Y],[X,[X,[X,[X,Y]]]]] - \frac{11}{12096}[[X,Y],[Y,[X,[X,[X,Y]]]]] \\
&\quad - \frac{3}{4480}[[X,Y],[Y,[Y,[X,[X,Y]]]]] - \frac{1}{10080}[[X,Y],[Y,[Y,[Y,[X,Y]]]]] \\
&\quad - \frac{1}{4032}[[X,Y],[[X,Y],[X,[X,Y]]]] - \frac{1}{6720}[[X,Y],[[X,Y],[Y,[X,Y]]]] \\
&\quad - \frac{19}{80640}[[X,[X,Y]],[X,[X,[X,Y]]]] - \frac{1}{10080}[[X,[X,Y]],[Y,[X,[X,Y]]]] \\
&\quad + \frac{17}{40320}[[X,[X,Y]],[Y,[Y,[X,Y]]]] - \frac{53}{60480}[[Y,[X,Y]],[X,[X,[X,Y]]]] \\
&\quad - \frac{19}{13440}[[Y,[X,Y]],[Y,[X,[X,Y]]]] - \frac{1}{5040}[[Y,[X,Y]],[Y,[Y,[X,Y]]]].
\end{aligned}
$$

On the other hand, expressing the BCH series in terms of right-nested commutators presents several advantages, especially when the series is considered in some particular physical settings: as we have seen in subsection 3.4.1, there are problems whose structure leads in a natural way to consider the BCH series of two operators X, Y satisfying $[Y,[Y,[X,Y]]] = 0$. Although in this case it is still possible to construct a generalized Hall basis [240], it is much simpler to identify the non-vanishing terms when using right-nested commutators. On the other hand, and contrary to the Hall basis, there is not a straightforward procedure to construct a set of independent right-nested commutators generating each homogeneous subspace of $\mathcal{L}(X,Y)$. In spite of these difficulties, in [14] compact expressions for Φ_m are obtained up to $m = 10$ in terms of right-nested commutators containing fewer terms than in the classical Hall basis. They can also be found at the previous website.

With the previous notation for right-nested commutators, the first terms for the BCH series (A.37) are

$$
\begin{aligned}
\Phi_1(X,Y) &= X + Y, \\
\Phi_2(X,Y) &= \frac{1}{2}[X\,Y], \\
\Phi_3(X,Y) &= \frac{1}{12}[X^2\,Y] - \frac{1}{12}[YXY], \\
\Phi_4(X,Y) &= -\frac{1}{24}[XYXY], \\
\Phi_5(X,Y) &= -\frac{1}{720}[X^4\,Y] - \frac{1}{120}[XYX^2Y] - \frac{1}{360}[XY^2XY] + \frac{1}{360}[YX^3Y] \\
&\quad + \frac{1}{120}[Y^2X^2Y] + \frac{1}{720}[Y^3XY], \\
\Phi_6(X,Y) &= \frac{1}{1440}\left(-2\,[X^2Y^2XY] + 6\,[XY^2X^2Y] + [XY^3XY] + [YX^4Y]\right),
\end{aligned}
$$

whereas for the symmetric BCH formula we have

$$
\begin{aligned}
W_1(X,Y) &= X + Y, \\
W_3(X,Y) &= -\frac{1}{24}[X^2\,Y] - \frac{1}{12}[YXY], \\
W_5(X,Y) &= \frac{7}{5760}[X^4\,Y] + \frac{1}{480}[XYX^2Y] - \frac{1}{360}[XY^2XY] + \frac{1}{360}[YX^3Y] \\
&\quad + \frac{1}{120}[Y^2X^2Y] + \frac{1}{720}[Y^3XY], \\
W_7(X,Y) &= -\frac{31}{967680}[X^6Y] - \frac{19}{80640}[X^2YX^3Y] - \frac{1}{10080}[X^2Y^2X^2Y] \\
&\quad + \frac{23}{161280}[XYX^4Y] - \frac{1}{20160}[XYXYX^2Y] - \frac{1}{6720}[XYXY^2XY] \\
&\quad + \frac{13}{60480}[XY^2X^3Y] + \frac{19}{40320}[XY^3X^2Y] + \frac{1}{10080}[XY^4XY] \\
&\quad - \frac{1}{10080}[YX^5Y] - \frac{1}{1260}[YXYX^3Y] - \frac{1}{1680}[YXY^2X^2Y] \\
&\quad + \frac{1}{3360}[Y^2X^4Y] - \frac{1}{3360}[Y^2XYX^2Y] - \frac{1}{2520}[Y^2XY^2XY] \\
&\quad + \frac{1}{7560}[Y^3X^3Y] + \frac{1}{10080}[Y^4X^2Y] - \frac{1}{30240}[Y^5XY].
\end{aligned}
$$

Bibliography

[1] A. Abad, R. Barrio, F. Blesa, and M. Rodríguez, *TIDES, a Taylor series Integrator for Differential EquationS*, ACM Trans. Math. Softw., 39 (2012), pp. 1–28.

[2] R. Abraham and J. E. Marsden, *Foundations of Mechanics*, Addison-Wesley, Second ed., 1978.

[3] M. Abramowitz and I. Stegun, *Handbook of Mathematical Functions*, Dover, 1965.

[4] A. Agrachev and R. Gamkrelidze, *The shuffle product and symmetric groups*, in Differential Equations, Dynamical Systems, and Control Science, K. Elworthy, W. Everitt, and E. Lee, eds., Marcel Dekker, 1994, pp. 365–382.

[5] M. Aichinger and E. Krotscheck, *A fast configuration space method for solving local Kohn–Sham equations*, Comput. Mater. Sci., 34 (2005), pp. 188–212.

[6] E. Akhmatskaya, M. Fernández-Pendás, T. Radivojević, and J. M. Sanz-Serna, *Adaptive splitting integrators for enhancing sampling efficiency of modified Hamiltonian Monte Carlo methods in molecular simulations*, Langmuir, 33 (2017), pp. 11530–11542.

[7] I. Alonso-Mallo, B. Cano, and N. Reguera, *Avoiding order reduction when integrating linear initial boundary value problems with exponential splitting methods*, IMA J. Numer. Anal., 38 (2018), pp. 1294–1323.

[8] I. Alonso-Mallo, B. Cano, and N. Reguera, *Avoiding order reduction when integrating reaction-diffusion boundary value problems with exponential splitting methods*, J. Comput. Appl. Math., 357 (2019), pp. 228–250.

[9] A. Alvermann and H. Fehske, *High-order commutator-free exponential time-propagation of driven quantum systems*, J. Comput. Phys., 230 (2011), pp. 5930–5956.

[10] D. An, D. Fang, and L. Lin, *Time-dependent Hamiltonian simulation of highly oscillatory dynamics and superconvergence for the Schrödinger equation*, Quantum, 6 (2022), p. 690.

[11] C. Andrieu, N. de Freitas, A. Doucet, and M. I. Jordan, *An introduction to MCMC for machine learning*, Machine Learning, 50 (2003), pp. 5–43.

[12] M. Antoñana, J. Makazaga, and A. Murua, *Efficient implementation of symplectic implicit Runge–Kutta schemes with simplified Newton iterations*, Numer. Algor., 78 (2018), pp. 63–86.

[13] A. Arnal, F. Casas, and C. Chiralt, *A general formula for the Magnus expansion in terms of iterated integrals of right-nested commutators*, J. Phys. Commun., 2 (2018), p. 035024.

[14] A. Arnal, F. Casas, and C. Chiralt, *A note on the Baker–Campbell–Hausdorff series in terms of right-nested commutators*, Mediterr. J. Math., 18 (2021), p. 53.

[15] V. I. Arnold, *Geometrical Methods in Theory of Ordinary Differential Equations*, Springer, Second ed., 1988.

[16] V. I. Arnold, *Mathematical Methods of Classical Mechanics*, Springer-Verlag, Second ed., 1989.

[17] V. I. Arnold, V. Kozlov, and A.I.Neishtadt, *Mathematical Aspects of Classical and Celestial Mechanics*, Springer, 3rd ed., 2006.

[18] J. Auer, E. Krotscheck, and S. A. Chin, *A fourth-order real-space algorithm for solving local Schrödinger equations*, J. Chem. Phys., 115 (2001), pp. 6841–6846.

[19] W. Auzinger, H. Hofstätter, and O. Koch, *Non-existence of generalized splitting methods with positive coefficients of order higher than fou*, Appl. Math. Letters, 97 (2019), pp. 48–52.

[20] P. Bader, *Geometric Integrators for Schödinger Equations*, PhD thesis, Universitat Politècnica de València, 2014.

[21] P. Bader, S. Blanes, and F. Casas, *Solving the Schrödinger eigenvalue problem by the imaginary time propagation technique using splitting methods with complex coefficients*, J. Chem. Phys., (2013), p. 124117.

[22] P. Bader, A. Iserles, K. Kropielnicka, and P. Singh, *Efficient methods for linear Schrödinger equation in the semiclassical regime with time-dependent potential*, Proc. R. Soc. A, 472 (2016), p. 20150733.

[23] P. Bader, D. McLaren, G. Quispel, and M. Webb, *Volume preservation by Runge–Kutta methods*, Appl. Numer. Math., 109 (2016), pp. 123–137.

[24] N. Balakrishnan, C. Kalyanaraman, and N. Sathyamurthy, *Time-dependent quantum mechanical approach to reactive scattering and related processes*, Phys. Rep., 280 (1997), pp. 79–144.

[25] A. Bandrauk, E. Dehghanian, and H. Lu, *Complex integration steps in decomposition of quantum exponential evolution operators*, Chem. Phys. Lett., 419 (2006), pp. 346–350.

[26] A. Bandrauk and H. Shen, *Improved exponential split operator method for solving the time-dependent Schrödinger equation*, Chem. Phys. Lett., 176 (1991), pp. 428–432.

[27] R. Barrio, *Taylor Series Method*, in Encyclopedia of Applied and Computational Mathematics, B. Engquist, ed., Springer, 2015.

[28] R. Barrio, M. Rodríguez, A. Abad, and F. Blesa, *Breaking the limits: The Taylor series method*, Appl. Math. Comput., 217 (2011), pp. 7940–7954.

[29] G. Benettin and A. Giorgilli, *On the Hamiltonian interpolation of near to the identity symplectic mappings with applications to symplectic integration algorithms*, J. Statist. Phys., 74 (1994), pp. 1117–1143.

[30] J. Bernier, S. Blanes, F. Casas, and A. Escorihuela-Tomàs, *Symmetric-conjugate splitting methods for linear unitary problems*, BIT Numer. Math., 63 (2023), p. 58.

[31] J. Bernier, S. Blanes, F. Casas, and A. Escorihuela-Tomàs, *On alternating-conjugate splitting methods*, tech. rep., arXiv:2503.08453, 2025.

[32] D. W. Berry, A. M. Childs, R. Cleve, R. Kothari, and R. D. Somma, *Simulating Hamiltonian dynamics with a truncated Taylor series*, Phys. Rev. Lett., 114 (2015), p. 090502.

[33] S. Blanes, *Novel parallel in time integrators for ODEs*, Appl. Math. Letters, 122 (2021), p. 107542.

[34] S. Blanes and C. J. Budd, *Adaptive geometric integrators for Hamiltonian problems with approximate scale invariance*, SIAM J. Sci. Comput., 26 (2005), pp. 1089–1113.

[35] S. Blanes, M. P. Calvo, F. Casas, and J. M. Sanz-Serna, *Symmetrically processed splitting integrators for enhanced Hamiltonian Monte Carlo sampling*, SIAM J. Sci. Comput., 43 (2021), pp. A3357–A3371.

[36] S. Blanes and F. Casas, *On the necessity of negative coefficients for operator splitting schemes of order higher than two*, Appl. Numer. Math., 54 (2005), pp. 23–37.

[37] S. Blanes and F. Casas, *Splitting methods for non-autonomous separable dynamical systems*, J. Phys. A: Math. Gen., 39 (2006), pp. 5405–5423.

[38] S. Blanes, F. Casas, P. Chartier, and A. Escorihuela-Tomàs, *On symmetric-conjugate composition methods in the numerical integration of differential equations*, Math. Comput., 91 (2022), pp. 1739–1761.

[39] S. Blanes, F. Casas, P. Chartier, and A. Murua, *Optimized high-order splitting methods for some classes of parabolic equations*, Math. Comput., 82 (2013), pp. 1559–1576.

[40] S. Blanes, F. Casas, and A. Escorihuela-Tomàs, *Applying splitting methods with complex coefficients to the numerical integration of unitary problems*, J. Comput. Dyn., 9 (2022), pp. 85–101.

[41] S. Blanes, F. Casas, A. Farrés, J. Laskar, J. Makazaga, and A. Murua, *New families of symplectic splitting methods for numerical integration in dynamical astronomy*, Appl. Numer. Math., 68 (2013), pp. 58–72.

[42] S. Blanes, F. Casas, C. González, and M. Thalhammer, *Convergence analysis of high-order commutator-free quasi-Magnus integrators for nonautonomous linear Schrödinger equations*, IMA J. Numer. Anal., 41 (2021), pp. 594–617.

[43] S. Blanes, F. Casas, C. González, and M. Thalhammer, *Efficient spliting methods based on modified potentials: numerical integration of linear parabolic problems and imaginary time propagation of the Schrödinger equation*, Commun. Comput. Phys., 33 (2023), pp. 937–961.

[44] S. Blanes, F. Casas, C. González, and M. Thalhammer, *Generalisation of splitting methods based on modified potentials to nonlinear evolution equations of parabolic and Schrödinger type*, Comput. Phys. Comm., 295 (2024), p. 109007.

[45] S. Blanes, F. Casas, and A. Murua, *On the numerical integration of ordinary differential equations by processed methods*, SIAM J. Numer. Anal., 42 (2004), pp. 531–552.

[46] S. Blanes, F. Casas, and A. Murua, *Composition methods for differential equations with processing*, SIAM J. Sci. Comput., 27 (2006), pp. 1817–1843.

[47] S. Blanes, F. Casas, and A. Murua, *Symplectic splitting operator methods tailored for the time-dependent Schrödinger equation*, J. Chem. Phys., 124 (2006), p. 234105.

[48] S. Blanes, F. Casas, and A. Murua, *On the linear stability of splitting methods*, Found. Comp. Math., 8 (2008), pp. 357–393.

[49] S. Blanes, F. Casas, and A. Murua, *Splitting and composition methods in the numerical integration of differential equations*, Bol. Soc. Esp. Mat. Apl., 45 (2008), pp. 89–145.

[50] S. Blanes, F. Casas, and A. Murua, *Error analysis of splitting methods for the time dependent Schrödinger equation*, SIAM J. Sci. Comput., 33 (2011), pp. 1525–1548.

[51] S. Blanes, F. Casas, and A. Murua, *An efficient algorithm based on splitting for the time integration of the Schrödinger equation*, J. Comput. Phys., 303 (2015), pp. 396–412.

[52] S. Blanes, F. Casas, and A. Murua, *Splitting methods for differential equations*, Acta Numerica, 33 (2024), pp. 1–161.

[53] S. Blanes, F. Casas, J. Oteo, and J. Ros, *Magnus and Fer expansions for matrix differential equations: the convergence problem*, J. Phys. A: Math. Gen., 22 (1998), pp. 259–268.

[54] S. Blanes, F. Casas, J. A. Oteo, and J. Ros, *The Magnus expansion and some of its applications*, Phys. Rep., 470 (2009), pp. 151–238.

[55] S. Blanes, F. Casas, and J. Ros, *Extrapolation of symplectic integrators*, Celest. Mech. & Dyn. Astr., 75 (1999), pp. 149–161.

[56] S. Blanes, F. Casas, and J. Ros, *Symplectic integrators with processing: a general study*, SIAM J. Sci. Comput., 21 (1999), pp. 711–727.

[57] S. Blanes, F. Casas, and J. Ros, *High-order Runge–Kutta–Nyström geometric methods with processing*, Appl. Numer. Math., 39 (2001), pp. 245–259.

[58] S. Blanes, F. Casas, and J. M. Sanz-Serna, *Numerical integrators for the Hybrid Monte Carlo method*, SIAM J. Sci. Comput., 36 (2014), pp. A1556–A1580.

[59] S. Blanes, F. Casas, and L. Shaw, *Generalized extrapolation methods based on compositions of a basic 2nd-order scheme*, Appl. Math. Comput., 473 (2024), p. 128663.

[60] S. Blanes, F. Casas, and M. Thalhammer, *High-order commutator-free quasi-Magnus exponential integrators for non-autonomous linear evolution equations*, Comput. Phys. Comm., 220 (2017), pp. 243–262.

[61] S. Blanes, F. Casas, and M. Thalhammer, *Convergence analysis of high-order commutator-free quasi-Magnus integrators for nonautonomous linear evolution equations*, IMA J. Numer. Anal., 38 (2018), pp. 743–778.

[62] S. Blanes and A. Iserles, *Explicit adaptive symplectic integrators for solving Hamiltonian systems*, Celest. Mech. & Dyn. Astr., 114 (2012), pp. 297–317.

[63] S. Blanes, A. Iserles, and S. MacNamara, *Positivity-preserving methods for ordinary differential equations*, ESAIM Math. Model. Numer. Anal., 56 (2022), pp. 1843–1870.

[64] S. Blanes and P. C. Moan, *Fourth- and sixth-order commutator-free Magnus integrators for linear and non-linear dynamical systems*, Appl. Numer. Math., 56 (2006), pp. 1519–1537.

[65] S. Blanes and P. C. Moan, *Practical symplectic partitioned Runge–Kutta and Runge–Kutta–Nyström methods*, J. Comput. Appl. Math., 142 (2002), pp. 313–330.

[66] P. B. Bochev and C. Scovel, *On quadratic invariants and symplectic structure*, BIT, 34 (1994), pp. 337–345.

[67] C. Bolley and M. Crouzeix, *Conservation de la positivité lors de la discrétisation des problèmes d'évolution paraboliques*, RAIRO: Anal. Numér., 12 (1978), pp. 237–245.

[68] A. Bonfiglioli and R. Fulci, *Topics in Noncommutative Algebra. The Theorem of Campbell, Baker, Hausdorff and Dynkin*, vol. 2034 of Lecture Notes in Mathematics, Springer, 2012.

[69] M. Born, *The Mechanics of the Atom*, Frederick Ungar Publ., 1960.

[70] Y. Borns-Weil, D. Fang, and J. Zhang, *Discrete superconvergence analysis for quantum Magnus algorithms of unbounded Hamiltonian simulation*, tech. rep., arXiv:2502.20255, 2025.

[71] J. L. Bosse, A. M. Childs, C. Derby, F. M. Gambetta, A. Montanaro, and R. A. Santos, *Efficient and practical Hamiltonian simulation from time-dependent product formulas*, Nature Communications, 16 (2025), p. 2673.

[72] N. Bou-Rabee and J. M. Sanz-Serna, *Geometric integrators and the Hamiltonian Monte Carlo method*, Acta Numerica, 27 (2018), pp. 113–206.

[73] N. Bourbaki, *Lie Groups and Lie Algebras*, Chapters 1–3, Springer, 1989.

[74] J. P. Boyd, *Chebyshev and Fourier Spectral Methods*, Dover, Second ed., 2001.

[75] T. Bridges and S. Reich, *Multi-symplectic integrators: numerical schemes for Hamiltonian PDEs that conserve symplecticity*, Phys. Lett. A, 284 (2001), pp. 184–193.

[76] S. Brooks, A. Gelman, G. L. Jones, and X.-L. Meng, eds., *Handbook of Markov Chain Monte Carlo*, CRC Press, 2011.

[77] J. C. Butcher, *The effective order of Runge–Kutta methods*, in Proceedings of the Conference on the Numerical Solution of Differential Equations, J. L. Morris, ed., vol. 109 of Lecture Notes in Mathematics, Springer, 1969, pp. 133–139.

[78] J. C. Butcher, *An algebraic theory of integration methods*, Math. Comput., 26 (1972), pp. 79–106.

[79] J. C. Butcher, *The Numerical Analysis of Ordinary Differential Equations*, John Wiley, 1987.

[80] M. P. Calvo, P. Chartier, A. Murua, and J. M. Sanz-Serna, *Numerical stroboscopic averaging for ODEs and DAEs*, Appl. Numer. Math., 61 (2011), pp. 1077–1095.

[81] M. P. Calvo and E. Hairer, *Accurate long-term integration of dynamical systems*, Appl. Numer. Math., 18 (1995), pp. 95–105.

[82] M. P. Calvo, A. Iserles, and A. Zanna, *Runge-Kutta methods for orthogonal and isospectral flows*, Appl. Numer. Math., 22 (1996), pp. 153–163.

[83] M. P. Calvo, A. Murua, and J. M. Sanz-Serna, *Modified equations for ODEs*, in Chaotic Numerics, P. Kloeden and K. Palmer, eds., vol. 172 of Contemporary Mathematics, American Mathematical Society, 1994, pp. 63–74.

[84] M. P. Calvo and J. M. Sanz-Serna, *The development of variable-step symplectic integrators, with application to the two-body problem*, SIAM J. Sci. Comput., 14 (1993), pp. 936–952.

[85] M. P. Calvo and J. M. Sanz-Serna, *Heterogeneous multiscale methods for mechanical systems with vibrations*, SIAM J. Sci. Comput., 32 (2010), pp. 2029–2046.

[86] M. P. Calvo, J. M. Sanz-Serna, and B. Zhu, *High-order stroboscopic averaging methods for highly oscillatory dealy problems*, Appl. Numer. Math., 152 (2020), pp. 466–472.

[87] M. Campostrini and P. Rossi, *A comparison of numerical algorithms for dynamical fermions*, Nuclear Phys. B, 329 (1990), pp. 753–764.

[88] J. Candy and W. Rozmus, *A symplectic integration algorithm for separable hamiltonian functions*, J. Comput. Phys., 92 (1991), pp. 230–256.

[89] C. Canuto, M. Y. Hussaini, A. Quarteroni, and T. Z. Zang, *Spectral Methods. Fundamentals in Single Domains*, Springer, 2006.

[90] P. A. M. Casares, M. J. Zini, and J. M. Arrazola, *Quantum simulation of time-dependent Hamiltonians via commutator-free quasi-Magnus operators*, Quantum, 8 (2024), p. 1567.

[91] F. Casas, *Sufficient conditions for the convergence of the Magnus expansion*, J. Phys. A: Math. Theor., 40 (2007), pp. 15001–15017.

[92] F. Casas, A. Escorihuela-Tomàs, and P. A. M. Casares, *Approximating exponentials of commutators by optimized product formulas*, Quantum Information Processing, 24 (2025), p. 47.

[93] F. Casas and A. Iserles, *Explicit Magnus expansions for nonlinear equations*, J. Phys. A: Math. Gen., 39 (2006), pp. 5445–5461.

[94] F. Casas and A. Murua, *An efficient algorithm for computing the Baker–Campbell–Hausdorff series and some of its applications*, J. Math. Phys., 50 (2009), p. 033513.

[95] F. Casas and B. Owren, *Cost efficient Lie group integrators in the RKMK class*, BIT, 43 (2003), pp. 723–742.

[96] F. Castella, P. Chartier, S. Descombes, and G. Vilmart, *Splitting methods with complex times for parabolic equations*, BIT Numer. Math., 49 (2009), pp. 487–508.

[97] E. Celledoni, V. Grimm, R. I. McLachlan, D. I. McLaren, D. O'Neale, B. Owren, and G. R. W. Quispel, *Preserving energy resp. dissipation in numerical PDEs using the "Average Vector Field" method*, J. Comput. Phys., 231 (2012), pp. 6770–6789.

[98] E. CELLEDONI, R. I. MCLACHLAN, B. OWREN, AND G. R. W. QUISPEL, *Energy-preserving integrators and the structure of B-series*, Found. Comput. Math., 10 (2010), pp. 673–693.

[99] D. M. CEPERLEY, *Path integrals in the theory of condensed helium*, Rev. Mod. Phys., 67 (1995), pp. 279–355.

[100] J. CHAMBERS, *A hybrid symplectic integrator that permits close encounters between massive bodies*, Mon. Not. R. Astron. Soc., 304 (1999), pp. 793–799.

[101] J. E. CHAMBERS, *Symplectic integrators with complex time steps*, Astron. J., 126 (2003), pp. 1119–1126.

[102] R. CHAN AND A. MURUA, *Extrapolation of symplectic methods for Hamiltonian problems*, Appl. Numer. Math., 34 (2000), pp. 189–205.

[103] P. J. CHANNELL AND J. C. SCOVEL, *Symplectic integration of Hamiltonian systems*, Nonlinearity, 3 (1990), pp. 231–259.

[104] P. CHARTIER, E. FAOU, AND A. MURUA, *An algebraic approach to invariant preserving integrators: the case of quadratic and Hamiltonian invariants*, Numer. Math., 103 (2006), pp. 575–590.

[105] Y.-A. CHEN, A. M. CHILDS, M. HAFEZI, Z. JIANG, H. KIM, AND Y. XU, *Efficient product formulas for commutators and applications to quantum simulations*, Phys. Rev. Res., 4 (2022), p. 013191.

[106] A. M. CHILDS AND Y. SU, *Nearly optimal lattice simulation by product formulas*, Phys. Rev. Lett., 123 (2019), p. 050503.

[107] A. M. CHILDS, Y. SU, M. C. TRAN, N. WIEBE, AND S. ZHU, *Theory of Trotter error with commutator scaling*, Phys. Rev. X, 11 (2021), p. 011020.

[108] A. M. CHILDS AND N. WIEBE, *Hamiltonian simulation using linear combinations of unitary operators*, Quantum Information & Computation, 12 (2012), pp. 901–924.

[109] S. A. CHIN, *Symplectic integrators from composite operator factorizations*, Phys. Lett. A, 226 (1997), pp. 344–348.

[110] S. A. CHIN, *Structure of positive decomposition of exponential operators*, Phys. Rev. E, 71 (2005), p. 016703.

[111] S. A. CHIN, *Multi-product splitting and Runge–Kutta–Nyström integrators*, Celest. Mech. & Dyn. Astr., 106 (2010), pp. 391–406.

[112] S. A. CHIN, *Modern light on ancient feud: Robert Hooke and Newton's graphical method*, Historia Mathematica, 60 (2022), pp. 1–14.

[113] S. A. CHIN AND J. GEISER, *Multi-product operator splitting as a general method of solving autonomous and nonautonomous equations*, IMA J. Numer. Anal., 31 (2011), pp. 1552–1577.

[114] E. CODDINGTON AND N. LEVINSON, *Theory of Ordinary Differential Equations*, McGraw Hill, 1955.

[115] D. COHEN AND L. GAUCKLER, *One-stage exponential integrators for nonlinear Schrödinger equations over long times*, BIT Numer. Math., 52 (2012), pp. 877–903.

[116] D. Cohen, E. Hairer, and C. Lubich, *Modulated Fourier expansions of highly oscillatory differential equations*, Found. Comput. Math., 3 (2003), pp. 327–345.

[117] D. Cohen, E. Hairer, and C. Lubich, *Numerical energy conservation for multi-frequency oscillatory differential equations*, BIT Numer. Math., 45 (2005), pp. 287–305.

[118] D. Cohen, T. Matsuo, and X. Raynaud, *A multi-symplectic numerical integrator for the two-component Camassa Holm equation*, J. Nonlinear Math. Phys., 21 (2014), pp. 442–453.

[119] D. Cohen and J. Schweitzer, *High order numerical methods for highly oscillatory problem*, Math. Model. Numer. Anal., 49 (2015), pp. 695–711.

[120] M. Creutz and A. Gocksch, *Higher-order hybrid Monte Carlo algorithms*, Phys. Rev. Lett., 63 (1989), pp. 9–12.

[121] P. E. Crouch and R. Grossman, *Numerical integration of ordinary differential equations on manifolds*, J. Nonlinear Sci., 3 (1993), pp. 1–33.

[122] J. M. A. Danby, *Fundamentals of Celestial Mechanics*, Willmann-Bell, 1988.

[123] R. de Vogelaere, *Methods of integration which preserve the contact transformation property of the hamiltonian equations.* Report No. 4, Dept. Math., Univ. of Notre Dame, Notre Dame, Ind., 1956.

[124] A. Debussche and E. Faou, *Modified energy for split-step methods applied to the linear Schrödinger equation*, SIAM J. Numer. Anal., 47 (2009), pp. 3705–3719.

[125] K. Dekker and J. Verwer, *Stability of Runge–Kutta Methods for Stiff Nonlinear Differential Equations*, North-Holland, Amsterdam, 1984.

[126] J. C. del Valle and K. Kropielnicka, *Second order exponential splittings in the presence of unbounded and time-dependent operators: construction and convergence*, SIAM J. Numer. Anal., 63 (2025), pp. 288–305.

[127] A. J. Dragt, *Lie Methods for Nonlinear Dynamics with Applications to Accelerator Physics*, tech. rep., University of Maryland, 2020.

[128] S. Duane, A. D. Kennedy, B. J. Pendleton, and D. Roweth, *Hybrid Monte-Carlo*, Phys. Lett. B, 195 (1987), pp. 216–222.

[129] E. Dynkin, *Calculation of the coefficients in the Campbell–Hausdorff series*, in Selected Papers of E.B. Dynkin with Commentary, E. Dynkin, A. Yushkevich, G. Seitz, and A. Onishchik, eds., American Mathematical Society, 2000, pp. 31–35.

[130] E. B. Dynkin, *Evaluation of the coefficients of the Campbell–Hausdorff formula*, Dokl. Akad. Nauk. SSSR, 57 (1947), pp. 323–326.

[131] E. Weinan and B. Engquist, *The heterogeneous multiscale methods*, Comm. Math. Sci., 1 (2003), pp. 87–132.

[132] L. Einkemmer, M. Moccaldi, and A. Ostermann, *Efficient boundary corrected Strang splitting*, Appl. Math. Comput., 332 (2018), pp. 76–89.

[133] L. Einkemmer and A. Ostermann, *Overcoming order reduction in diffusion-reaction splitting. Part I: Dirichlet boundary conditions*, SIAM J. Sci. Comput., 37 (2015), pp. A1577–A1592.

[134] L. Einkemmer and A. Ostermann, *Overcoming order reduction in diffusion-reaction splitting. Part II: oblique boundary conditions*, SIAM J. Sci. Comput., 38 (2016), pp. A3741–A3757.

[135] K.-J. Engel and R. Nagel, *A Short Course on Operator Semigroups*, Springer, 2006.

[136] B. Engquist, A. Fokas, and A. Iserles, eds., *Highly Oscillatory Problems*, Cambridge University Press, 2009.

[137] B. Engquist and R. Tsai, *Heterogeneous multiscale methods for stiff ordinary differential equations*, Math. Comput., 74 (2005), pp. 1707–1742.

[138] D. Fang, D. Liu, and R. Sarkar, *Time-dependent Hamiltonian simulation via Magnus expansion: algorithm and superconvergence*, Commun. Math. Phys. 406 (2025), p. 128.

[139] E. Faou, *Geometric Numerical Integration and Schrödinger Equations*, European Mathematical Society, 2012.

[140] M. D. Feit, J. A. Fleck, and A. Steiger, *Solution of the Schrödinger equation by a spectral method*, J. Comp. Phys., 47 (1982), pp. 412–433.

[141] K. Feng, *Difference schemes for Hamiltonian formalism and symplectic geometry*, J. Comput. Math., 4 (1986), pp. 279–289.

[142] K. Feng and M. Qin, *Symplectic Geometric Algorithms for Hamiltonian Systems*, Zheijang Publ. United Group - Springer, 2010.

[143] K. Feng and M. Z. Qin, *The symplectic methods for the computation of Hamiltonian equations*, in Numerical Methods for Partial Differential Equations, Y. Zhu and B. Gao, eds., vol. 1297 of Lecture Notes in Mathematics, Springer, 1987, pp. 1–37.

[144] K. Feng and Z.-J. Shang, *Volume-preserving algorithms for source-free dynamical systems*, Numer. Math., 71 (1995), pp. 451–463.

[145] K. Feng, H. M. Wu, M. Z. Qin, and D. L. Wang, *Construction of canonical difference schemes for Hamiltonian formalism via generating functions*, J. Comput. Math., 7 (1989), pp. 71–96.

[146] M. Fernández-Pendás, E. Akhmatskaya, and J. M. Sanz-Serna, *Adaptive multi-stage integrators for optimal energy conservation in molecular simulations*, J. Comput. Phys., 327 (2016), pp. 434–449.

[147] R. P. Feynman, *Simulating Physics with computers*, Int. J. Theor. Phys., 21 (1982), pp. 467–488.

[148] J. A. Fleck, J. R. Morris, and M. D. Feit, *Time-dependent propagation of high energy laser beams through the atmosphere*, Appl. Phys. A: Materials Science & Processing, 10 (1976), pp. 129–160.

[149] W. Fleming, *Functions of Several Variables*, Springer Verlag, 2nd ed., 1977.

[150] W. Folkner, J. Williams, D. Boggs, R. Park, and P. Kuchynka, *The planetary and lunar ephemerides de430 and de431*, IPN Progress Report, 42 (2014).

[151] B. Fornberg, *A Practical Guide to Pseudospectral Methods*, Cambridge University Press, 1998.

[152] D. Furihata and T. Matsuo, *Discrete Variational Derivative Method. A Structure-Preserving Numerical Method for Partial Differential Equations*, CRC Press, 2011.

[153] B. García-Archilla, J. M. Sanz-Serna, and R. D. Skeel, *Long-time-steps methods for oscillatory differential equations*, SIAM J. Sci. Comput., 20 (1999), pp. 930–963.

[154] L. Gauckler, E. Hairer, and C. Lubich, *Energy separation in oscillatory Hamiltonian systems without any non-resonance condition*, Comm. Math. Phys., 321 (2013), pp. 803–815.

[155] L. Gauckler and C. Lubich, *Splitting integrators for nonlinear schrödinger equations over long times*, Found. Comput. Math., 10 (2010), pp. 141–169.

[156] Z. Ge and J. Marsden, *Lie-Poisson Hamiltonian-Jacobi theory and Lie-Poisson integrators*, Phys. Lett., 133 (1988), pp. 134–139.

[157] R. Glowinski, S. J. Osher, and W. Yin, eds., *Splitting Methods in Communication, Imaging, Science, and Engineering*, Springer, 2016.

[158] D. Goldman and T. Kaper, *nth-order operator splitting schemes and non-reversible systems*, SIAM J. Numer. Anal., 33 (1996), pp. 349–367.

[159] H. Goldstein, *Classical Mechanics*, Addison Wesley, Second ed., 1980.

[160] H. H. Goldstine, *A History of Numerical Analysis from the 16th through the 19th Century*, Springer, 1977.

[161] V. V. Gorbatsevich, A. L. Onishchik, and E. B. Vinberg, *Foundations of Lie Theory and Lie Transformation Groups*, Springer, 1997.

[162] D. Gottlieb and S. A. Orszag, *Numerical Analysis of Spectral Methods: Theory and Applications*, SIAM, 1977.

[163] S. Gray and D. E. Manolopoulos, *Symplectic integrators tailored to the time-dependent Schrödinger equation*, J. Chem. Phys., 104 (1996), pp. 7099–7112.

[164] S. Gray and J. M. Verosky, *Classical Hamiltonian structures in wave packet dynamics*, J. Chem. Phys., 100 (1994), pp. 5011–5022.

[165] D. F. Griffiths and J. M. Sanz-Serna, *On the scope of the method of modified equations*, SIAM J. Sci. Statist. Comput., 7 (1986), pp. 994–1008.

[166] J. Guckenheimer and P. Holmes, *Nonlinear Oscillations, Dynamical Systems, and Bifurcations of Vector Fields*, Springer, 1983.

[167] E. Hairer, *Backward analysis of numerical integrators and symplectic methods*, Ann. Numer. Math, 1 (1994), pp. 107–132.

[168] E. Hairer and M. Hairer, GniCodes–Matlab programs for geometric numerical integration, in Frontiers in Numerical Analysis, J. F. Blowey, A. W. Craig and T. Shardlow, eds., Springer, 2003, pp. 199-240.

[169] E. Hairer, C. Lubich, and G. Wanner, *Geometric numerical integration illustrated by the Störmer–Verlet method*, Acta Numerica, 12 (2003), pp. 399–450.

[170] E. Hairer, C. Lubich, and G. Wanner, *Geometric Numerical Integration. Structure-Preserving Algorithms for Ordinary Differential Equations*, Springer-Verlag, Second ed., 2006.

[171] E. Hairer, S. P. Nørsett, and G. Wanner, *Solving Ordinary Differential Equations I, Nonstiff Problems*, Springer-Verlag, Second revised ed., 1993.

[172] E. Hairer and G. Söderling, *Explicit, time reversible, adaptive step size control*, SIAM J. Sci. Comput., 26 (2005), pp. 1838–1851.

[173] E. Hairer and G. Wanner, *On the Butcher group and general multi-value methods*, Computing, 13 (1974), pp. 1–15.

[174] E. Hairer and G. Wanner, *Solving Ordinary Differential Equations II, Stiff and Differential-Algebraic Problems*, Springer, 2nd ed., 1996.

[175] M. Hajiketabi and F. Casas, *Numerical integrators based on the Magnus expansion for nonlinear dynamical systems*, Appl. Math. Comput., 369 (2020), p. 124844.

[176] R. W. Hamming, *Numerical Methods for Scientists and Engineers*, Dover, 1986.

[177] E. Hansen and A. Ostermann, *Exponential splitting for unbounded operators*, Math. Comput., 78 (2009), pp. 1485–1496.

[178] E. Hansen and A. Ostermann, *High order splitting methods for analytic semigroups exist*, BIT Numer. Math., 49 (2009), pp. 527–542.

[179] R. H. Hardin and F. D. Tappert, *Applications of the split-step Fourier method to the numerical solution of nonlinear and variable coefficient wave equations*, SIAM Rev., 15 (1973), p. 423.

[180] W. K. Hastings, *Monte-Carlo methods using Markov chains and their applications*, Biometrika, 57 (1970), pp. 97–109.

[181] P. Henrici, *Discrete Variable Methods in Ordinary Differential Equations*, John Wiley & Sons, 1962.

[182] J. S. Hesthaven, S. Gottlieb, and D. Gottlieb, *Spectral Methods for Time-Dependent Problems*, Cambridge University Press, 2007.

[183] M. W. Hirsch, S. Smale, and R. Devaney, *Differential Equations, Dynamical Systems, and an Introduction to Chaos*, Academic Press, 2013.

[184] M. Hochbruck and C. Lubich, *On Magnus integrators for time-dependent Schrödinger equations*, SIAM J. Numer. Anal., 41 (2003), pp. 945–963.

[185] M. Hochbruck and A. Ostermann, *Exponential integrators*, Acta Numerica, 19 (2010), pp. 209–286.

[186] H. Hofstätter and O. Koch, *Non-satisfiability of a positivity condition for commutator-free exponential integrators of order higher than four*, Numer. Math., 141 (2019), pp. 681–691.

[187] M. Holmes, *Introduction to Numerical Methods in Differential Equations*, Springer, 2007.

[188] W. Huang and B. Leimkuhler, *The adaptive Verlet method*, SIAM J. Sci. Comput., 18 (1997), pp. 239–256.

[189] W. Hundsdorfer and J. G. Verwer, *A note on splitting errors for advection-reaction equations*, Appl. Numer. Math., 18 (1995), pp. 191–199.

[190] W. Hundsdorfer and J. G. Verwer, *Numerical Solution of Time-Dependent Advection-Diffusion-Reaction Equations*, Springer, 2003.

[191] T. N. Ikeda, A. Abrar, I. L. Chuang, and S. Sugiura, *Minimum Trotterization formulas for a time-dependent Hamiltonian*, Quantum, 7 (2023), p. 1168.

[192] A. Iserles, *On the global error of discretization methods for highly-oscillatory ordinary differential equations*, BIT, 42 (2002), pp. 561–599.

[193] A. Iserles, *A First Course in the Numerical Analysis of Differential Equations*, Cambridge University Press, 2nd ed., 2009.

[194] A. Iserles, H. Z. Munthe-Kaas, S. P. Nørsett, and A. Zanna, *Lie-group methods*, Acta Numerica, 9 (2000), pp. 215–365.

[195] A. Iserles and S. P. Nørsett, *On the solution of linear differential equations in Lie groups*, Phil. Trans. Royal Soc. A, 357 (1999), pp. 983–1019.

[196] A. Iserles and G. R. W. Quispel, *Why geometric integration?*, in Discrete Mechanics, Geometric Integration and Lie–Butcher Series, K. Ebrahimi-Fard and M. B. Liñán, eds., Springer, 2018, pp. 1–28.

[197] T. Jahnke and C. Lubich, *Error bounds for exponential operator splittings*, BIT, 40 (2000), pp. 735–744.

[198] A. Jorba and M. Zou, *A software package for the numerical integration of ODEs by means of high-order Taylor methods*, Exp. Math., 14 (2005), pp. 99–117.

[199] W. Kahan and R. C. Li, *Composition constants for raising the order of unconventional schemes for ordinary differential equations*, Math. Comput., 66 (1997), pp. 1089–1099.

[200] W. O. Kermack and A. G. McKendrick, *A contribution to the mathematical theory of epidemics*, Proc. R. Soc. Lond. A, 115 (1927), pp. 700–721.

[201] E. Kieri, *Stiff convergence of force-gradient operator splitting methods*, Appl. Numer. Math., 94 (2015), pp. 33–45.

[202] U. Kirchgraber, *An ODE-solver based on the method of averaging*, Numer. Math., 53 (1988), pp. 621–652.

[203] S. Klarsfeld and J. A. Oteo, *Recursive generation of higher-order terms in the Magnus expansion*, Phys. Rev. A, 39 (1989), pp. 3270–3273.

[204] A. Knapp, *Lie Groups Beyond an Introduction*, Birkhäuser, Second ed., 2005.

[205] P.-V. Koseleff, *Calcul formel pour les méthodes de Lie en mécanique hamiltonienne*, PhD thesis, École Polytechnique, 1993.

[206] D. Kosloff and R. Kosloff, *A Fourier method solution for the time dependent Schrödinger equation as a tool in molecular dynamics*, J. Comp. Phys., 52 (1983), pp. 35–53.

[207] J. P. Labute, *Groups and Lie algebras: the Magnus theory*, in The Mathematical Legacy of Wilhelm Magnus. Groups, Geometry and Special Functions, W. Abikoff, J. S. Birman, and K. Kuiken, eds., 1994, pp. 397–406.

[208] J. D. Lambert, *Computational Methods in Ordinary Differential Equations*, John Wiley & Sons, 1973.

[209] J. D. Lambert, *The initial value problem for ordinary differential equations*, in The State of the Art in Numerical Analysis, D. Jacobs, ed., Academic Press, 1976, pp. 451–500.

[210] D. P. Landau and K. Binder, eds., *A Guide to Monte Carlo Simulations in Statistical Physics*, Cambridge University Press, 2005.

[211] F. M. Lasagni, *Canonical Runge–Kutta methods*, ZAMP, 39 (1988), pp. 952–953.

[212] J. Laskar, *A numerical experiment on the chaotic behaviour of the solar system*, Nature, 338 (1989), pp. 237–238.

[213] J. Laskar and P. Robutel, *High order symplectic integrators for perturbed Hamiltonian systems*, Celest. Mech. and Dyn. Astro., 80 (2001), pp. 39–62.

[214] J. Laskar, P. Robutel, F. Joutel, M. Gastineau, A. C. M. Correia, and B. Levrard, *A long-term numerical solution for the insolation quantities of the Earth*, Astron. Astrophys., 428 (2004), pp. 261–285.

[215] C. Leforestier, R. H. Bisseling, C. Cerjan, M. D. Feit, R. Friesner, A. Guldberg, A. Hammerich, G. Jolicard, W. Karrlein, H.-D. Meyer, N. Lipkin, O. Roncero, and R. Kosloff, *A comparison of different propagation schemes for the time dependent Schrödinger equation*, J. Comp. Phys., 94 (1991), pp. 59–80.

[216] L. Lehtovaara, J. Toivanen, and J. Eloranta, *Solution of time-independent Schrödinger equation by the imaginary time propagation method*, J. Comput. Phys., 221 (2007), pp. 148–157.

[217] B. Leimkuhler and C. Matthews, *Molecular Dynamics*, Springer, 2015.

[218] B. Leimkuhler and S. Reich, *Simulating Hamiltonian Dynamics*, Cambridge University Press, 2004.

[219] M. Leok, *Spectral variational integrators*, in Encyclopedia of Applied and Computational Mathematics, B. Engquist, ed., Springer, 2015.

[220] M. Leok and T. Shingel, *General techniques for constructing variational integrators*, Front. Math. China, 7 (2012), pp. 273–303.

[221] M. Leok and J. Zhang, *Discrete Hamiltonian variational integrators*, IMA J. Numer. Anal., 31 (2011), pp. 1497–1532.

[222] A. Lew, J. E. Marsden, M. Ortiz, and M. West, *Variational time integrators*, Int. J. Numer. Meth. Engng, 60 (2004), pp. 153–212.

[223] J. S. Liu, *Monte Carlo Strategies for Scientific Computing*, Springer, 2nd ed., 2008.

[224] S. Lloyd, *Universal quantum simulators*, Science, 273 (1996), pp. 1073–1078.

[225] M. A. López-Marcos, J. M. Sanz-Serna, and R. D. Skeel, *Cheap enhancement of symplectic integrators*, in Proceedings of the Dundee Conference on Numerical Analysis, D. Griffiths and G. A. Watson, eds., Longman, 1996, pp. 107–122.

[226] M. A. López-Marcos, J. M. Sanz-Serna, and R. D. Skeel, *An explicit symplectic integrator with maximal stability interval*, in Numerical Analysis, A.R. Mitchel 75th Birthday Volume, D. Griffiths and G. Watson, eds., Singapore, 1996, World Scientific, pp. 163–176.

[227] M. A. López-Marcos, J. M. Sanz-Serna, and R. D. Skeel, *Explicit symplectic integrators using Hessian-vector products*, SIAM J. Sci. Comput., 18 (1997), pp. 223–238.

[228] C. Lubich, *From Quantum to Classical Molecular Dynamics: Reduced Models and Numerical Analysis*, European Mathematical Society, 2008.

[229] W. Magnus, *On the exponential solution of differential equations for a linear operator*, Comm. Pure and Appl. Math., VII (1954), pp. 649–673.

[230] J. Makazaga and A. Murua, *A new class of symplectic integration schemes based on generating functions*, Numer. Math., 113 (2009), pp. 631–642.

[231] F. Mandl and G. Shaw, *Quantum Field Theory*, John Wiley & Sons, 1984.

[232] G. I. Marchuk, *Splitting and alternating direction methods*, in Handook of Numerical Analysis, P. G. Ciarlet and J. L. Lions, eds., vol. 1, North-Holland, Amsterdam, 1990, pp. 197–462.

[233] J. Marsden, G. Patrick, and W. Shadwick, eds., *Integration Algorithms and Classical Mechanics*, American Mathematical Society, 1996.

[234] J. E. Marsden and T. S. Ratiu, *Introduction to Mechanics and Symmetry*, Springer-Verlag, 1994.

[235] J. E. Marsden and M. West, *Discrete mechanics and variational integrators*, Acta Numerica, 10 (2001), pp. 357–514.

[236] R. I. McLachlan, *Composition methods in the presence of small parameters*, BIT, 35 (1995), pp. 258–268.

[237] R. I. McLachlan, *On the numerical integration of ODE's by symmetric composition methods*, SIAM J. Sci. Comput., 16 (1995), pp. 151–168.

[238] R. I. McLachlan, *More on symplectic correctors*, in Integration Algorithms and Classical Mechanics, J. E. Marsden, G. W. Patrick, and W. F. Shadwick, eds., vol. 10 of Fields Institute Communications, American Mathematical Society, 1996, pp. 141–149.

[239] R. I. McLachlan and S. K. Gray, *Optimal stability polynomials for splitting methods, with applications to the time-dependent Schrödinger equation*, Appl. Numer. Math., 25 (1997), pp. 275–286.

[240] R. I. McLachlan and A. Murua, *The Lie algebra of classical mechanics*, J. Comput. Dyn., 6 (2019), pp. 198–213.

[241] R. I. McLachlan and G. R. W. Quispel, *Six lectures on the Geometric Integration of ODEs*, in Foundations of Computational Mathematics, R. A. DeVore, A. Iserles, and E. Süli, eds., Cambridge University Press, 2001, pp. 155–210.

[242] R. I. McLachlan, G. R. W. Quispel, and N. Robidoux, *Geometric integration using discrete gradients*, Phil. Trans. Royal Soc. A, 357 (1999), pp. 1021–1046.

[243] R. I. McLachlan and R. Quispel, *Splitting methods*, Acta Numerica, 11 (2002), pp. 341–434.

[244] R. I. McLachlan and R. Quispel, *Geometric integrators for ODEs*, J. Phys. A: Math. Gen., 39 (2006), pp. 5251–5285.

[245] R. I. McLachlan, B. N. Ryland, and Y. Sun, *High order multisymplectic Runge–Kutta methods*, SIAM J. Sci. Comput., 36 (2014), pp. A2199–A2226.

[246] L. Meirovich, *Methods of Analytical Dynamics*, McGraw-Hill, 1988.

[247] A. Messiah, *Quantum Mechanics*, Dover, 1999.

[248] N. Metropolis, A. W. Rosenbluth, M. N. Rosenbluth, A. H. Teller, and E. Teller, *Equations of state calculations by fast computing machines*, J. Chem. Phys., 21 (1953), pp. 1087–1092.

[249] B. Mielnik and J. Plebański, *Combinatorial approach to Baker–Campbell–Hausdorff exponents*, Ann. Inst. Henri Poincaré, XII (1970), pp. 215–254.

[250] S. Miesbach and H. Pesch, *Symplectic phase flow approximation for the numerical integration of canonical systems*, Numer. Math., 61 (1992), pp. 501–521.

[251] A. Miessen, P. J. Ollitrault, F. Tacchino, and I. Tavernelli, *Quantum algorithms for quantum dynamics*, Nat. Comput. Sci., 3 (2023), pp. 25–37.

[252] S. Mikkola, *Practical symplectic methods with time transformation for the few-body problem'*, Celest. Mech., 67 (1997), pp. 145–165.

[253] P. C. Moan, *On backward error analysis and Nekhoroshev stability in the numerical analysis of conservative systems of ODEs*, PhD thesis, University of Cambridge, 2002.

[254] P. C. Moan, *On the KAM and Nekhorosev theorems for symplectic integrators and implications for error growth*, Nonlinearity, 17 (2004), pp. 67–83.

[255] P. C. Moan and J. Niesen, *Convergence of the Magnus series*, Found. Comput. Math., 8 (2008), pp. 291–301.

[256] P. C. Moan and J. A. Oteo, *Convergence of the exponential Lie series*, J. Math. Phys., 42 (2001), pp. 501–508.

[257] H. Munthe-Kaas, *Lie-Butcher theory for Runge–Kutta methods*, BIT, 35 (1995), pp. 572–587.

[258] H. Munthe-Kaas, *Runge–Kutta methods on Lie groups*, BIT, 38 (1998), pp. 92–111.

[259] H. Munthe-Kaas, *High order Runge–Kutta methods on manifolds*, Appl. Numer. Math., 29 (1999), pp. 115–127.

[260] H. Munthe-Kaas and B. Owren, *Computations in a free Lie algebra*, Phil. Trans. Royal Soc. A, 357 (1999), pp. 957–981.

[261] A. Murua, *Métodos simplécticos desarrollables en P-series*, PhD thesis, Universidad de Valladolid, 1994.

[262] A. Murua, *On order conditions for partitioned symplectic methods*, SIAM J. Numer. Anal., 34 (1997), pp. 2204–2211.

[263] A. Murua, *The Hopf algebra of rooted trees, free Lie algebras, and Lie series*, Found. Comp. Math., 6 (2006), pp. 387–426.

[264] A. Murua, *B-series*, in Encyclopedia of Applied and Computational Mathematics, B. Engquist, ed., Springer, 2015.

[265] R. M. Neal, *MCMC using Hamiltonian Dynamics*, in Handbook of Markov Chain Monte Carlo, S. Brooks, A. Gelman, G. L. Jones, and X.-L. Meng, eds., CRC Press, 2011, pp. 113–174.

[266] M. A. Nielsen and I. L. Chuang, *Quantum Computation and Quantum Information*, Cambridge University Press, 10th anniversary ed., 2010.

[267] P. J. Olver, *Applications of Lie Groups to Differential Equations*, Springer-Verlag, Second ed., 1993.

[268] I. P. Omelyan, I. M. Mryglod, and R. Folk, *On the construction of high order force gradient algorithms for integration of motion in classical and quantum systems*, Phys. Rev. E, 66 (2002), p. 026701.

[269] I. P. Omelyan, I. M. Mryglod, and R. Folk, *Symplectic analytically integrable decomposition algorithms: classification, derivation, and application to molecular dynamics, quantum and celestial mechanics simulations*, Comput. Phys. Comm., 151 (2003), pp. 272–314.

[270] B. Owren and A. Marthinsen, *Runge–Kutta methods adapted to manifolds and based on rigid frames*, BIT, 39 (1999), pp. 116–142.

[271] L. Pars, *A Treatise on Analytical Dynamics*, Ox Bow Press, 1979.

[272] J. Partington, *Linear Operators and Linear Systems. An Analytical Approach to Control Theory*, Cambridge University Press, 2004.

[273] A. Pazy, *Semigroups of Linear Operators and Applications to Partial Differential Equations*, Springer, 1983.

[274] I. Percival and D. Richards, *Introduction to Dynamics*, Cambridge University Press, 1982.

[275] M. Postnikov, *Lie Groups and Lie Algebras. Semester V of Lectures in Geometry*, URSS Publishers, 1994.

[276] T. Prosen and I. Pizorn, *High order non-unitary split-step decomposition of unitary operators*, J. Phys. A: Math. Gen., 39 (2006), pp. 5957–5964.

[277] G. R. W. Quispel, *Volume-preserving integrators*, Phys. Lett. A, 206 (1995), pp. 26–30.

[278] G. R. W. Quispel and D. McLaren, *A new class of energy-preserving numerical integration methods*, J. Phys. A: Math. Theor., 41 (2008), p. 045206.

[279] S. Reich, *Symplectic integration of constrained Hamiltonian systems by composition methods*, SIAM J. Numer. Anal., 33 (1996), pp. 475–491.

[280] S. Reich, *Backward error analysis for numerical integrators*, SIAM J. Numer. Anal., 36 (1999), pp. 1549–1570.

[281] S. Reich, *Multi-symplectic Runge–Kutta collocation methods for Hamiltonian wave equations*, J. Comput. Phys., 157 (2000), pp. 473–499.

[282] C. P. Robert and G. Casella, *Monte Carlo Statistical Methods*, Springer-Verlag, 1999.

[283] W. Rossmann, *Lie Groups. An Introduction Through Linear Groups*, Oxford University Press, 2002.

[284] G. Rowlands, *A numerical algorithm for Hamiltonian systems*, J. Comput. Phys., 97 (1991), pp. 235–239.

[285] R. Ruth, *A canonical integration technique*, IEEE Trans. Nucl. Sci., 30 (1983), pp. 26–69.

[286] J. M. Sanz-Serna, *Geometric integration*, in The State of the Art in Numerical Analysis (York, 1996), vol. 63 of Inst. Math. Appl. Conf. Ser. New Ser., New York, 1997, Oxford Univ. Press, pp. 121–143.

[287] J. M. SANZ-SERNA, *Runge–Kutta schemes for Hamiltonian systems*, BIT, 28 (1988), pp. 877–883.

[288] J. M. SANZ-SERNA, *Symplectic integrators for Hamiltonian problems: an overview*, Acta Numerica, 1 (1992), pp. 243–286.

[289] J. M. SANZ-SERNA, *Backward error analysis of symplectic integrators*, in Integration Algorithms and Classical Mechanics, J. E. Marsden, G. W. Patrick, and W. F. Shadwick, eds., Fields Institute Communications, American Mathematical Society, 1996, pp. 193–205.

[290] J. M. SANZ-SERNA, *Mollified impulse methods for highly-oscillatory differential equations*, SIAM J. Numer. Anal., 46 (2008), pp. 1040–1059.

[291] J. M. SANZ-SERNA, *Symplectic Runge–Kutta schemes for adjoint equations, automatic differentiation, optimal control, and more*, SIAM Review, 58 (2016), pp. 3–33.

[292] J. M. SANZ-SERNA AND L. ABIA, *Order conditions for canonical Runge–Kutta schemes*, SIAM J. Numer. Anal., 28 (1991), pp. 1081–1096.

[293] J. M. SANZ-SERNA AND M. P. CALVO, *Numerical Hamiltonian Problems*, Chapman & Hall, 1994.

[294] J. M. SANZ-SERNA AND A. PORTILLO, *Classical numerical integrators for wave-packet dynamics*, J. Chem. Phys., 104 (1996), pp. 2349–2355.

[295] T. SCHLICK, *Molecular Modelling and Simulation: An Interdisciplinary Guide*, Springer, Second ed., 2010.

[296] J.-P. SERRE, *Lie Algebras and Lie Groups*, no. 1500 in Lecture Notes in Mathematics, Springer-Verlag, Second ed., 1992.

[297] Z. SHANG, *KAM theorem of symplectic algorithms for Hamiltonian systems*, Numer. Math., 83 (1999), pp. 477–496.

[298] Q. SHENG, *Solving linear partial differential equations by exponential splitting*, IMA J. Numer. Anal., 9 (1989), pp. 199–212.

[299] Y. SHI, Y. SUN, Y. WANG, AND J. LIU, *Study of adaptive symplectic methods for simulating charged particle dynamics*, J. Comput. Dyn., 6 (2019), pp. 429–448.

[300] R. D. SKEEL AND J. J. BIESIADECKI, *Symplectic integration with variable stepsize*, Annals of Numerical Mathematics, 1 (1994), pp. 191–198.

[301] R. D. SKEEL AND J. L. CIEŚLIŃSKI, *On the famous unpublished preprint "methods of integration which preserve the contact transformation property of the Hamilton equations*, Tech. Rep. arXiv:2003.12268, 2020.

[302] G. SÖDERLING, L. JAY, AND M. CALVO, *Stiffness 1952–2012: Sixty years in search of a definition*, BIT, 55 (2015), pp. 531–558.

[303] M. SOFRONIOU AND G. SPALETTA, *Derivation of symmetric composition constants for symmetric integrators*, Optim. Method. Softw., 20 (2005), pp. 597–613.

[304] J. STOER AND R. BULIRSCH, *Introduction to Numerical Analysis*, Springer–Verlag, 1980.

[305] A. STUART AND A. HUMPHRIES, *Dynamical Systems and Numerical Analysis*, Cambridge University Press, 1996.

[306] E. Süli and D. Mayers, *An Introduction to Numerical Analysis*, Cambridge University Press, 2003.

[307] Y. B. Suris, *Preservation of symplectic structure in the numerical solution of Hamiltonian systems*, in Numerical Solution of Differential Equations, S. S. Filippov, ed., 1988, pp. 148–160. In Russian.

[308] Y. B. Suris, *Hamiltonian methods of Runge–Kutta type and their variational interpretation*, Math. Model., 2 (1990), pp. 78–87.

[309] G. Sussman and J. Wisdom, *Chaotic evolution of the Solar System*, Science, 257 (1992), pp. 56–62.

[310] M. Suzuki, *Fractal decomposition of exponential operators with applications to many-body theories and Monte Carlo simulations*, Phys. Lett. A, 146 (1990), pp. 319–323.

[311] M. Suzuki, *General theory of fractal path integrals with applications to many-body theories and statistical physics*, J. Math. Phys., 32 (1991), pp. 400–407.

[312] M. Suzuki, *Hybrid exponential product formulas for unbounded operators with possible applications to Monte Carlo simulations*, Phys. Lett. A, 201 (1995), pp. 425–428.

[313] M. Thalhammer, *High-order exponential operator splitting methods for time-dependent Schrödinger equations*, SIAM J. Numer. Anal., 46 (2008), pp. 2022–2038.

[314] M. Thalhammer, *Convergence analysis of high-order time-splitting pseudo-spectral methods for nonlinear Schrödinger equations*, SIAM J. Numer. Anal., 50 (2012), pp. 3231–3258.

[315] M. Thalhammer and J. Abhau, *A numerical study of adaptive space and time discretisations for Gross–Pitaevskii equations*, J. Comput. Phys., 231 (2012), pp. 6665–6681.

[316] W. Thirring, *Classical Dynamical Systems*, Springer-Verlag, 1992.

[317] L. N. Trefethen, *Spectral Methods in MATLAB*, SIAM, 2000.

[318] L. N. Trefethen and D. Bau, *Numerical Linear Algebra*, SIAM, 1997.

[319] W. Tucker, *A rigorous ODE solver and Smale's 14th problem*, Found. Comput. Math., 2 (2002), pp. 53–117.

[320] V. S. Varadarajan, *Lie Groups, Lie Algebras, and Their Representations*, Springer-Verlag, 1984.

[321] V. Vinokurov, *Logarithm of the solution of a linear differential equation, the Hausdorff formula and conservation laws*, Soviet Math. Dokl., 44 (1992), pp. 200–205.

[322] G. Wanner, *Kepler, Newton and numerical analysis*, Acta Numerica, (2010), pp. 561–598.

[323] J. Wei, *Note on global validity of the Baker–Hausdorff and Magnus theorems*, J. Math. Phys., 4 (1963), pp. 1337–1341.

[324] T. Weissert, *The Genesis of Simulation in Dynamics. Pursuing the Fermi–Pasta–Ulam Problem*, Springer, 1997.

[325] M. WEST, *Variational Integrators*, PhD thesis, California Institute of Technology, 2004.

[326] E. WHITTAKER, *A Treatise on the Analytical Dynamics of Particles and Rigid Bodies*, Cambridge University Press, 4th ed., 1964.

[327] S. WIGGINS, *Global Bifurcations and Chaos*, Springer-Verlag, 1988.

[328] J. WISDOM AND M. HOLMAN, *Symplectic maps for the N-body problem*, Astron. J., 102 (1991), pp. 1528–1538.

[329] J. WISDOM, M. HOLMAN, AND J. TOUMA, *Symplectic correctors*, in Integration Algorithms and Classical Mechanics, J. Marsden, G. Patrick, and W. Shadwick, eds., vol. 10 of Fields Institute Communications, American Mathematical Society, 1996, pp. 217–244.

[330] Y. H. WU, *The generating function for the solution of ODE's and its discrete methods*, Computers Math. Applic., 15 (1988), pp. 1041–1050.

[331] N. N. YANENKO, *The Method of Fractional Steps*, Springer, 1971.

[332] H. YOSHIDA, *Construction of higher order symplectic integrators*, Phys. Lett. A, 150 (1990), pp. 262-268.

[333] K. YOSIDA, *Functional Analysis*, Springer-Verlag, Third ed., 1971.

[334] B. ZHU, L. JI, A. ZHU, AND Y. TANG, *Poisson integrators based on splitting method for Poisson systems*, Commun. Comput. Phys., 32 (2022), pp. 1129–1155.

[335] S. ZHUK, N. F. ROBERTSON, AND S. BRAVY, *Trotter error bounds and dynamic multi-product formulas for Hamiltonian simulation*, Phys. Rev. Res., 6 (2024), p. 033309.

Index

so that $x_{n+1} \approx x(t_{n+1})$ with $t_{n+1} = (n+1)h)$. This is the so-called *Lie–Trotter scheme.* When the matrices commute, the sequence produces the exact solution. Otherwise, a direct calculation shows that

$$\mathrm{e}^{h(F_1+F_2)} - \mathrm{e}^{hF_2}\mathrm{e}^{hF_1} = \frac{1}{2}h^2[F_1, F_2] + \mathcal{O}(h^3) \qquad \text{as} \qquad h \to 0,$$

and so the previous approximation is only of first order of accuracy. Another version of the method is possible, of course, by reversing the order of F_1 and F_2, namely

$$x_{n+1} = \mathrm{e}^{hF_1}\,\mathrm{e}^{hF_2}\,x_n, \qquad n \geq 0 \tag{3.7}$$

has the same order of accuracy and properties as (3.6). If we have $m > 2$ matrices in (3.5), the Lie–Trotter scheme reads

$$x_{n+1} = \mathrm{e}^{hF_m}\,\cdots\,\mathrm{e}^{hF_2}\,\mathrm{e}^{hF_1}\,x_n$$

(or any other permutation of the matrices F_j).

If we instead consider a symmetrized version of (3.6), then

$$\mathrm{e}^{h(F_1+F_2)} - \mathrm{e}^{\frac{1}{2}hF_1}\mathrm{e}^{hF_2}\mathrm{e}^{\frac{1}{2}hF_1} = Ch^3 + \mathcal{O}(h^4), \tag{3.8}$$

where the constant C can be obtained either by comparing Taylor series or by applying the Baker–Campbell–Hausdorff (BCH) formula (see Appendix A.4) as $C = \frac{1}{24}([F_1, [F_1, F_2]] + 2[F_2, [F_1, F_2]])$. Therefore, the sequence

$$x_{n+1} = \mathrm{e}^{\frac{1}{2}hF_1}\mathrm{e}^{hF_2}\mathrm{e}^{\frac{1}{2}hF_1}\,x_n, \qquad n \geq 0 \tag{3.9}$$

produces a second-order approximation for the solution of (3.4). This corresponds to the *Strang* (splitting) *scheme.* If the roles of F_1 and F_2 are interchanged, one has another version of the Strang splitting scheme.

Simple generalizations to the case $m > 2$ include, in particular, the product

$$\mathrm{e}^{\frac{1}{2}hF_1}\,\mathrm{e}^{\frac{1}{2}hF_2}\,\cdots\,\mathrm{e}^{hF_m}\,\cdots\,\mathrm{e}^{\frac{1}{2}hF_2}\,\mathrm{e}^{\frac{1}{2}hF_1}.$$

Higher-order splitting methods could in principle be constructed by including more exponentials with their corresponding coefficients in a time step, namely

$$\Psi(h) = \mathrm{e}^{a_{s+1}hF_1}\,\mathrm{e}^{b_s hF_2}\,\mathrm{e}^{a_s hF_1}\,\cdots\,\mathrm{e}^{b_1 hF_2}\,\mathrm{e}^{a_1 hF_1}. \tag{3.10}$$

The number s (called the *number of stages*) as well as the coefficients a_j, b_j are chosen so that

$$\Psi(h) = \mathrm{e}^{h(F_1+F_2)} + \mathcal{O}(h^{r+1}) \qquad \text{as} \qquad h \to 0$$

for a given order r. In (3.10), the first and last exponentials correspond to F_1. This format is convenient for implementation, since the last exponential in one step can be concatenated with the first at the next step, thus reducing by one the number of evaluations. This corresponds to the well known FSAL

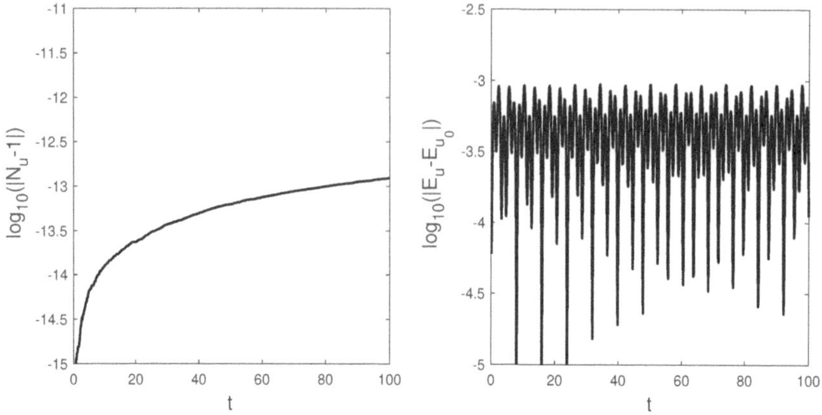

FIGURE 6.2: Error in norm (left) and in energy (right) for the Strang splitting (6.23) in the interval $t \in [0, 100]$ with time step size $h = 5/2^5$ for the Schrödinger equation with quadratic potential (6.28).

whose solution is given by

$$\begin{pmatrix} q(t) \\ p(t) \end{pmatrix} = O(tH) \begin{pmatrix} q_0 \\ p_0 \end{pmatrix}, \tag{6.30}$$

where

$$O(tH) = \begin{pmatrix} \cos(tH) & \sin(tH) \\ -\sin(tH) & \cos(tH) \end{pmatrix} \tag{6.31}$$

is an orthogonal and symplectic $2M \times 2M$ matrix. Computing $O(tH)$ exactly by diagonalizing the matrix H presents the same problems as its complex representation e^{-itH}, so that one proceeds in the usual step-by-step way by dividing the whole time interval into subintervals of length h and then approximating $O(hH)$ acting on the initial condition at each step $(q_n, p_n)^T$. In this respect, notice that if we introduce the nilpotent matrices A and B,

$$A = \begin{pmatrix} 0 & H \\ 0 & 0 \end{pmatrix}, \qquad B = \begin{pmatrix} 0 & 0 \\ -H & 0 \end{pmatrix}, \tag{6.32}$$

then

$$\mathrm{e}^{ha_kA} = \begin{pmatrix} I & a_khH \\ 0 & I \end{pmatrix}, \qquad \mathrm{e}^{hb_kB} = \begin{pmatrix} I & 0 \\ -b_khH & I \end{pmatrix},$$

and splitting methods of the form

$$\begin{pmatrix} q_{n+1} \\ p_{n+1} \end{pmatrix} = \mathrm{e}^{ha_{s+1}A}\, \mathrm{e}^{hb_sB}\, \mathrm{e}^{ha_sA} \cdots \mathrm{e}^{hb_1B}\, \mathrm{e}^{ha_1A} \begin{pmatrix} q_n \\ p_n \end{pmatrix} \tag{6.33}$$

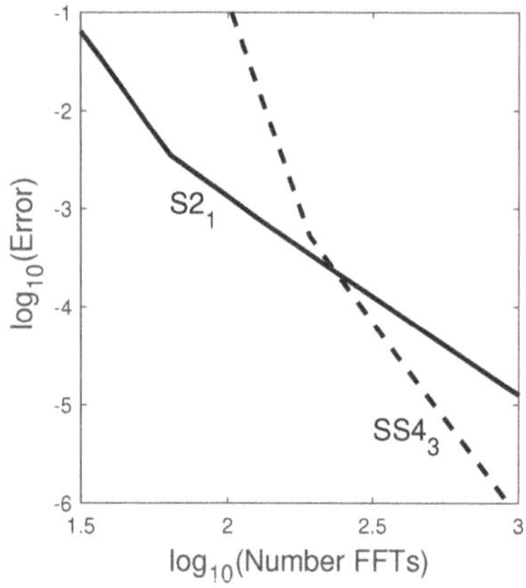

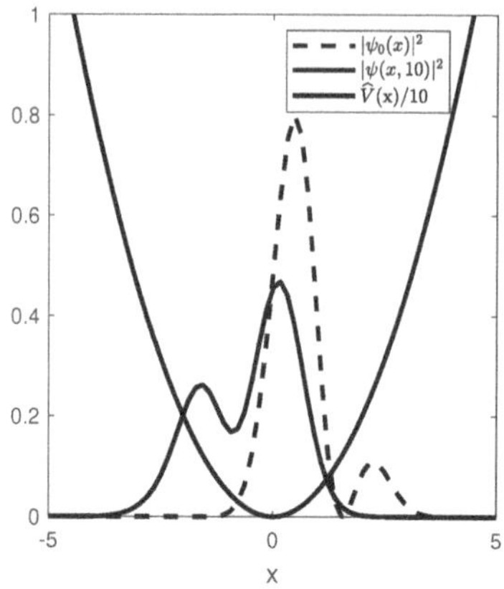

plane for some angle $0 < \theta < \pi/2$,

$$\Sigma_\theta = \{t \in \mathbb{C} : |\arg(t)| < \theta\},$$

so that $T(t)$ is analytic in t for all $t \in \Sigma_\theta$ [135, 273]. Thus, if

- L, A and B generate analytic semigroups on $\mathcal{X}$ (now a complex Banach space) in the sector Σ_θ, $0 < \theta < \pi/2$, with

$$\|\mathrm{e}^{tA}\| \leq \mathrm{e}^{\omega|t|}, \qquad \|\mathrm{e}^{tB}\| \leq \mathrm{e}^{\omega|t|}$$

  for $\omega \geq 0$ and all $t \in \Sigma_\theta$;

- $\|E_{r+1}\mathrm{e}^{t(A+B)}u_0\| \leq C$, where E_{r+1} is a composition of the operators A and B that consist of exactly $r+1$ factors,

then the splitting method of classical order r with coefficients $a_i, b_i \in \Sigma_\theta \subset \mathbb{C}$ satisfies

$$\|(\Psi(h)^n - \mathrm{e}^{nhL})u_0\| \leq Ch^r, \qquad 0 \leq nh < t_{\mathrm{f}}. \tag{6.48}$$

Since $\Psi(h)u_0$ is complex valued, this approach cannot be applied in principle when the operators A and B are real, i.e., for problems defined in a real Banach space $\mathcal{X}$. The most straightforward remedy consists of projecting the numerical solution after each time step on the real axis, i.e., computing the approximations $u_n \approx u(t_n)$, as $u_n = \mathrm{Re}(\Psi(h)u_{n-1})$. In that case one still has the bound (6.48) for the resulting integration scheme [178].

Although this is only valid in the linear case, similar results are observed in practice in the nonlinear heat equation $\partial_t u = \Delta u + F(u)$ with periodic boundary conditions, at least when F is analytic [96, 39], and so they constitute a strong motivation to study splitting methods with complex coefficients in general. This is precisely the subject of the following section.

6.4 Splitting and composition methods with complex coefficients

It has been known for a long time that, besides real solutions, the order conditions arising from splitting and composition methods (see section 3.3) also admit complex solutions [26, 310, 311, 312]. In fact, some of the resulting methods were explored in the context of Hamiltonian [101] and quantum mechanics [25, 276] before being used more recently in parabolic partial differential equations [96, 178] as a strategy to develop effective high-order methods.

We have already seen at the end of the previous section that a strategy to develop effective methods of order higher than two for systems evolving in a

For Product Safety Concerns and Information please contact our
EU representative GPSR@taylorandfrancis.com Taylor & Francis
Verlag GmbH, Kaufingerstraße 24, 80331 München, Germany